180th Anniversary of Ludwig Boltzmann

180th Anniversary of Ludwig Boltzmann

Guest Editors

Antonio M. Scarfone
Sergio Luiz E. F. Da Silva

MDPI

Basel • Beijing • Wuhan • Barcelona • Belgrade • Novi Sad • Cluj • Manchester

Guest Editors
Antonio M. Scarfone
Complex Systems Institute
Consiglio Nazionale delle Ricerche
Roma
Italy

Sergio Luiz E. F. Da Silva
Geoscience Institute
Fluminense Federal University
Niterói
Brazil

Editorial Office
MDPI AG
Grosspeteranlage 5
4052 Basel, Switzerland

This is a reprint of the Special Issue, published open access by the journal *Entropy* (ISSN 1099-4300), freely accessible at: https://www.mdpi.com/journal/entropy/special_issues/7GXB49GKLH.

For citation purposes, cite each article independently as indicated on the article page online and as indicated below:

Lastname, A.A.; Lastname, B.B. Article Title. *Journal Name* **Year**, *Volume Number*, Page Range.

ISBN 978-3-7258-4633-7 (Hbk)
ISBN 978-3-7258-4634-4 (PDF)
https://doi.org/10.3390/books978-3-7258-4634-4

Contents

About the Editors

Antonio M. Scarfone

Antonio M. Scarfone, working mainly in statistical mechanics, is a senior researcher in mathematical physics at the Complex Systems Institute of the National Research Council, Italy. Born in 1968, he studied physics at the University of Torino, earning his degree in 1996, and successively completed his Ph.D. in Physics at the Polytechnic of Torino in 2000. He has authored more than 100 scientific papers on statistical mechanics, kinetic theory, geometry information, nonlinear classical and quantum dynamics, and noncommutative algebras, all published in many international peer-reviewed journals. He is Editor-in-Chief of the Section Statistical Physics of *Entropy* and a member of the Editorial Board of several journals in mathematical physics, like *Advances in Mathematical Physics*, *Bulletin of Mathematical Analysis and Applications*, and belongs to the advisory panel of *Journal of Physics A*. *Lastly*, he is one of the main organizers of the SigmaPhi International Conference series that has been held every three years starting from 2005.

Sergio Luiz E. F. Da Silva

Sergio Luiz E. F. Da Silva is a Researcher affiliated with the Laboratory of Parallel Architectures for Signal Processing at the Federal University of Rio Grande do Norte (UFRN). He holds a PhD in Physics (2021), which he completed one year ahead of schedule, focusing on statistical physics and complex systems, and a Master's degree (2016) in Geophysics, both from the UFRN. He also obtained his degree in Physics from UFRN in 2013. His primary research interests are in the analysis of data-centered problems, especially in the theory of inverse problems. He is currently involved in projects aimed at improving seismic imaging techniques using wave equation-based approaches for complex geologic structures, overcoming challenges such as low resolution and non-Gaussian noise. These projects are conducted in collaboration with multinational companies and universities from the United States, Norway, France, and Brazil. Over the years, Dr. da Silva has collaborated with international institutions, including Politecnico di Torino, Italy, and Mines ParisTech, France, where he expanded his research horizons in geophysical methodologies. He was a member of the Organizing Committee of the SigmaPhi International Conference series in 2023. Dr. da Silva has published in reputable journals, contributing to the advancement of knowledge in inverse problems and related fields.

entropy

MDPI

Editorial

180th Anniversary of Ludwig Boltzmann

Antonio M. Scarfone [1] and Sérgio Luiz E. F. da Silva [1,2,*]

1 Istituto dei Sistemi Complessi—Consiglio Nazionale delle Ricerche (ISC-CNR), c/o Dipartimento di Scienza Applicata e Tecnologia del Politecnico di Torino, 10129 Torino, Italy; antoniomaria.scarfone@cnr.it
2 Laboratory of Parallel Architectures for Signal Processing, Federal University of Rio Grande do Norte, Natal 59078-970, Brazil
* Correspondence: sergio.silva.highres@imd.ufrn.br

Everybody who has devoted themselves in some way to the study of the scientific disciplines knows the giant contribution that Ludwig Boltzmann made to theoretical physics. Born in Vienna on 20 February 1844 into a world on the cusp of scientific revolution, Boltzmann showed an aptitude for mathematics and physics from a young age. Influenced by prominent scientists such as Christian Doppler and Josef Stefan, who followed him from his early studies at the University of Vienna, he completed his doctoral thesis in 1866 on the behavior of gases, a topic that would become central to most of his future works and which would mark him in the following century as one of the most important figures in the history of physics, alongside contemporaries such as James Clerk Maxwell and Josiah Willard Gibbs, establishing foundational principles that continue to shape the field of statistical physics to this day.

By the age of 25, he had already been appointed as a full professor of mathematical physics at the University of Graz, Austria. Widely regarded as one of Austria's greatest scientists, Boltzmann received invitations to occupy prestigious chairs in theoretical physics at leading academic centers, most notably in Berlin, Germany. In addition, he was awarded numerous honorary titles, reflecting the high esteem in which he was held by the scientific community.

His profound interest in thermodynamics shaped Boltzmann's early academic career, beginning the development of the kinetic theory of gas by introducing the idea of monads in a modern key and by highlighting the necessity of employing statistical methods in physics [1]. In 1872, by the age of only 28, he established the groundwork for contemporary statistical mechanics with a pair of influential papers: "Weitere studien über das wärmegleichgewicht unter gasmolekülen" (Further Studies on the Thermal Equilibrium of Gas Molecules) [2] and "Über die beziehung eines allgemeinen mechanischen satzes zum zweiten hauptsatz der wärmetheorie" (On the relationship of a general mechanical theorem to the second law of thermodynamics) [3]. In these papers, Ludwig Boltzmann introduced groundbreaking concepts such as the transport equation that bears his name and the statistical interpretation of entropy. His incorporation of probabilistic reasoning into physics revolutionized our comprehension of thermodynamics, entropy, and transport theory and provided insights into the second law of thermodynamics, which is central to modern physics, facilitating subsequent scientific advancements that continue to shape the various disciplines.

Boltzmann, by studying the approach to equilibrium in a system, introduces following the kinetic equation:

$$\frac{\partial}{\partial t} f(\boldsymbol{v}, t) + \boldsymbol{a} \cdot \nabla_v f(\boldsymbol{v}, t) = C[f(\boldsymbol{v}, t)] , \tag{1}$$

today named the Boltzmann equation, for a single-particle probability distribution function $f(v, t)$, where $C[f]$, the collisional term, accounts for the binary interactions among the monads of the system. In this way, he was able to introduce the so-called H-functional,

$$H[f] = \int f(v, t) \ln f(v, t) \, dv \tag{2}$$

which, together with the relation $dH/dt \leq 0$, constitutes his celebrated H-theorem.

Several years later, Josiah Willard Gibbs [4], a pioneer in the study of statistical mechanics, introduced the well-known expression

$$S[p] = -k_B \int p(x, v, t) \ln(p(x, v, t)) \, dx \, dv \, , \tag{3}$$

which shows an evident similarity with the H-functional (2), where $p(x, v, t)$ is the probability distribution function of the system in the phase space. The dimensional constant k_B has since been baptized in honor of Ludwing as the Boltzmann constant. This relation (3) is now known as Boltzmann–Gibbs entropy. It is used in statistical mechanics to measure how disordered or random a system with many parts is on a microscopic level. The legitimacy of this last statement finds validity in the following expression:

$$S = k_B \ln(W) \, , \tag{4}$$

written in this form by Max Planck [5] during his studies on black-body radiation. Although Boltzmann never explicitly wrote this last relation, it is nowadays known as the Boltzmann–Planck formula of entropy. Epigraphed on the Boltzmann tomb as a testimony to his enormous contribution to this field, it is a pillar in the field of thermostatistics.

However, Boltzmann's ideas, probably due to their extraordinary innovation, were not largely accepted and recognized as such initially by the wider scientific community. These have given rise to several controversial discussions with prominent personalities, such as Ernst Mach, and have also given rise to the birth of logical paradoxes like Maxwell's devil, the Poincare recurrence, and the Loschmidt paradox, to cite the most well known, which, in some way, exist in contrast to the prediction made by the H-theorem, and which have only recently found partial explanations.

Furthermore, Boltzmann's contributions to the understanding of heat, energy, and the behavior of gases in terms of statistical laws have surely revolutionized the way scientists today approach the behavior of matter. The principles he introduced have profoundly influenced not only physics but also fields such as biology, engineering, and even economics, where statistical approaches play a fundamental role in the understanding of many complex systems. A complete compilation of Ludwig Boltzmann's published works, including all his articles originally published in journals, can be found in ref. [6].

This Special Issue celebrates the 180th anniversary of Ludwig Boltzmann's birth by bringing together high-quality reviews and original research articles that examine the enduring influence of Boltzmann's concepts on statistical physics and related fields. Our aim is not only to acknowledge Boltzmann's foundational contributions but also to explore how his pioneering ideas continue to inspire advancements at the forefront of contemporary science and technology. The breadth of research presented here demonstrates the ongoing relevance of statistical mechanics in diverse domains.

The articles featured in this Special Issue illustrate the far-reaching impact of Boltzmann's work. Tamburrini, Davis, and Moya (2023) [7] introduce the Ehrenfest procedure as an efficient alternative to probability density functions for macroscopic properties in systems with multiple degrees of freedom, such as plasmas. This approach, rooted in the conjugate variable theorem and fluctuation–dissipation theorem, offers a computationally efficient method for studying adiabatic invariants in magnetized plasmas. Their study validates theoretical predictions through test particle simulations, demonstrating the utility of statistical mechanics in understanding dynamical systems far from equilibrium.

Kais (2024) [8] investigates quantum phase transitions in finite systems, challenging the conventional view that true criticality only emerges in the thermodynamic limit. Traditionally, finite-size scaling methods have been employed to extrapolate results to larger systems, but recent advances in ultra-cold atomic and molecular systems have raised new questions about the nature of phase transitions in genuinely finite systems. By developing a finite-size scaling approach tailored for such cases, this study examines quantum critical parameters in the context of chemical processes at ultra-cold temperatures. The recent observation of a quantum phase transition in a single trapped $^{171}Yb^+$ ion suggests that these transitions may play a fundamental role in chemical bond formation and dissociation, offering fresh perspectives on quantum chemistry.

Expanding on the statistical framework for microscopic systems, Alvarez-Estrada (2024) [9] extends a previous analysis of the quantum Wigner function and non-equilibrium equations for a microscopic particle in one spatial dimension. By deriving new solutions for the hierarchy equations and linking them to a Smoluchowski equation, this study develops a model describing molecular chain growth through atomic binding and catalyst activation in three-dimensional space. This work provides valuable insights into stochastic dynamics at the microscopic scale.

Rondoni and Di Florio (2024) [10] contribute to the fundamental understanding of the Boltzmann equation by analyzing conditions under which probability behaves analogously to mass. They emphasize that probability is the only suitable mathematical tool for describing small systems that deviate from the classical Boltzmann equation. Their study underscores the continued relevance of Boltzmann's insights in interpreting statistical mechanics applications.

From a conceptual perspective, Gimbel (2024) [11] explores the broader implications of Boltzmann's statistical approach to entropy. This study highlights how his 1877 formulation introduced quantization, reformulated entropy, and challenged deterministic interpretations of the laws of nature. Boltzmann's statistical perspective on the Second Law of Thermodynamics marked a paradigm shift, aligning with other scientific developments of the time, such as Darwin's theory of evolution and Hutton's geological theories. This paper also examines potential influences between Boltzmann and Darwin, noting Boltzmann's frequent references to evolutionary theory in his later writings.

Edrisi, Patwa, and Morales Escalante (2024) [12] examine Wehrl entropy for harmonic potential benchmark problems, deriving a steady-state analytical formula. However, their findings indicate a lack of theoretical results on absolute entropy monotonicity in the presence of friction. By employing a stochastic Monte Carlo numerical solver, this study provides numerical evidence suggesting the possibility of monotonic behavior, shedding light on entropy evolution in non-equilibrium systems.

In the domain of granular suspensions, Gómez González and Garzó (2024) [13] investigate non-Newtonian transport properties using inelastic Maxwell models and BGK-type kinetic approaches. Their study explicitly accounts for collisions between granular particles and a molecular gas, contrasting with traditional coarse-grained models. By employing both inelastic Maxwell models (IMM) and BGK-type kinetic models, they provide exact transport property computations and detailed velocity moment analyses. Notably, their findings highlight the phenomenon of discontinuous shear thickening (DST), demonstrating its dependence on mass ratios and other parameters.

Hunt, Sahimi, and Newman (2024) [14] apply percolation theory and ecological optimality principles to refine predictions of plant species richness based on net primary productivity. Their approach enhances species-energy theory by offering a more precise expression for primary productivity as a function of precipitation and potential evapotranspiration, improving our understanding of biodiversity patterns and non-climatic influences on ecosystem dynamics.

Addressing the dynamics of complex networks, van Elteren, Quax, and Sloot (2024) [15] identify two critical roles that nodes play in catalyzing abrupt transitions. Initiator nodes act as amplifiers of short-lived fluctuations, destabilizing neighboring nodes, while stabilizer nodes encode long-term memory, counteracting cascading effects. This study introduces a new framework for understanding and controlling endogenous metastability in networked systems, with implications for fields ranging from epidemiology to financial systems.

In the realm of materials science, Svyetlichnyy (2025) [16] simulates the powder bed fusion (PBF) process for functionally graded materials (FGM) fabrication using a combination of unity-based deposition models and lattice Boltzmann method (LBM) simulations. By introducing a diffusion model for material mixtures, specifically AISI 316L austenitic steel and 18Ni maraging 300 martensitic steel, this study demonstrates consistency with experimental data. The proposed approach offers a powerful tool for simulating, studying, and optimizing FGM production in PBF processes, reinforcing the practical applications of Boltzmann's methods in modern engineering.

Finally, Boguñá and Serrano (2025) [17] present a statistical mechanics framework for directed networks, conceptualizing them as ensembles of interacting fermions. Their approach systematically models network structures and dynamics, introducing novel perspectives and analytical tools for studying directed networks. By addressing reciprocity and other network characteristics, their work provides a principled method for investigating influence propagation and connectivity in complex systems.

We are honored to present this Special Issue as a tribute to Ludwig Boltzmann's scientific vision. The breadth and depth of the research included here reaffirm the foundational role of statistical mechanics, and we hope the articles included herein will inspire future investigations into the challenges and opportunities that lie ahead.

Conflicts of Interest: The authors declare no conflicts of interest.

References

1. Cercignani, C. *Ludwig Boltzmann: The Man Who Loved Atoms*; Oxford University Press: Oxford, UK, 1998.
2. Boltzmann, L. Weitere studien über das wärmegleichgewicht unter gasmolekülen. In *The Kinetic Theory of Gases*; World Scientific: Singapore, 2010; pp. 262–349.
3. Boltzmann, L. Über die beziehung eines allgemeinen mechanischen satzes zum zweiten hauptsatz der wärmetheorie. In *The Kinetic Theory of Gases*; World Scientific: Singapore, 2010; pp. 362–367.
4. Gibbs, J.W. *Elementary Principles in Statistical Mechanics*; Charles Scribner's Sons: New York, NY, USA, 1902.
5. Planck, M. Ueber das Gesetz der Energieverteilung im Normalspectrum. *Ann. Phys.* **1901**, *309*, 553.
6. Boltzmann, L.; Hasenöhrl, F. *Wissenschaftliche Abhandlungen*; Cambridge Library Collection—Physical Sciences; Cambridge University Press: Cambridge, UK, 2012.
7. Tamburrini, A.; Davis, S.; Moya, P.S. Evaluating the Adiabatic Invariants in Magnetized Plasmas Using a Classical Ehrenfest Theorem. *Entropy* **2023**, *25*, 1559. [CrossRef] [PubMed]
8. Kais, S. Walking with the Atoms in a Chemical Bond: A Perspective Using Quantum Phase Transition. *Entropy* **2024**, *26*, 230. [CrossRef] [PubMed]
9. Alvarez-Estrada, R.F. Non-Equilibrium Wigner Function and Application to Model of Catalyzed Polymerization. *Entropy* **2024**, *26*, 104. [CrossRef] [PubMed]
10. Rondoni, L.; Di Florio, V. Probability Turns Material: The Boltzmann Equation. *Entropy* **2024**, *26*, 171. [CrossRef] [PubMed]
11. Gimbel, S. It Ain't Necessarily So: Ludwig Boltzmann's Darwinian Notion of Entropy. *Entropy* **2024**, *26*, 238. [CrossRef] [PubMed]
12. Edrisi, A.; Patwa, H.; Morales Escalante, J.A. A Numerical Study of Quantum Entropy and Information in the Wigner–Fokker–Planck Equation for Open Quantum Systems. *Entropy* **2024**, *26*, 263. [CrossRef] [PubMed]

13. Gómez González, R.; Garzó, V. Exact Results for Non-Newtonian Transport Properties in Sheared Granular Suspensions: Inelastic Maxwell Models and BGK-Type Kinetic Model. *Entropy* **2024**, *26*, 265. [CrossRef] [PubMed]
14. Hunt, A.G.; Sahimi, M.; Newman, E.A. Species Richness Net Primary Productivity and the Water Balance Problem. *Entropy* **2024**, *26*, 641. [CrossRef] [PubMed]
15. van Elteren, C.; Quax, R.; Sloot, P.M.A. Cascades Towards Noise-Induced Transitions on Networks Revealed Using Information Flows. *Entropy* **2024**, *26*, 1050. [CrossRef] [PubMed]
16. Svyetlichnyy, D. Lattice Boltzmann Modeling of Additive Manufacturing of Functionally Graded Materials. *Entropy* **2025**, *27*, 20. [CrossRef] [PubMed]
17. Boguñá, M.; Serrano, M.Á. Statistical Mechanics of Directed Networks. *Entropy* **2025**, *27*, 86. [CrossRef] [PubMed]

entropy

MDPI

Article

Evaluating the Adiabatic Invariants in Magnetized Plasmas Using a Classical Ehrenfest Theorem

Abiam Tamburrini [1,*], Sergio Davis [2,3] and Pablo S. Moya [1,*]

1 Departamento de Física, Facultad de Ciencias, Universidad de Chile, Santiago 8370459, Chile
2 Research Center in the Intersection of Plasma Physics, Matter and Complexity (P^2mc), Comisión Chilena de Energía Nuclear, Casilla 188-D, Santiago 8320000, Chile; sergdavis@gmail.com
3 Departamento de Física, Facultad de Ciencias Exactas, Universidad Andres Bello, Santiago 8370136, Chile
* Correspondence: abiamtamburrini@gmail.com (A.T.); pablo.moya@uchile.cl (P.S.M.)

Abstract: In this article, we address the reliance on probability density functions to obtain macroscopic properties in systems with multiple degrees of freedom as plasmas, and the limitations of expensive techniques for solving Equations such as Vlasov's. We introduce the Ehrenfest procedure as an alternative tool that promises to address these challenges more efficiently. Based on the conjugate variable theorem and the well-known fluctuation-dissipation theorem, this procedure offers a less expensive way of deriving time evolution Equations for macroscopic properties in systems far from equilibrium. We investigate the application of the Ehrenfest procedure for the study of adiabatic invariants in magnetized plasmas. We consider charged particles trapped in a dipole magnetic field and apply the procedure to the study of adiabatic invariants in magnetized plasmas and derive Equations for the magnetic moment, longitudinal invariant, and magnetic flux. We validate our theoretical predictions using a test particle simulation, showing good agreement between theory and numerical results for these observables. Although we observed small differences due to time scales and simulation limitations, our research supports the utility of the Ehrenfest procedure for understanding and modeling the behavior of particles in magnetized plasmas. We conclude that this procedure provides a powerful tool for the study of dynamical systems and statistical mechanics out of equilibrium, and opens perspectives for applications in other systems with probabilistic continuity.

Keywords: Ehrenfest theorem; magnetized plasmas; non-equilibrium statistical mechanics

1. Introduction

Statistical mechanics provides a theoretical framework and a toolbox that gives us a deep understanding and detailed description of the collective behavior of particles in physical systems. One of the main advantages of statistical mechanics lies in its ability to relate observable macroscopic properties to the microscopic properties of individual particles. By using probabilistic and statistical methods, we can greatly simplify the analysis of complex systems, and obtain general rules applicable to a wide range of conditions. Through this connection, we can understand how macroscopic properties emerge from the behavior of particles at the microscopic level [1]. Furthermore, statistical mechanics offers a unified approach to analyze equilibrium and non-equilibrium systems. It allows us to study not only systems at thermodynamic equilibrium but also non-equilibrium systems [2–5]. This capability is especially relevant for modeling complex phenomena in nature [6–8]. Treating a non-equilibrium system is still an open task [9]. However, we have seen approximations based on the fundamental theory of statistical mechanics, such as the principle of maximum entropy [10], that, from first principles and information theory [11], we can infer probability distribution for the state of the system, which is most likely given certain restrictions and the information available. This approach allows us to describe the dynamics of non-equilibrium systems by providing a coherent picture of how they evolve over time, thus providing a clear prescription for constructing models in non-equilibrium statistical mechanics [12,13]. These models have been explored to understand the time evolution of dynamical systems [14–17], leaving us with general rules applicable to this type of system [18].

Although we can find plasma in an equilibrium state, usually when collisions are scarce, a plasma is typically a system out of equilibrium with a high number of degrees of freedom [19]. In the study of plasmas, there are various approaches from non-equilibrium statistical mechanics that allow an adequate description of these complex systems. Two prominent approaches are non-extensive statistical mechanics [20] and superstatistics [21], both aimed at addressing the features that emerge when plasma particles are far from thermodynamic equilibrium. Non-extensive statistical mechanics represents a theoretical framework that quantifies the distance from equilibrium in these systems through non-extensive entropy. This approximation generalizes the Boltzmann–Gibbs entropy, which has traditionally been used to describe systems in thermodynamic equilibrium. However, plasmas often have dynamic and structural properties that differ significantly from a classical steady state. In this sense, non-extensive statistical mechanics offers a more suitable way to characterize systems with non-trivial fluctuations and correlations, which is common in plasmas due to their turbulent behavior and non-linear transport properties [22–25]. On the other hand, superstatistics is another approach that is not based on generalizing the Boltzmann–Gibbs entropy, but instead proposes the existence of other parameters that follow a distribution different than Boltzmann [21,26,27]. These parameters can be considered effective temperatures in different cells of the plasma, where each cell has a temperature distribution that deviates from a global temperature in a possible thermodynamic equilibrium. In this view, the plasma particles do not experience a single, global temperature, but rather 'feel' different temperatures depending on the cell they belong to. This non-trivial characteristic of temperature in the plasma has important implications for the dynamics and macroscopic properties of the system. These advanced theoretical tools contribute to a better understanding of the physical processes in plasmas [28–31].

In general, from the point of view of statistical mechanics, these systems are described by partial differential equations (PDEs) for the time-dependent probability distribution function of the system. With this, it is possible to calculate the macroscopic properties of the system using the expectation value of its microscopic properties. The plasmas are described by Vlasov's PDE [32]. This equation comes from Liouville's theorem, where the Hamiltonian involves kinetic terms and the interaction of charged particles with electromagnetic fields [33]. There is no general analytical solution to this problem for arbitrary electromagnetic fields for the probability density function, and the numerical methods are extremely expensive in terms of computational cost, as they require the discretization of space and time with a grid size as small as the desired resolution, which depends of the highest frequency of modes of oscillation in the plasma, typically the frequency associated with the electrons [34].

To study the plasma macroscopic properties, we have developed new non-equilibrium statistical mechanics techniques, in particular, what has been called the 'Ehrenfest procedure' due to its resemblance to Ehrenfest's theorem in quantum mechanics [35]. The procedure allows the construction of particular differential equations associated with an arbitrary and time-dependent macroscopic property w. Using two fundamental theorems of statistical mechanics, the conjugate variable theorem (CVT) and the fluctuation–dissipation theorem (FDT) [18,36], it is possible to relate the derivatives in time and space of the probability density function with the derivatives in time and space of an arbitrary time-dependent macroscopic property. Therefore, employing this feature, we can eliminate the explicit dependence of the probability density function, ρ, on the original PDE, obtaining a particular PDE associated with an arbitrary macroscopic property w. Moreover, in Ref. [35], the procedure was applied to the Vlasov equation to formulate new equations for the general evolution of arbitrary fluctuations in collisionless plasmas.

On the other hand, it is well known that in classical mechanics, whenever there is a system with periodic movements, the action integral $\oint pdq$, with p and q the generalized variables in the phase space, taken over a period, is a constant of motion [37]. If under small perturbations in the system this quantity remains constant, then we call it adiabatic invariant. These quantities are important in plasma physics and help us simplify the answer in cases involving complex motion. The study of adiabatic invariants allows us to identify and characterize properties and relations that remain constant during the temporal evolution of a system. It is possible to predict significant quantities and magnitudes of systems in different stages [38]. It allows studying conditions in which the system is stable or maintains an equilibrium state [39]. At the particle scale, it allows the characterization of their movement and the understanding of their dynamics in environments where they interact with electromagnetic fields, such as particles trapped in the Earth's magnetic field [40], giving rise to Earth's radiation belts. In particular, for particles trapped in a dipole magnetic field, there are three invariants associated with three types of movement, namely: gyromotion around magnetic field lines, bouncing between two magnetic mirror points, and orbital motion around the Earth due to the curvature drift [41–43]. In this work, we make use of the Ehrenfest procedure applied to the Vlasov equation to study the adiabatic invariants in a magnetized collisionless plasma.

We have organized this document as follows. First, in Section 2, we show the three adiabatic invariants associated with the three motions of the particles under a magnetic field, and we present the expression that the Ehrenfest procedure determines for the time evolution of each of these quantities. Then, in Section 3, we present the numerical result obtained from a test particle simulation and show the correlation between the analytical expression shown in Section 2 and the numerical result. Finally, in Section 4, we summarize our findings and present the main conclusions.

2. Ehrenfest Procedure in Adiabatic Invariants

As already mentioned, in a previous study [35], it was shown that by applying the Ehrenfest procedure to the Vlasov equation, it is possible to derive a dynamical PDE for an arbitrary observable, equivalent to the Vlasov equation. Following Ref. [35], we define the Ehrenfest operator given by

$$\mathrm{Eh}[\bullet] = \left\langle \frac{\partial \bullet}{\partial t} + \mathbf{v} \cdot \partial_{\mathbf{r}} \bullet + \frac{\mathbf{F}}{m} \cdot \partial_{\mathbf{v}} \bullet \right\rangle, \tag{1}$$

where the brackets correspond to the expectation value, defined as follows:

$$\mathcal{W}(t) = \langle w(\mathbf{r}, \mathbf{v}, t) \rangle = \int d\mathbf{r} d\mathbf{v}\, \rho(\mathbf{r}, \mathbf{v}) w(\mathbf{r}, \mathbf{v}, t) \tag{2}$$

Therefore, the PDE for the arbitrary observable corresponds to the equation that follows the form

$$\frac{\partial}{\partial t}\langle w \rangle = \mathrm{Eh}[w]\,, \tag{3}$$

namely,

$$\frac{\partial}{\partial t}\langle w \rangle = \left\langle \frac{\partial w}{\partial t} \right\rangle + \left\langle \mathbf{v} \cdot \partial_{\mathbf{r}} w \right\rangle + \left\langle \frac{\mathbf{F}}{m} \cdot \partial_{\mathbf{v}} w \right\rangle, \tag{4}$$

where the notation Eh[w] represents the expression derived from the Ehrenfest procedure on an arbitrary observable, equal to the temporal evolution of this observable w.

Here, we are interested in the study of adiabatic invariants of particles trapped in a magnetic field. For example, in the case of the Earth's radiation belt, there are three main motions of the particles due to their interaction with the approximate dipole magnetic field of the Earth [44]. Given this approximation, it is clear to see that due to the curvature of the field, there will be different drifts that give rise to these movements [37]. The invariants, their associated motion, and the Ehrenfest procedure applied to each of them are described below.

- First Invariant.

  The first invariant of magnetic moment is associated with the Larmor gyration, and it is given by the expression,

$$\mu = \frac{mv_{\perp}^2}{2B}. \tag{5}$$

  Considering the magnetic moment as an observable $\mu = w$, we replace the right hand of Equation (5) in Equation (4):

$$\frac{\partial}{\partial t}\langle \mu \rangle = \mathrm{Eh}[\mu]\,.$$

  Therefore, we obtain

$$\frac{\partial}{\partial t}\langle \mu \rangle = -\left[\left\langle \frac{mv_{\perp}^2}{2}\frac{\partial}{\partial t}\left(\frac{1}{B}\right)\right\rangle + \left\langle \frac{mv_{\perp}^2}{2}\mathbf{v}\cdot\nabla_r\left(\frac{1}{B}\right)\right\rangle\right], \tag{6}$$

  where Equation (6) corresponds to the Ehrenfest expression for the time variation of the magnetic moment. The first addend of this expression refers to the temporal variation of the magnetic field. If the magnetic field has fast variations, it can cause direct variation in the magnetic moment of a particle. If the change in the magnetic field is fast enough, it could even induce electric currents, which in turn would affect its magnetic moment. The second addend refers to the existence of drifts. Given a curvature in the magnetic field, the Larmor orbits can expand or contract accordingly, and the particles may lose or gain energy in the transverse direction, affecting the average magnetic moment.

- Second Invariant.

  The second adiabatic invariant, known as the longitudinal invariant, is associated with the periodic motion along the magnetic field line for particles trapped between two magnetic mirror points. It is possible to study this motion indirectly through the pitch angle between the particle velocity and the magnetic field direction, given by the expression

$$\cos(\theta) = \frac{\mathbf{B}\cdot\mathbf{v}}{|B||v|}. \tag{7}$$

  Considering the pitch angle as an observable $\cos(\theta) = w$, we replace the right hand of Equation (7) in Equation (4):

$$\frac{\partial}{\partial t}\big\langle \cos(\theta)\big\rangle = \mathrm{Eh}[\cos(\theta)] = \big\langle \mathbf{v}\cdot(\hat{\mathbf{v}}\cdot\nabla_{\mathbf{r}})\hat{\mathbf{B}}\big\rangle. \tag{8}$$

  Thus, Equation (8) shows that when a charged particle moves in a nonuniform magnetic field, it experiences a magnetic force that acts perpendicular to its velocity and the magnetic field. We can demonstrate this non-uniformity in Equation (8), given by the gradient of the projected field on the direction of movement of the particle. This force can change the direction of motion of the particle, which in turn affects the angle between the velocity and the magnetic field lines, that is, the pitch angle.

- Third Invariant.

  Finally, the third adiabatic invariant, called magnetic flux invariant, is associated with the nonuniform **B** field drift, which if it is not time-dependent, should expect close trajectories; therefore, we can study it indirectly through the average (squared) radial distance of a particle radio particles. Then, making the observable $w = r^2$, we replace in Equation (4), obtaining

$$\frac{\partial}{\partial t}\langle r^2\rangle = \mathrm{Eh}[\mathrm{r}^2] = \langle \mathbf{v}\cdot\mathbf{r}\rangle. \tag{9}$$

In this case, it is more direct to see that the variation of the average radial distance of each particle will depend on whether it moves in the radial direction. For this reason, the variation of the third invariant depends directly on the projection of the velocity of the particles in the radial direction.
The set of equations that form Equations (6), (8), and (9) corresponds to the analytical results for the temporal evolution of the quantities that allow us to study the movements to which the three adiabatic invariants are associated in this system. To verify these three expressions obtained using the Ehrenfest procedure, both sides of Equations (6), (8), and (9) will be compared using the data extracted from test particle simulations.

3. Numerical Results for Testing the Ehrenfest Procedure in Adiabatic Invariants

In order to verify the theoretical expressions derived in Section 2, we developed a test particle simulation [45] code written in the C++ programming language. In this kind of simulation, the particles follow the trajectories prescribed by the electromagnetic field, do not interact with each other, as in particle-in-cell simulations [34,46,47], and do not influence the dynamics of electromagnetic fields. Our objective was to simulate the outer Earth's radiation belt conditions. The particles were trapped by the Earth's magnetic field, which was considered a dipole magnetic field of the form

$$\mathbf{B}(\mathbf{r}) = \frac{3(\mathbf{m}\cdot\mathbf{r}) - r^2\mathbf{m}}{r^5}, \tag{10}$$

where **r** is particle position, and **m** is the magnetic moment that, in our case, was considered in the $\hat{z}$ direction. In setting the initial conditions, we adopted a Maxwellian distribution comprising 10^5 particles. The velocities were stochastically assigned according to this distribution, employing the Box–Muller method and employing a thermal velocity that aligns with energy levels on the order of keV, normalized with respect to the rest energy of electrons. The particles were injected in a small cube near the magnetic equator at a radial distance of $4.5 \pm 0.5\ R_E$, with R_E being the Earth's radius. The simulation was run for 1.4×10^6 time steps $\Delta\tau = 0.1$, where $\tau = c\,t/R_E$ represents the normalized time, and c is the speed of light. To obtain the velocity and position of the particle, the Boris algorithm was used, which is based on Liouville's theorem, and preserves energy by construction without being symplectic [48]. The first result corresponds to the particle trajectories, where we notice that the particles follow the expected behavior given the conditions. In Figure 1, we can see the evolution of the particle distribution in four different moments during the simulation. Three of them ($\tau = 1$, 100, and 500) correspond to the first moments of the simulation in which the radiation belt is still forming, and the last one ($\tau = 20{,}000$) corresponds to a later stage in which the belt has been completed.

Since the spatial particle distribution reaches a stationary state, that is, once the radiation belt has been formed after approximately 10,000 time steps, and the particles follow their periodic motion, we build the time series for each quantity. On the one hand, we build the time derivative of the observable, and on the other, we build the time series for the quantities that the Ehrenfest procedure indicates to determine the time evolution of the observable; that is, using the simulation we construct and compare the left and right sides of the Equations (6), (8), and (9), and to construct the expectation values of the different quantities, we calculate the arithmetic average over all the particles for a fixed time, given by the discrete data that we can extract from the simulation. In the left column of Figure 2, from top to bottom, time series corresponding to both sides of Equations (6), (8), and (9) are plotted in red and blue. In addition, the right column of the same figure shows scatter plots of the same quantities (magnetic moment, pitch angle, and average radial distance), so we compare them with each other with respect to the trend line that represents Slope 1 (black dashed line). We calculate the Pearson correlation coefficient between quantities and plot both sides of the Equations (6), (8), and (9) to be able to see the level of correlation between both sides of the Ehrenfest equation for each adiabatic invariant. Finally, we made a linear fit to each data set and calculated the slope of the fit in each case, which is shown in Figure 2 along with the Pearson correlation coefficient.

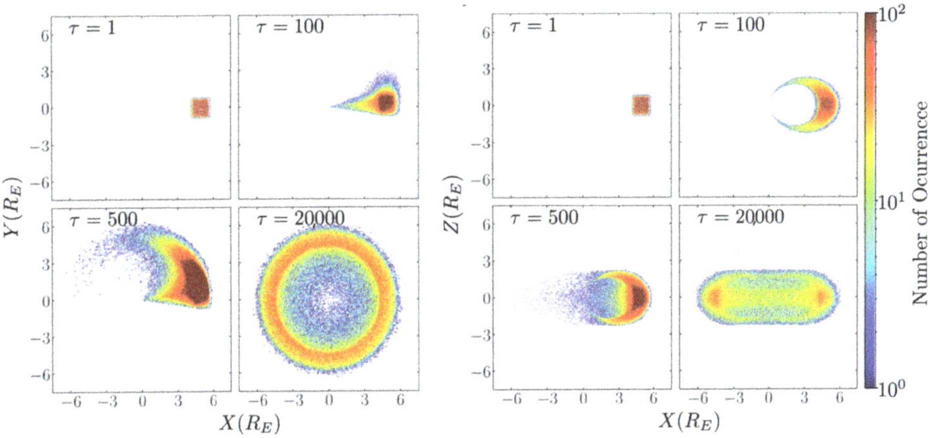

Figure 1. Particle trajectories in the $\hat{x}$-$\hat{y}$ plane in the left panel and $\hat{x}$-$\hat{z}$ plane in the right panel , in four different time steps: $\tau = 1$, $\tau = 100$, $\tau = 500$, and $\tau = 20{,}000$.

In the complete dynamics of the system, we have that the three adiabatic invariants are on different time scales. For example, the first invariant is on the order of the motion of the cyclotron around the magnetic field line, given by the Larmor frequency, and the flux invariant is given by the precession of particles around the Earth. This means that the equations of motion must be solved over a wide range of time scales.

Since the time series are constructed based on different parameters and fluctuate on different time scales, the data of each series were normalized before being treated for construction, This allows the order of fluctuation of each time series to be the same, eliminating the error caused by the centered derivative by having to make a finite difference between quantities that are in the order of 10^{-25}, such as the magnetic moment. The time series show small differences that are best evident in the lower window of each time series, where we approach a specific time interval.

We can see that the three invariants fluctuate on different time scales, and each quantity presents a different noise in its curve. For example, for the magnetic moment, it is possible to distinguish its distinctive oscillation period with respect to the equilibrium value being at least 10 times smaller than the other two quantities. On the other hand, for the average radius of the orbit, it is possible to distinguish a period of oscillation around the equilibrium value, but at the same time, it contains noise in the curve, which occurs to a lesser extent for the pitch angle. From this second invariant, we can see that the fluctuation follows the same behavior and, as expected, presents a better Pearson correlation coefficient of 0.961 compared to the correlation coefficient of 0.941 obtained for the magnetic moment.

We also note that by the construction of the second invariant, this is already normalized and is where we find the best slope $m = 0.958$, and as expected, it is better than the slope of the magnetic moment $m = 0.856$, which, in principle, fluctuates in the order previously mentioned. Finally, for the third adiabatic invariant, we note that although the curve contains more noise, this expression presents the best Pearson correlation coefficient of

0.980, and in the zoom window, we distinguish that the time series also follows the same behavior.

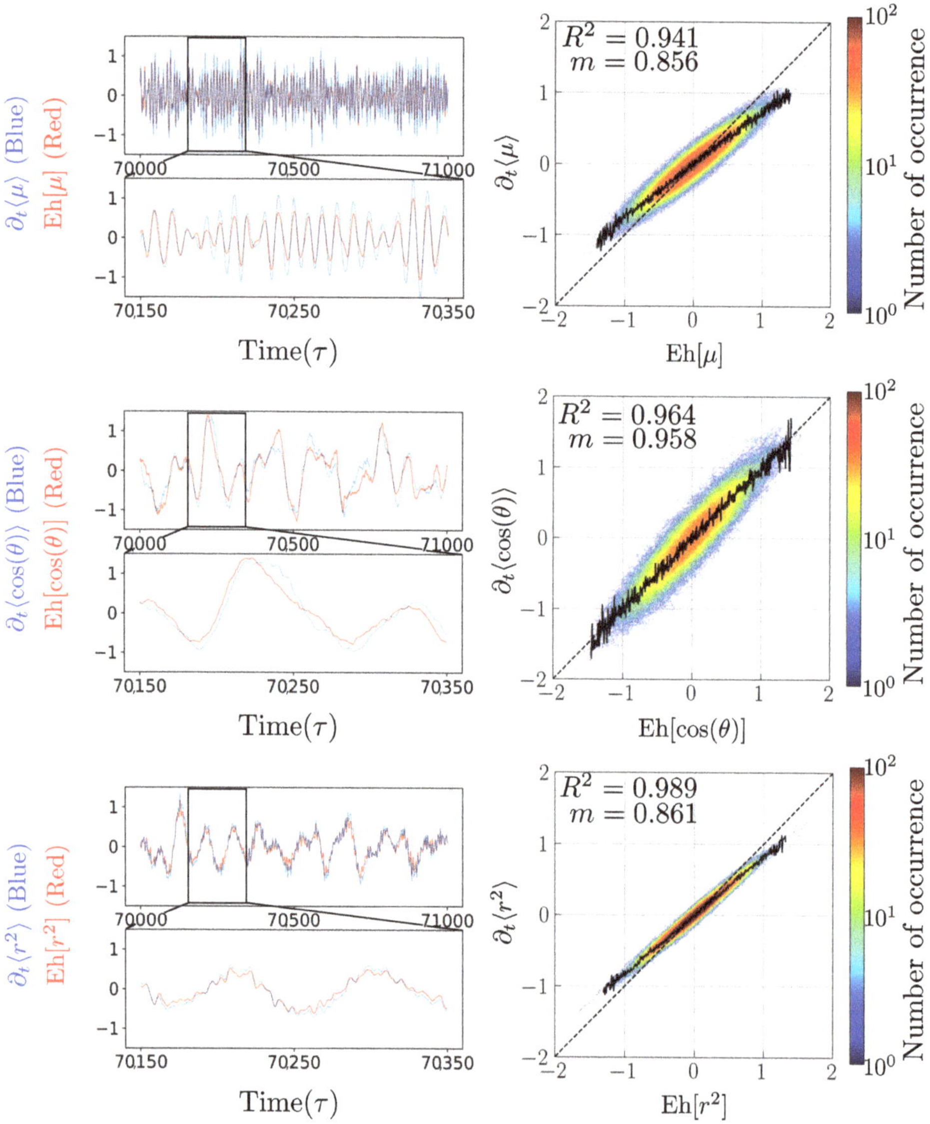

Figure 2. Application of the Ehrenfest procedure to particles trapped in a dipole magnetic field. Left: time series of the left-hand (blue) and right-hand (red) sides of Equation (4) using data extracted from the simulation. From top to bottom, each panel corresponds to the Ehrenfest procedure applied to the magnetic moment, pitch angle, and radial distance, respectively. Right: correlations between left- and right-hand sides of each relation. In each scatter plot, the solid black line represents the maximum in each column, and black dashed lines represent the tendency and identity, respectively.

Although the analytical expression suggests that the relationship between the temporal evolution of the adiabatic invariant should be 1 to 1 with the expression determined by the Ehrenfest procedure, it is expected to obtain small differences between the time series that represent both sides of the Equations obtained using the Ehrenfest procedure. This is because it only approximates the behavior of a Vlasov system since, by using the numerical simulation, we discretized the space. The numerical methods are in fact finite differences; that is, we are working only with a section of the phase space and not with the entire phase space as a Vlasov system would follow. In Figure 2, these differences and errors are manifested by less than one correlation coefficient and also in that the slopes of the fits for each quantity are below the 1-to-1 relationship suggested by the analytical expression. Despite this, we notice in the time series windows that each quantity follows the same behavior and presents a very good correlation and a slope very close to Slope 1.

4. Conclusions

In this article, we have explored the applicability of the Ehrenfest procedure for the study of adiabatic invariants in magnetized plasmas. The reliance on probability density functions to obtain macroscopic properties in systems with a high number of degrees of freedom and the expensive techniques, both theoretical (such as canonical transformations in the Vlasov equation) and computational to solve the problem, motivated us to propose a procedure that would allow obtaining information about the system in a less expensive way than conventional methods, that is, the Ehrenfest procedure. This procedure is based on general rules such as the conjugate variable theorem [18], valid for states arbitrarily far from equilibrium, covering specific cases such as hypervirial [49] relations in statistical mechanics and the expression of temperature in the microcanonical ensemble, as derived by Rugh and generalized by Rickayzen [50,51]. Its validity remains independent of considerations about ergodicity or a priori equality of probabilities for microstates in a thermodynamic system, even when the existence of such a system is uncertain. Instead, it holds as long as we base our reasoning on incomplete information, given in the form of expected values. This theorem can serve as an alternative tool for the estimation of parameters in probability distributions that present sufficient statistics, seen from the perspective of frequentist statistics.

We have demonstrated the application of our 'Ehrenfest procedure' to the Vlasov equation for collisionless plasmas. We use the general relation in equation (4), which gives the time evolution of any macroscopic property in a Vlasov system. From Ehrenfest's general relation, we have derived the partial differential equation for the study, both direct and indirect, of the three adiabatic invariants: magnetic moment, longitudinal invariant, and magnetic flux, which we show in Equations (5), (7), and (9). We compare the theory with the numerical simulations, and the results suggest a good agreement between the theory and the numerical simulations for the magnetic moment, the average radius of the orbit, and the pitch angle, which suggests that the procedure is suitable for these observables. In addition, the increase in the statistics of the problem since the simulation approaches a Vlasov system, and we only have a part of the phase space, together with the fact that the numerical methods used are of finite difference, can give rise to differences and errors in the observed correlations, which explains the fluctuations and noises present in the curves of the different quantities analyzed. Due to the aforementioned, with this procedure, we were able to demonstrate the equality given by the analytical expression in the numerical simulation data up to the simulation precision. It is also worth mentioning that although Boris's algorithm is based on construction on Liouville's theorem, it is not symplectic; therefore, it does not preserve the volume in the phase space. Thus, although it is expected that there is some numerical error, we observe that the analytical expressions are mostly confirmed by the simulations up to numerical precision.

We have observed an increase in the Pearson correlation coefficient and the slope of the linear fit as we increase the statistics collected from the problem. Specifically, by increasing the number of particles in the simulation by an order of magnitude, the correlation coefficient increases significantly, as shown in Figure 3, where we also see that it seems to stabilize between 10^4 and 10^5. Here, we present the case of 10^5 particles, as it is the smallest number that leads to meaningful results without the use of unnecessary numerically expensive simulations. In summary, we have shown that the Ehrenfest procedure is a general useful tool, applicable to any system following a continuity equation for probability, such as the Vlasov equation in collisionless plasmas. Given these advantages, applications seem to come naturally. For example, in a plasma modeled by the Fokker–Planck equation, it would be possible to have expressions for observables with dependencies on the drift and diffusion coefficient. It remains for future work to show that this procedure is useful to analyze the case in which the magnetic field intensity is suddenly altered in this system of trapped particles, simulating a geomagnetic storm. In such a case, it should be possible to violate the third adiabatic invariant and expect the particles to undergo radial diffusion [44] that, in principle, may be estimated using this procedure. In general, this procedure allows one to quickly obtain equations of dynamic evolution for particular properties and, therefore, seems to be a powerful addition to the study of out-of-equilibrium plasmas or dynamical systems in general.

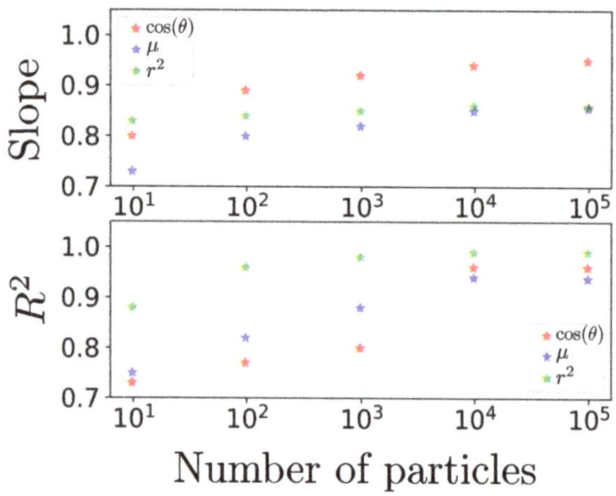

Figure 3. The upper panel shows the change in the slope of the linear fit for each quantity as we increase the particle statistics of the problem in steps of an order of magnitude. The lower panel shows the variation of the Pearson correlation coefficient by increasing the number of particles. The error associated with each data is on the order of 0.02% on average; therefore, the error bars are not distinguishable on the scale of the figure.

Author Contributions: Conceptualization, A.T., S.D. and P.S.M.; methodology, A.T., S.D., and P.S.M.; formal analysis, A.T., S.D. and P.S.M.; investigation, A.T., S.D. and P.S.M.; writing—original draft preparation, A.T., S.D. and P.S.M.; writing—review and editing, A.T., S.D. and P.S.M. All authors contributed equally to this manuscript. All authors have read and agreed to the published version of the manuscript.

Funding: A. Tamburrini is grateful to Agencia Nacional de Investigación y Desarrollo (ANID, Chile) for the National Doctoral Scholarship No. 21210407. S. Davis thanks the support of ANID FONDECYT 1220651, Grant. P. S. Moya is thankful for the support of the Research Vice-Rectory of the University of Chile (VID) through Grant ENL08/23.

Institutional Review Board Statement: Not applicable.

Data Availability Statement: The data presented in this study are openly available in Zenodo at https://doi.org/10.5281/zenodo.8411207.

Acknowledgments: We thank Iván Gallo and Javier Silva for useful discussions.

Conflicts of Interest: The authors declare no conflict of interest.

Abbreviations

The following abbreviations are used in this manuscript:

PDE	partial differential equations
CVT	conjugate variable theorem
FDT	fluctuation–dissipation theorem

References

1. McQuarrie, D.A. *Statistical Mechanics*; Chemistry Series; Harper & Row: New York, NY, USA, 1975.
2. Espinoza, C.; Stepanova, M.; Moya, P.; Antonova, E.; Valdivia, J. Ion and Electron κ Distribution Functions Along the Plasma Sheet. *Geophys. Res. Lett.* **2018**, *45*, 6362–6370. [CrossRef]
3. Eyelade, A.V.; Stepanova, M.; Espinoza, C.M.; Moya, P.S. On the Relation between Kappa Distribution Functions and the Plasma Beta Parameter in the Earth's Magnetosphere: THEMIS Observations. *Astrophys. J. Suppl. Ser.* **2021**, *253*, 34. [CrossRef]
4. Ross, D. Equilibrium free energies from non-equilibrium trajectories with relaxation fluctuation spectroscopy. *Nat. Phys.* **2018**, *14*, 842–847. [CrossRef]
5. Accardi, L. Dynamical detailed balance and local kms condition for non-equilibrium states. *Int. J. Mod. Phys. B* **2002**, *18*, 435–467. [CrossRef]
6. Jensen, H. Complexity, collective effects, and modeling of ecosystems: Formation, function, and stability. *Ann. N. Y. Acad. Sci.* **2007**, *1195*. [CrossRef]
7. Chave, J. Scale and Scaling in Ecological and Economic Systems. *Environ. Resour. Econ.* **2003**, *26*, 527–557. [CrossRef]
8. Clark, J.; Kiwi, M.; Torres, F.; Rogan, J.; Valdivia, J.A. Generalization of the Ehrenfest urn model to a complex network. *Phys. Rev. E* **2015**, *92*, 012103. [CrossRef]
9. Capitelli, M.; Armenise, I.; Bruno, D.; Cacciatore, M.; Celiberto, R.; Colonna, G.; Pascale, O.D.; Diomede, P.; Esposito, F.; Gorse, C.; et al. Non-equilibrium plasma kinetics: A state-to-state approach. *Plasma Sources Sci. Technol.* **2007**, *16*, S30. [CrossRef]
10. Jaynes, E.T. Information Theory and Statistical Mechanics. *Phys. Rev.* **1957**, *106*, 620–630. [CrossRef]
11. Jaynes, E.T. Information Theory and Statistical Mechanics. II. *Phys. Rev.* **1957**, *108*, 171–190. [CrossRef]
12. Belkin, A.; Hubler, A.; Bezryadin, A. Self-Assembled Wiggling Nano-Structures and the Principle of Maximum Entropy Production. *Sci. Rep.* **2015**, *5*, 8323. [CrossRef] [PubMed]
13. Auletta, G.; Rondoni, L.; Vulpiani, A. On the relevance of the maximum entropy principle in non-equilibrium statistical mechanics. *Eur. Phys. J. Spec. Top.* **2017**, *226*, 2327–2343. [CrossRef]
14. Dewar, R. Information theory explanation of the fluctuation theorem, maximum entropy production and self-organized criticality in non-equilibrium stationary states. *J. Phys. Math. Gen.* **2003**, *36*, 631. [CrossRef]
15. Pressé, S.; Ghosh, K.; Lee, J.; Dill, K.A. Principles of maximum entropy and maximum caliber in statistical physics. *Rev. Mod. Phys.* **2013**, *85*, 1115–1141. [CrossRef]
16. Ourabah, K.; Gougam, L.A.; Tribeche, M. Nonthermal and suprathermal distributions as a consequence of superstatistics. *Phys. Rev. E* **2015**, *91*, 12133. [CrossRef]
17. Ourabah, K. Demystifying the success of empirical distributions in space plasmas. *Phys. Rev. Res.* **2020**, *2*, 23121. [CrossRef]
18. Davis, S.; Gutierrez, G. Conjugate variables in continuous maximum-entropy inference. *Phys. Rev. E* **2012**, *86*, 051136. [CrossRef]
19. Ichimaru, S. *Statistical Plasma Physics, Volume I: Basic Principles*; Frontiers in Physics; Avalon Publishing: New York, NY, USA, 2004.
20. Tsallis, C. *Introduction to Nonextensive Statistical Mechanics: Approaching a Complex World*; Springer: Cham, Switzerland, 2009; Volume 1.
21. Beck, C.; Cohen, E.G. Superstatistics. *Phys. A Stat. Mech. Its Appl.* **2003**, *322*, 267–275. [CrossRef]
22. Viñas, A.F.; Moya, P.S.; Navarro, R.; Araneda, J.A. The role of higher-order modes on the electromagnetic whistler-cyclotron wave fluctuations of thermal and non-thermal plasmas. *Phys. Plasmas* **2014**, *21*, 012902. [CrossRef]
23. Livadiotis, G. Derivation of the entropic formula for the statistical mechanics of space plasmas. *Nonlinear Process. Geophys.* **2017**, *25*, 77–88. [CrossRef]
24. Yoon, P. Non-equilibrium statistical mechanical approach to the formation of non-Maxwellian electron distribution in space. *Eur. Phys. J. Spec. Top.* **2020**, *229*, 819–840. [CrossRef]
25. Moya, P.S.; Lazar, M.; Poedts, S. Toward a general quasi-linear approach for the instabilities of bi-Kappa plasmas. Whistler instability. *Plasma Phys. Control. Fusion* **2021**, *63*, 25011. [CrossRef]
26. Sánchez, E.; González-Navarrete, M.; Caamaño-Carrillo, C. Bivariate superstatistics: An application to statistical plasma physics. *Eur. Phys. J. B* **2021**, *94*, 55. [CrossRef]
27. Beck, C. Superstatistics: Theory and applications. *Contin. Mech. Thermodyn.* **2004**, *16*, 293–304. [CrossRef]
28. Davis, S.; Avaria, G.; Bora, B.; Jain, J.; Moreno, J.; Pavez, C.; Soto, L. Single-particle velocity distributions of collisionless, steady-state plasmas must follow superstatistics. *Phys. Rev. E* **2019**, *100*, 023205. [CrossRef]
29. Livadiotis, G.; McComas, D.J. Invariant kappa distribution in space plasmas out of equilibrium. *Astrophys. J.* **2011**, *741*, 88. [CrossRef]

30. Sattin, F. Derivation of Tsallis statistics from dynamical equations for a granular gas. *J. Phys. Math. Gen.* **2003**, *36*, 1583–1591. [CrossRef]
31. Beck, C. Dynamical Foundations of Nonextensive Statistical Mechanics. *Phys. Rev. Lett.* **2001**, *87*, 180601. [CrossRef]
32. Bellan, P.M. *Fundamentals of Plasma Physics*; Cambridge University Press: Cambridge, UK, 2008.
33. Soto, R. *Kinetic Theory and Transport Phenomena*; Oxford University Press: Oxford, UK, 2016; Volume 25.
34. Birdsall, C.K.; Langdon, A.B. *Plasma Physics via Computer Simulation*; CRC press: Boca Raton, FL, USA, 2004.
35. González, D.; Tamburrini, A.; Davis, S.; Jain, J.; Gutiérrez, G. Expectation values of general observables in the Vlasov formalism. In *Proceedings of the Journal of Physics: Conference Series, Proceedings of the XX Chilean Physics Symposium, Santiago, Chile, 30 November–2 December 2016*; IOP Publishing: Bristol, UK, 2018; Volume 1043, p. 012008.
36. Kubo, R. The fluctuation-dissipation theorem. *Rep. Prog. Phys.* **1966**, *29*, 255. [CrossRef]
37. Chen, F.F. *Introduction to Plasma Physics and Controlled Fusion*; Springer: Cham, Switzerland, 1984; Volume 1.
38. Kulinskii, V.L.; Glavatskiy, K.S. Thermodynamics without ergodicity. *arXiv* **2018**, arXiv:1811.04591.
39. Schmidt, M. Power functional theory for Brownian dynamics. *J. Chem. Phys.* **2013**, *138*, 214101. [CrossRef] [PubMed]
40. Hastie, R. Adiabatic invariants and the equilibrium of magnetically trapped particles. *Ann. Phys.* **1967**, *41*, 302–338. [CrossRef]
41. Bhattacharjee, A.; Schreiber, G.M.; Taylor, J.B. Geometric phase, rotational transforms, and adiabatic invariants in toroidal magnetic fields. *Phys. Fluids B Plasma Phys.* **1992**, *4*, 2737–2739.
42. Zhao, H.; Baker, D.N.; Jaynes, A.N.; Li, X.; Elkington, S.R.; Kanekal, S.G.; Spence, H.E.; Boyd, A.J.; Huang, C.L.; Forsyth, C. On the relation between radiation belt electrons and solar wind parameters/geomagnetic indices: Dependence on the first adiabatic invariant and L*. *J. Geophys. Res. Space Phys.* **2017**, *122*, 1624–1642. [CrossRef]
43. Subbotin, D.A.; Shprits, Y.Y. Three-dimensional radiation belt simulations in terms of adiabatic invariants using a single numerical grid. *J. Geophys. Res. Space Phys.* **2012**, *117*. [CrossRef]
44. Kivelson, M.G.; Russell, C.T. *Introduction to Space Physics*; Cambridge university press: Cambridge, UK, 1995.
45. Marchand, R. Test-particle simulation of space plasmas. *Commun. Comput. Phys.* **2010**, *8*, 471. [CrossRef]
46. Büchner, J. *Space and Astrophysical Plasma Simulation*; Springer: Cham, Switzerland, 2023.
47. Matsumoto, H.; Omura, Y. International School for Space Simulations. In *Computer Space Plasma Physics: Simulation Techniques and Software*; Terra Scientific Pub. Co.: Tokyo, Japan, 1993.
48. Qin, H.; Zhang, S.; Xiao, J.; Liu, J.; Sun, Y.; Tang, W.M. Why is Boris algorithm so good? *Phys. Plasmas* **2013**, *20*, 084503. [CrossRef]
49. Gray, C.G.; Gubbins, K.E. *Theory of Molecular Fluids: Fundamentals*; Oxford University Press: Oxford, UK, 1984.
50. Rugh, H.H. Dynamical Approach to Temperature. *Phys. Rev. Lett.* **1997**, *78*, 772–774. [CrossRef]
51. Rickayzen, G.; Powles, J.G. Temperature in the classical microcanonical ensemble. *J. Chem. Phys.* **2001**, *114*, 4333–4334. [CrossRef]

entropy

MDPI

Article

Non-Equilibrium Wigner Function and Application to Model of Catalyzed Polymerization

Ramon F. Alvarez-Estrada

Departamento de Física Teórica, Universidad Complutense de Madrid, 28040 Madrid, Spain; rfa@ucm.es

Abstract: The quantum Wigner function and non-equilibrium equation for a microscopic particle in one spatial dimension ($1D$) subject to a potential and a heat bath at thermal equilibrium are considered by non-trivially extending a previous analysis. The non-equilibrium equation yields a general hierarchy for suitable non-equilibrium moments. A new non-trivial solution of the hierarchy combining the continued fractions and infinite series thereof is obtained and analyzed. In a short thermal wavelength regime (keeping quantum features adequate for chemical reactions), the hierarchy is approximated by a three-term one. For long times, in turn, the three-term hierarchy is replaced by a Smoluchovski equation. By extending that $1D$ analysis, a new model of the growth (polymerization) of a molecular chain (template or *te*) by binding an individual unit (an atom) and activation by a catalyst is developed in three spatial dimensions ($3D$). The atom, *te*, and catalyst move randomly as solutions in a fluid at rest in thermal equilibrium. Classical statistical mechanics describe the *te* and catalyst approximately. Atoms and bindings are treated quantum-mechanically. A mixed non-equilibrium quantum–classical Wigner–Liouville function and dynamical equations for the atom and for the *te* and catalyst, respectively, are employed. By integrating over the degrees of freedom of *te* and with the catalyst assumed to be near equilibrium, an approximate Smoluchowski equation is obtained for the unit. The mean first passage time (MFPT) for the atom to become bound to the *te*, facilitated by the catalyst, is considered. The resulting MFPT is consistent with the Arrhenius formula for rate constants in chemical reactions.

Keywords: non-equilibrium Wigner function and hierarchy for moments; short thermal wavelength and long-time regimes; approximate Smoluchovski equation; catalyzed polymerization

1. Introduction

Non-equilibrium quantum statistical mechanics has its own scientific importance [1–17], and its applications make it even more important. Part of the latter, of paramount relevance, are chemical reactions [18–23]: in them, typically, atoms/molecules in an initial state are not the same as those in the final one, while on the other hand, those processes occur in the presence or inside a statistical medium (for instance, a fluid). Here, chemical reactions can be understood in a broad sense and include, namely, biochemical processes [23]. The very formation and breaking up of bound states with discretized binding energies in chemical reactions unavoidably requires quantum mechanics and statistical mechanical concepts, even if other features in those processes can be accounted for in terms of classical physics for that purpose. For further knowledge regarding thermodynamics and statistical mechanics, see, for instance [24–28].

Before proceeding further, it seems adequate to remind the reader of part of the previous work that has motivated the present one. For that purpose, see [29,30] and references therein.

The non-equilibrium dynamics of a closed classical gas composed of a large number of identical non-relativistic particles was described by the classical Liouville distribution f and equation, which depend on the time t and on the positions and momenta of all particles with some suitable initial condition at $t = 0$. It was assumed that only the inner part of the gas was off-equilibrium while the remainder of it was at equilibrium at absolute temperature T. The standard Boltzmann equilibrium distribution f_{eq} with temperature T, being Gaussian in all momenta, generated a family of orthogonal polynomials in the latter: the standard Hermite polynomials. The latter, upon integrating on the momenta, enabled the introduction of non-equilibrium moments (depending on t and particle positions). The Liouville equation gave rise to an infinite three-term hierarchy for those classical moments. The hierarchy was entirely different from the standard Bogoliubov–Born–Green–Kirkwood–Yvon (BBGKY) hierarchy [1,3,4]. The hierarchy was solved in terms of suitable infinite continued fractions of operators. Those continued fractions suggested and allowed for the implementation of a long-time approximation. After further approximations, the lowest non-equilibrium moment was shown to satisfy a Smoluchovsky equation, which is formally similar to the one characterizing the so-called Rouse model in polymer dynamics [31]. An approximate approach to thermal equilibrium over a long time can then be established.

Attempts at extensions of those procedures to quantum processes faced two difficulties. First, it was not warranted that, contrary to the classical situation, quantum–mechanical distributions be non-negative in certain limited spatial domains. This, in turn, would, in general, prevent a direct use of the corresponding equilibrium distribution to generate, in the standard way, a family of orthogonal polynomials in momenta and, so, non-equilibrium moments. Such difficulty was bypassed by invoking a suitable generalization of the theory of orthogonal polynomials, as will be illustrated later in Section 2.3 of this work. The second difficulty was (and still continues partly to be) that, after having implemented that solution of the first difficulty, the resulting hierarchy for the non-equilibrium moments is not, in general, a three-term one: this demands proper analysis.

The present work presents: (1) a new study of non-equilibrium quantum statistical mechanics aimed at providing a new (at least, partial) solution to the second difficulty (Section 2) and (2) one possible application to a certain chemical reaction: namely, the one playing a key role in the polymerization of a molecular chain [32] (Sections 3 and 4). In so doing, previous analysis will be generalized [30,33] in a non-trivial way.

Section 2 presents: a one-dimensional ($1D$) non-equilibrium quantum Wigner function (W) and dynamical equation, the introduction of a family of orthogonal polynomials generated by the equilibrium Wigner distribution (W_{eq}), a general n-term recurrence relation for non-equilibrium moments of the corresponding W (determined, in turn, by those orthogonal polynomials), a formal solution by combining continued operator fractions and series expansions thereof, an approximate three-term recurrence relation for short thermal wavelengths (still in the quantum regime), long-time approximations, and an irreversible Smoluchowski equation.

Section 3 treats a three-dimensional ($3D$) model for the addition (polymerization) of one single atom to a freely jointed molecular chain acting as a template (*te*) activated by a catalyst in a fluid at equilibrium at a given temperature.

Section 4 studies the model in Section 3 through $3D$ non-equilibrium quantum–classical Wigner–Liouville functions, a dynamical equation for the individual atom, the *te*, and the catalyst, and successive approximations (assuming that the *te* and catalyst are at thermal equilibrium) so as to yield a standard $3D$ non-equilibrium Wigner equation. At this stage, the $1D$ developments in Section 2 (short thermal wavelength and long-time approximations in Section 2) are extended to the above standard $3D$ equation, thereby yielding a Smoluchowski equation for the single atom, and a mean first passage time (MFPT) equation for the atom to become attached to the *te* chain.

Successive approximations are made for conditions suitable for chemical reactions.

Section 5 offers the conclusions and some discussions.

This work is a contribution to "180th Anniversary of Ludwig Boltzmann", a Special Issue of Entropy.

2. One-Dimensional Wigner Function and Equation

2.1. General Aspects

We consider a simplified model of a microscopic non-relativistic particle of mass m in one spatial dimension (x) in a finite interval Ω ($= (-L/2, +L/2)$), which is large on the microscopic scale although possibly small on the macroscopic one, by omitting unnecessary details. The particle is subject to a real (time-independent and velocity-independent) potential $V = V(x)$ in $-L/2 < x < L/2$ and to a heat bath (HB) at absolute temperature T. The quantum particle Hamiltonian is $H = -(\hbar^2/2m)(\partial^2/\partial x^2) + V$, with $\hbar$ being Planck's constant. For definiteness, the wave functions may be assumed to fulfill Dirichlet boundary conditions in $\mid x \mid= L/2$.

The conditions on the potential $V = V(x)$ are:

$V(x) = V(-x)$ (for simplicity, although strictly unnecessary).
$V(x)$ is attractive (< 0) in the interval $-a < x < a$ ($0 < a < L/2$); it is very small (with an arbitrary sign) in $a <\mid x \mid< L/2$ and vanishes fast as $\mid x \mid\to L/2$. Specifically, $V(x)$ is finite everywhere, and its magnitude $|V|$ is appreciable only in the limited interval $-a < x < a$; a is understood to be the range of V.
$V(x)$ and all $d^n V(x)/dx^n$ for $n = 1, 2, 3, \ldots$ are continuous everywhere.

In general, H has both a denumerably infinite number of discrete states with non-negative energy (an almost continuous spectrum: L being large but finite at the microscopic scale) and, in principle, a finite discrete spectrum ($j =$ d) with energies $E_d < 0$. With j being a general label, let $\varphi_j = \varphi_j(x)$ generically denote a suitably normalized eigenfunction of H with corresponding eigenvalue E_j.

As a simplifying assumption, $V(x)$ does give rise to only one bound state (bound spectrum).

The denumerably infinite discrete (almost continuous) spectrum E_j of H has a small spacing, approaches a continuous spectrum more the larger L is, and would become a continuous one (sweeping the continuous positive real axis $0 < E_j < +\infty$) if $L^{-1} \to 0$. We shall always denote it by CS, even if the small L^{-1} remains > 0. The eigenfunctions corresponding to CS are $\varphi_j = \varphi_k(x)$, with $j \equiv k$ being an almost continuous wavevector and eigenvalues $E_j = E_k = \hbar^2 k^2/(2m) \geq 0$. The CS eigenfunctions are normalized through: $(\varphi_k, \varphi_{k'}) = \int dx \varphi_k^* \varphi_{k'} = \delta_{k,k'}$ (a Kronecker delta). Also, by taking into account the bound state: $(\varphi_{\mathrm{d}}, \varphi_{\mathrm{d}}) = \int dx \varphi_{\mathrm{d}}^* \varphi_d = 1$ (normalized) and $(\varphi_{\mathrm{d}}, \varphi_k) = \int dx \varphi_{\mathrm{d}}^* \varphi_k = 0$. Integrations are performed over Ω. Hence, φ_{d} and all CS φ_k span two separate Hilbert subspaces $\mathcal{H}_{\mathrm{d}}$ and $\mathcal{H}_{\mathrm{CS}}$. See [34–36].

The time (t) evolution of the quantum particle is given by the general density operator $\rho = \rho(t)$ (a statistical mixture of quantum states) for $t > 0$. It fulfills the (t-reversible) operator equation $\partial\rho/\partial t = (i\hbar)^{-1}[H, \rho]$ ($[H, \rho] = H\rho - \rho H$ being the commutator) with initial condition $\rho(t = 0) = \rho_{\mathrm{in}}$. The Hermitian and positive-definite linear operators $\rho(t)$ and ρ_{in} act on the Hilbert space spanned by the set of all eigenfunctions of H. See [34].

Let $\beta = (k_{\mathrm{B}} T)^{-1}$, with k_{B} being Boltzmann's constant. We shall introduce the fixed and physically relevant x-independent momentum and thermal wavelength:

$$q_{\mathrm{eq}} = (2m/\beta)^{1/2}, \lambda_{th} = \hbar/q_{\mathrm{eq}} \tag{1}$$

We now resort to the non-equilibrium (reversible) Wigner function [1,2,4,6–8,37] at t, which reads formally:

$$W(x, q; t) = \frac{1}{\pi\hbar} \int dx' \exp\left[\frac{2i}{\hbar} x' q\right] \langle x - x' | \rho(t) | x + x' \rangle, \tag{2}$$

The initial Wigner function W_{in} is given by Equation (2) by using $\rho(t=0)$. The equilibrium density operator $\rho_{\text{eq}} = \exp[-\beta H]$ determines the equilibrium Wigner function $W_{\text{eq}}(x,q)$ formally as:

$$W_{\text{eq}}(x,q) = \frac{1}{\pi\hbar}\int dx' \exp\left(\frac{i2qx'}{\hbar}\right)\langle x - x'|\rho_{\text{eq}}|x + x'\rangle. \tag{3}$$

A more specific analysis is required since Ω is not strictly an infinite interval. However, as L is large, we shall approximate spatial integrals as those for an infinite length interval when such approximations are harmless unless some specific discussion is required. Strictly speaking, as Ω is not an infinite interval, the q' are discretized momenta and $\int dq'$ (which will appear later) should be interpreted as a series. However, as L is large, we shall disregard the small spacings in q' and understand $\int dq'$, as the notation indicates, as an integral (q' thus varying continuously) instead of as a sum. This remark and interpretation will apply and be understood whenever integrations over momenta occur later. We shall accept that all integrals (or all series) over momenta converge for large values of the latter: explicit expressions and computations will support this assumption. Summarizing: as L is large and unless otherwise stated, we shall approximate using practical calculations (without writing it explicitly) spatial integrals by those in $-\infty < L < +\infty$ and series over momenta by integrations over them in $-\infty < q < +\infty$.

Integrability properties of the Wigner function hold and will not be considered here for brevity: they have been treated previously in [29,30] and, in particular, in references therein.

In principle, $\sum_j$ will denote sums over all eigenfunctions in $j = \text{d}$ and $j = \text{CS}$; that is, the former includes the contribution of both the single (bound state) discrete eigenfunction plus that of an infinite summation over the whole CS ones. Since L is large, we approximate for the CS: $\sum_j \to (L/(2\pi))\int dk$ as $L^{-1} \to 0$. Therefore, with the latter understanding for a large-interval Ω, W_{eq} is approximated as:

$$W_{\text{eq}}(x,q) = \frac{1}{(\pi\hbar)}\int dx' \exp\left(\frac{i2qx}{\hbar}\right)\sum_j \exp[-\beta E_j]\varphi_j(x-x')\varphi_j^*(x+x'). \tag{4}$$

While at $T=0$ there is no transition between the bound state and the CS ones; such a transition is indeed possible for $T>0$ (due to the HB) and can play a key role as $k_{\text{B}}T$ approaches $|E_d|$. We emphasize that the contribution of the bound state becomes more negligible the higher the temperature is (quasi-classical regime) and even disappears in the full classical regime. On the other hand, there seems to be no compelling reason for not using the quasi-classical or even classical formula as rough or zeroth-order approximations in regions where the contributions due to the bound states are negligible.

For $t>0$, the exact (t-reversible) dissipationless quantum master equation for the general off-equilibrium Wigner function W [6,37] is:

$$\frac{\partial W(x,q;t)}{\partial t} = -\frac{q}{m}\frac{\partial W(x,q;t)}{\partial x} + M_Q W, \tag{5}$$

$$\begin{aligned} M_Q W &= \int dq' W(x,q';t)\int \frac{idx'}{\pi\hbar^2}[V(x+x') - V(x-x')]\exp\left[\frac{i2(q-q')x'}{\hbar}\right] \\ &= \frac{dV}{dx}\frac{\partial W}{\partial q} - \frac{\hbar^2}{3!2^2}\frac{d^3V}{dx^3}\frac{\partial^3 W}{\partial q^3} + \frac{\hbar^4}{5!2^4}\frac{d^5V}{dx^5}\frac{\partial^5 W}{\partial q^5} - \cdots, \end{aligned} \tag{6}$$

with initial condition W_{in}. W is real: this can be directly established through either Equation (2) or Equation (6) by taking complex conjugates and changing $x' \to x'' = -x'$.

If $\hbar \to 0$, then the above equation becomes the classical Liouville equation [1,4,6].

$$\frac{\partial W(x,q;t)}{\partial t} = -\frac{q}{m}\frac{\partial W(x,q;t)}{\partial x} + \frac{dV}{dx}\frac{\partial W}{\partial q} \tag{7}$$

2.2. Equilibrium Wigner Function near the Classical Limit

Recall that in the classical case, the equilibrium (or Boltzmann's) canonical distribution describing the thermal equilibrium of a classical particle with an HB is proportional to a Gaussian in q: that is, $\exp\left[-\beta q^2/(2m)\right]$, with q now being the classical momentum. In the high temperature or small β (quasiclassical) regime, Wigner [6] obtained successive approximations for $W_{\rm eq}(x,q)$ as a power series in $\hbar$. Equations (22) and (25) in [6] (the leading terms in that expansion for $W_{\rm eq}(x,q)$) are directly recast as:

$$W_{\rm eq}(x,q) = c_0 f_{\rm eq}[1 + c_1 + c_2\frac{\partial^2 V}{\partial x^2}H_2(q/q_{eq})] \tag{8}$$

with

$$f_{\rm eq} = \exp[-(q^2/q_{eq}^2 + \beta V)] \tag{9}$$

$$c_1 = \hbar^2[-\frac{\beta^2}{12m}\frac{\partial^2 V}{\partial x^2} + \frac{\beta^3}{24m}(\frac{\partial V}{\partial x})^2] \tag{10}$$

$$c_2 = \frac{(\beta\hbar)^2}{48m} \tag{11}$$

where $c_0 = 2\pi\hbar$. H_2 denotes the standard Hermite polynomial of second order [38]. We shall refer to the contributions associated with c_1 and c_2 as quantum contributions in the quasiclassical regime. Actually, Wigner's series in [6], after factoring out $c_0 f_{\rm eq}$, turns out to also be a power series in β and q so that the contributions in his Equation (25) contain only up to quadratic powers in q. The latter have been rewritten in terms of H_2 in (8) in order to facilitate an eventual comparison with $W_{\rm eq}$ if desired.

In the classical case ($\hbar = 0$) with attractive V in some domain, $f_{\rm eq}$ contains the factor $\exp[-\beta V]$, which is > 1 and can be certainly large in the regions where $V < 0$. Recall that in the classical case, in the latter domains there are no bound states but a continuum of classical orbits. At this point, one may wonder how such a continuum of classical orbits can become a discrete set of bound states if one resorts back to the quantum (statistical) regime of interest here. The above quasiclassical expressions do provide some hint for that. In fact, for $W_{\rm eq}$ as given in by the small β (or, formally, small $\hbar$) Equation (8), by integrating over y (with $y = q/q_{eq}$) and doing some algebra, one has that $\int dy W_{eq}$ tends to:

$$c_0\pi^{1/2}\exp[-\beta V].\left[1 - \hbar^2\frac{\beta^2}{12m}\frac{\partial^2 V}{\partial x^2} + \frac{\hbar^2\beta^3}{24m}(\frac{\partial V}{\partial x})^2\right] \tag{12}$$

We state that the crucial term $-\hbar^2\frac{\beta^2}{12m}\frac{\partial^2 V}{\partial x^2}$ is a qualitative signal of how a continuum of classical orbits can become a discrete set of bound states even if it does not provide a quantitative mechanism. In fact, the attractive V, being finite and having minima (with values < 0) in certain domains, have $\frac{\partial^2 V}{\partial x^2} > 0$ at those minima. Then, the leading correction $-\hbar^2\frac{\beta^2}{12m}\frac{\partial^2 V}{\partial x^2}$ tends to reduce the importance of $V(< 0)$ in those domains and contributes to restrict a continuum of classical bounded orbits to a discrete set of quantum bound states. The remaining term $\frac{\hbar^2\beta^3}{24m}(\frac{\partial V}{\partial x})^2$ (> 0) would tend to reinforce $\exp[-\beta V]$ (with $-V(> 0)$) but at a smaller amount due to the additional factor β and due to the fact that it is small and vanishes at the minimum ($\frac{\partial V}{\partial x} = 0$). We shall regard this qualitative behavior as an indication that even the lowest quantum corrections to the classical $f_{\rm re,eq}$ contain certain, even if small and partial, signals of quantum effects.

2.3. W_{eq} as a Quasi-Definite Functional of Momenta

Notice that neither $W_{\rm eq}$ nor W can be warranted to be nonnegative in general [8,37,39,40]. However, by invoking a suitable extension [41] of the theory of orthogonal polynomials, it is natural: (i) to accept that $W_{\rm eq}$ is a quasi-definite functional of momenta (as one can justify through examples in certain cases: see references in [29]) and (ii) to invoke that $W_{\rm eq}$ can be regarded as a generating function to construct recurrently an infinite family of orthogonal polynomials in q.

Let $y = q/q_{eq}$. We introduce (unnormalized) orthogonal polynomials $H_{Q,n} = H_{Q,n}(y)$ in y determined by W_{eq}, which acts as a (in general, non-Gaussian) weight function. The term n is a non-negative integer. $H_{Q,n}(y)$ also depends parametrically on x, although such a dependence is not displayed explicitly. Under the assumptions about V in Section 2.1, $W_{\rm eq}$ is an even function of q: this is immediately confirmed, in particular just by looking at the classical and quasiclassical expressions for it given in Section 3.2. Then, $H_{Q,n}$ is even or odd in y for even or odd n, respectively. We choose $H_{Q,0} = H_{Q,0}(y) = 1$. We also choose: $H_{Q,1} = H_{Q,1}(y) = y$, $H_{Q,2} = H_{Q,2}(y) = y^2 + \epsilon_{2,0}$, $H_{Q,3} = H_{Q,3}(y) = y^3 + \epsilon_{3,1}y$, and so on. In general, the $H_{Q,n}$ terms are constructed recurrently as follows. We impose for $n \neq n'$ and any x (left unintegrated) that

$$\int dy W_{\rm eq}(x,y) H_{\rm Q,n}(y) H_{\rm Q,n'}(y) = 0, \tag{13}$$

where

$$H_{Q,n}(y) = y^n + \sum_j \epsilon_{n,n-j} y^{n-j} + \cdots \tag{14}$$

with $0 \le j \le n$ and $n - j = 2, 4, \ldots$. Here, $\epsilon_{n,n-j}$ are dimensionless and y-independent (though x-dependent, in general). One has $\epsilon_{n,n-j} = 0$ if j is odd so that $H_{Q,n}(-y) = (-1)H_{Q,n}(y)$.

The nonvanishing $\epsilon_{n,n-2}$ for low-order $n = 2,3,4,5$ (j even) are

$$\epsilon_{2,0} = -\langle y^2 \rangle, \quad \epsilon_{3,1} = -\frac{\langle y^4 \rangle}{\langle y^2 \rangle}, \tag{15}$$

$$\epsilon_{4,2} = \frac{\langle y^2 \rangle \langle y^4 \rangle - \langle y^6 \rangle}{\langle y^4 \rangle - \langle y^2 \rangle^2}, \quad \epsilon_{4,0} = \frac{\langle y^2 \rangle \langle y^6 \rangle - \langle y^4 \rangle^2}{\langle y^4 \rangle - \langle y^2 \rangle^2} \tag{16}$$

$$\epsilon_{5,3} = \frac{\langle y^4 \rangle \langle y^6 \rangle - \langle y^2 \rangle \langle y^8 \rangle}{\langle y^2 \rangle \langle y^6 \rangle - \langle y^4 \rangle^2}, \tag{17}$$

$$\langle y^n \rangle = \frac{\int dy W_{\rm eq}(x,q) y^n}{\int dy W_{\rm eq}(x,q)}. \tag{18}$$

They fulfill the following exact quantum identities:

$$(\epsilon_{3,1} - \epsilon_{4,2})\epsilon_{2,0} + \epsilon_{4,0} = 0 \tag{19}$$

$$-(\epsilon_{4,2} - \epsilon_{5,3})\epsilon_{3,1} + \epsilon_{4,0} - \epsilon_{5,1} = 0 \tag{20}$$

$$\epsilon_{5,1} = \frac{\langle y^4 \rangle \langle y^8 \rangle - \langle y^6 \rangle^2}{\langle y^2 \rangle \langle y^6 \rangle - \langle y^4 \rangle^2} \tag{21}$$

There is an infinite number of identities among higher ϵ values, which lie outside our scope here. In the strict quantum regime, $H_{Q,n}(y)$ are different from the standard Hermite polynomials. The procedure for successively constructing $H_{Q,n}(y)$ based upon (13) and (14) becomes increasingly cumbersome as n increases, even if it is conceptually straightforward: notice that possible recurrence relations among $H_{Q,n}(y)$ are still lacking so far. In the classical limit ($\hbar \to 0$), with $W_{\rm eq}(x,q)$ approximated by the classical Boltzmann distribution $c_0 f_{\rm eq}$, the orthogonal polynomials $H_{Q,n}(y)$ are equal to $2^{-n} H_n(y)$, with $H_n(y)$ being the standard Hermite polynomials [38]. Then, for the latter, the computation of all coefficients in the (classical) counterpart of (14) boils down to that of $\langle y^n \rangle$ and the latter to that of Gaussian integrals. In the classical limit, one finds:

$$\epsilon_{2,0} = -\frac{1}{2}, \epsilon_{3,1} = -\frac{3}{2}, \epsilon_{4,0} = \frac{3}{4}, \epsilon_{4,2} = -3, \epsilon_{5,3} = -5, \epsilon_{5,1} = \frac{15}{4} \tag{22}$$

which are x-independent. The term $\epsilon_{2,0}$ in the quantum regime will play an important role: it will be studied in Section 2.7.

2.4. Non-Equilibrium Moments and Hierarchy

We shall proceed to the non-equilibrium Equations (5) and (6). We shall use the (unnormalized) polynomials in y $(= q/q_{\text{eq}})$ $H_{Q,n} = H_{Q,n}(y)$ $(n = 0, 1, 2, 3, \ldots)$ orthogonalized in y (for fixed x) by using the equilibrium distribution W_{eq} as a weight function. The actual $H_{Q,n}(y)$ lead to defining the non-equilibrium moments $(n = 0, 1, 2, \ldots)$:

$$W_n = W_n(x;t) = \int dy H_{Q,n}(y) W \tag{23}$$

The initial moments $W_{\text{in},n}$ for W_n are obtained by replacing W with W_{in} in Equation (23). The transformation of the one-dimensional Equations (5) and (6) into a linear hierarchy for the non-equilibrium moments W_n will play an important role in this work. It can be carried out through direct computations and cancellations employing Equations (13) and (23), which are increasingly cumbersome as n increases. The general (t-reversible) hierarchy implied by Equations (5) and (6) for any n reads:

$$\frac{\partial W_n}{\partial t} = -M_{n,n+1} W_{n+1} - \sum_{n'=1}^{n} M_{n,n-n'} W_{n-n'} \tag{24}$$

The M's are t-independent operator coefficients, which are characterized below. In general, the quantum hierarchy is not a three-term one. The following operator coefficients for any n will play a key role:

$$M_{n,n+1} W_{n+1} \equiv \frac{q_{\text{eq}}}{m} \frac{\partial W_{n+1}}{\partial x} \tag{25}$$

$$M_{n,n-1} W_{n-1} = -\frac{q_{\text{eq}}}{m} \left[(\epsilon_{n+1,n-1} - \epsilon_{n,n-2}) \frac{\partial W_{n-1}}{\partial x} \right] + \frac{n}{q_{\text{eq}}} \frac{\partial V}{\partial x} W_{n-1} - \frac{q_{\text{eq}}}{m} \frac{\partial \epsilon_{n,n-2}}{\partial x} W_{n-1}. \tag{26}$$

Thus, in the exact non-equilibrium quantum hierarchy (24), the contributions from W_{n+1} always have the same structure $(-(q_{\text{eq}}/m)\partial W_{n+1}/\partial x)$ for any n. Then $M_{n,n+1}$ is n-independent.

It is very convenient to perform a Laplace transform of the general hierarchy (24) so as to replace each W_n with its Laplace transform $\tilde{W}_n = \tilde{W}_n(s) = \int_0^{+\infty} dt W_n(t) \exp(-st)$.

Then, the above hierarchy (24) becomes the Laplace-transformed quantum hierarchy:

$$s\tilde{W}_n = W_{\text{in},n} - M_{n,n+1} \tilde{W}_{n+1} - \sum_{n'=1}^{n} M_{n,n-n'} \tilde{W}_{n-n'} \tag{27}$$

with the same t-independent operators (the M's) as in (24). Notice that $M_{n,n'=0} = 0$ for any n except for $n = 1$, and $M_{n,n-n'} = 0$ if $n - n'$ is even. For simplicity and without any essential loss of generality, we shall assume $W_{\text{in},n} = 0$ for $, n > 0$ while only $W_{\text{in},n=0} \neq 0$.

In detail, the lowest equations in the Laplace-transformed quantum hierarchy (27) are:

$$s\tilde{W}_0 = W_{\text{in},0} - M_{0,1}\tilde{W}_1 \quad (28)$$
$$s\tilde{W}_1 = -M_{1,2}\tilde{W}_2 - M_{1,0}\tilde{W}_0 \quad (29)$$
$$s\tilde{W}_2 = -M_{2,3}\tilde{W}_3 - M_{2,1}\tilde{W}_1 \quad (30)$$
$$s\tilde{W}_3 = -M_{3,4}\tilde{W}_4 - M_{3,2}\tilde{W}_2 \quad (31)$$
$$s\tilde{W}_4 = -M_{4,5}\tilde{W}_5 - M_{4,3}\tilde{W}_3 - M_{4,1}\tilde{W}_1 \quad (32)$$
$$s\tilde{W}_5 = -M_{5,6}\tilde{W}_6 - M_{5,4}\tilde{W}_4 - M_{5,2}\tilde{W}_2 \quad (33)$$
$$s\tilde{W}_6 = -M_{6,7}\tilde{W}_7 - M_{6,5}\tilde{W}_5 - M_{6,3}\tilde{W}_3 - M_{6,1}\tilde{W}_1 \quad (34)$$

and so on for higher values of the integer n. A key feature of the non-equilibrium hierarchies (24) and (27) is that all operator coefficients (M) in the former are expressed in terms of V and of quantities computed out of the equilibrium Wigner function W_{eq}. The $M_{n,n'}$ for $n = 0, 1, 2, 3, 4$ are identified upon comparing Equation (27) and Equations (28)–(32). Specifically, the operator coefficients (M) in the above equations follow directly from:

$$s\tilde{W}_0 = W_{\text{in},0} - \frac{q_{\text{eq}}}{m}\frac{\partial \tilde{W}_1}{\partial x}, \quad (35)$$

$$s\tilde{W}_1 = -\frac{q_{\text{eq}}}{m}\frac{\partial \tilde{W}_2}{\partial x} + \frac{q_{eq}}{m}\frac{\partial}{\partial x}((\epsilon_{2,0})\tilde{W}_0) + \frac{1}{q_{\text{eq}}}\frac{dV}{dx}\tilde{W}_0, \quad (36)$$

$$s\tilde{W}_2 = -\frac{q_{\text{eq}}}{m}\frac{\partial \tilde{W}_3}{\partial x} + \frac{q_{\text{eq}}}{m}\frac{\partial}{\partial x}((\epsilon_{3,1} - \epsilon_{2,0})\tilde{W}_1) + \frac{q_{\text{eq}}}{m}\frac{d\epsilon_{2,0}}{dx}\tilde{W}_1 + \frac{2}{q_{\text{eq}}}\frac{dV}{dx}\tilde{W}_1, \quad (37)$$

$$s\tilde{W}_3 = -\frac{q_{\text{eq}}}{m}\frac{\partial \tilde{W}_4}{\partial x} + \frac{q_{\text{eq}}}{m}\frac{\partial}{\partial x}((\epsilon_{4,2} - \epsilon_{3,1})\tilde{W}_2) + \frac{q_{\text{eq}}}{m}\frac{d\epsilon_{3,1}}{dx}\tilde{W}_2 + \frac{3}{q_{\text{eq}}}\frac{dV}{dx}\tilde{W}_2, \quad (38)$$

$$s\tilde{W}_4 = -\frac{q_{\text{eq}}}{m}\frac{\partial \tilde{W}_5}{\partial x} + \frac{q_{\text{eq}}}{m}\frac{\partial}{\partial x}((\epsilon_{5,3} - \epsilon_{4,2})\tilde{W}_3) + \frac{q_{\text{eq}}}{m}\frac{d\epsilon_{4,2}}{dx}\tilde{W}_3 + \frac{4}{q_{\text{eq}}}\frac{dV}{dx}\tilde{W}_3 + \frac{\hbar^2}{2^2 q_{\text{eq}}^3}\frac{d^3V}{dx^3}\left(-6 + \frac{\epsilon_{4,2}}{\epsilon_{2,0}}\right)\tilde{W}_1 \quad (39)$$

The $\epsilon_{n,n-2}$ in the first four Equations (35)–(38) do contain inside them (even if not in an explicit way) quantum effects arising from W_{eq}.

The derivations of the successive equations in the non-equilibrium recurrence starting from the equation for $s\tilde{W}_3$ onward are increasingly difficult, as they involve exact cancellations in order to arrive at consistent equations.

Thus, Equation (38) for $s\tilde{W}_3$ makes use of the identity in (19) and it involves the exact cancellation of the contribution of $\tilde{W}_0$.

Also, Equation (39) for $s\tilde{W}_4$ employs the identity in (20) and involves other exact cancellations in such a way that the contribution of $\tilde{W}_1$ is explicitly proportional to the quantum correction $\frac{\hbar^2}{2^2 q_{\text{eq}}^3}\frac{d^3V}{dx^3}$ (of order $\hbar^2$) as it stands. We emphasize that such a contribution multiplying $\tilde{W}_1$ in Equation (39) for $s\tilde{W}_4$ has an explicit quantum origin. In that aspect, (39) does differ from (35)–(38). The reason for that difference is that the quantum corrections in Equations (5) and (6) manifest themselves explicitly only at order $\hbar^2$ and then, in turn, in the equations in the hierarchy at orders $n \geq 4$.

The very fact that the full quantum Equation (39) for $n = 4$ does contain a term of order $\hbar^2$ in $\tilde{W}_1$ confirms that the quantum hierarchy (24) is not a three-term hierarchy.

$\tilde{W}_{n'}$ ($0 < n' \leq n - 1$) do carry n-dependent coefficients, which increase with n. In particular (and leaving aside other contributions), the equation for $s\tilde{W}_5$ can be shown to contain in its right-hand-side $(\hbar^2/q_{\text{eq}}^3)(\partial^3 V/\partial x^3)\tilde{W}_2$ as the highest spatial derivative of V, while that for $s\tilde{W}_6$ contains $(\hbar^2/q_{\text{eq}}^3)(\partial^3 V/\partial x^3)\tilde{W}_3$ and $(\hbar^4/q_{\text{eq}}^5)(\partial^5 V/\partial x^5)\tilde{W}_1$ and so on.

We stress that the complicated structure of the general hierarchy (24) is a genuine consequence of quantum mechanics. If $\hbar \to 0$, the hierarchy for the classical limit of $\tilde{W}_n$ becomes the following three-term one for Equation (7):

$$\frac{\partial W_n}{\partial t} = -M_{n,n+1}W_{n+1} - M_{n,n-1}W_{n-1} \tag{40}$$

$$M_{n,n+1} \equiv \frac{q_{\text{eq}}}{m}\frac{\partial}{\partial x} \tag{41}$$

$$M_{n,n-1} = -\frac{q_{\text{eq}}}{m}\left[(\epsilon_{n+1,n-1} - \epsilon_{n,n-2})\frac{\partial}{\partial x}\right] + \frac{n}{q_{\text{eq}}}\frac{\partial V}{\partial x}. \tag{42}$$

with $M_{n,n-n'} = 0$ for $n' \neq 1$ and the same $M_{n,n+1}$. Specifically, one has:

$$M_{n,n-1} = \frac{q_{\text{eq}}n}{m}[\frac{1}{2}\frac{\partial}{\partial x} + \frac{m}{q_{\text{eq}}^2}\frac{\partial V}{\partial x}] \tag{43}$$

Notice the crucial simplification for $\hbar \to 0$: ϵ become independent on x.

2.5. Formal Solution of General Hierarchy

It is methodologically important and useful to obtain a formal solution of the general quantum hierarchy (24) that is not a three-term one (that is, without proceeding to the classical regime). For that purpose, one solves the hierarchy (24) for $n > 0$ recurrently. Then, all W_n for $n > 0$ can be expressed through suitable linear operators in terms of W_0 and of the initial condition $W_{\text{in},0}$ (assuming $W_{\text{in},n} = 0$ for $n > 0$). In particular, it will be important to get W_1 as a linear functional of W_0.

In practice, it is convenient to operate in terms of Laplace transforms by employing (27). The solution of the hierarchy proceeds specifically as follows: (i) for suitable $n_0 \geq 2$, in the equation of the hierarchy (27) yielding $\tilde{W}_{n_0}(s)$, one omits $\tilde{W}_{n_0+1}(s)$ and solves formally the resulting equation for $\tilde{W}_{n_0}(s)$ in terms of those $\tilde{W}_{n_0-n}(s)$ for $n > 0$ appearing in the right-hand-side of that equation; (ii) one proceeds to the equation of the hierarchy yielding $\tilde{W}_{n_0-1}(s)$, reshuffles into it the above expression for $\tilde{W}_{n_0}(s)$, and solves for $\tilde{W}_{n_0-1}(s)$; (iii) one proceeds by iteration to the equations for $\tilde{W}_{n_0-n}(s)$ for $n > 1$ and so on until one arrives at $\tilde{W}_1(s)$ in terms of $\tilde{W}_0(s)$. Then, one repeats the above procedure for $\tilde{W}_{n_0+1}(s)$ and infers by induction what the formal structure of the solution is as $n_0 \to +\infty$. One finds (s-dependences being understood for brevity as suitable):

$$\tilde{W}_1(s) = D_1(s)[-M_{1,0}]\tilde{W}_0(s). \tag{44}$$

$$D_1 = \frac{1}{sI - M_{1,2}D_2(M_{2,1} + M_{2,3}D_3M_{3,4}D_4M_{4,1} + M_{2,3}D_3M_{3,4}D_4M_{4,5}D_5M_{5,6}D_6M_{6,1} + \ldots)}, \tag{45}$$

$$D_2 = \frac{1}{sI - M_{2,3}D_3(M_{3,2} + M_{3,4}D_4M_{4,5}D_5M_{5,2} + M_{3,4}D_4M_{4,5}D_5M_{5,6}D_6M_{6,7}D_7M_{7,2} + \ldots)}, \tag{46}$$

$$D_3 = \frac{1}{sI - M_{3,4}D_4(M_{4,3} + M_{4,5}D_5M_{5,6}D_6M_{6,3} + M_{4,5}D_5M_{5,6}D_6M_{6,7}D_7M_{7,8}D_8M_{8,3} + \ldots)} \tag{47}$$

and so on. The D vales are continued-fraction linear operators. I is the unit operator. In a compact form:

$$D_n = D_n(s) - \frac{1}{sI - M_{n,n+1}D_{n+1}(M_{n+1,n} + N_{n+1,n})} \tag{48}$$

$$N_{n+1,n} = N_{n+1,n}(s) = M_{n+1,n+2}D_{n+2}M_{n+2,n+3}D_{n+3}M_{n+3.n} + \ldots \tag{49}$$

with the understanding that if the infinite continued fraction related to D_n in (48) is approximated by a finite continued fraction by cutting it off at some n_0, the infinite series for $N_{n+1,n}$ is approximated by a finite sum that is cut off at n_0; thereby, its last term contains $M_{n_0.n}$.

Then, one has the exact equation for $\tilde{W}_0(s)$:

$$s\tilde{W}_0(s) = W_{\mathrm{in},n} - M_{0,1}D_1[-M_{1,0}]\tilde{W}_0(s) \tag{50}$$

For previous approaches to stochastic equations that make use of continued fractions, see [42,43]. As a general reference on ordinary continued fractions, see [44]. Notice the genuine cyclic structure (determined by the structure of the hierarchy) in successive terms in the denominators in D_n. For instance, in D_1, the contribution $M_{2,3}D_3M_{3,4}D_4M_{4,1}$ in the series $N_{2,1}$ indeed achieves a cyclic structure in the indices of M due precisely to $M_{4,1} = -\frac{\hbar^2}{2^2 q_{\mathrm{eq}}^3}\frac{\partial^3 V}{\partial x^3}\left(-6 + \frac{\epsilon_{4,2}}{\epsilon_{2,0}}\right)$, which is a quantum correction depending explicitly on $\hbar$ (and also implicitly on it through $\frac{\epsilon_{4,2}}{\epsilon_{2,0}}$). In the strict classical limit, $\epsilon_{2,0} = -(1/2)$ and $\epsilon_{4,2} = -3$, and that quantum correction disappears. And there is similarly with $M_{6,1}$ in $M_{2,3}D_3M_{3,4}D_4M_{4,5}D_5M_{5,6}D_6M_{6,1}$ and so on.

There are other approaches (different from the one pursued here, to the best of the present authors' knowledge) to the analysis of general linear recurrence relations: see, for instance [45].

2.6. Convergence: A Qualitative Analysis

A detailed mathematical study lies outside the scope of this work, as we shall limit ourselves to indicate arguments justifying convergence. We shall consider first the case $\hbar \to 0$. Then, one has the three-term hierarchy in Equations (40), (41) and (43) for the Liouville equation. Equation (48) still holds with the series $N_{n+1,n} \to 0$ so that

$$D_n = D_n(s) = \frac{1}{sI - M_{n,n+1}D_{n+1}M_{n+1,n}} \tag{51}$$

with $M_{n,n+1}$ and $M_{n+1,n}$ being given in Equations (41) and (43). We shall introduce the n-independent operators

$$F = M_{n,n+1}[(-M_{n+1,n})/(n+1)] \tag{52}$$

$$a_+ = \frac{1}{F^{1/4}}M_{n,n+1}\frac{1}{F^{1/4}}, a_- = \frac{1}{F^{1/4}}[(-M_{n+1,n})/(n+1)]\frac{1}{F^{1/4}} \tag{53}$$

Let $s = 0$ in Equation (48) and iterate it indefinitely. By using Equations (52) and (53), a direct formal computation yields:

$$D_n(0) = 2^{-1/2}\frac{\Gamma((n/a+-2)+1/2)}{\Gamma((n/2)+1)}F^{-1/4}AF^{-1/4}, A = \frac{1}{a_-Aa_+} \tag{54}$$

where Γ is the standard Gamma function [46] and the operators F and A being n-independent. The following formal and short discussion (simplifying unimportant dependences for it) may be adequate. Let us assume the operator $F = -\frac{\partial}{\partial x}(a\frac{\partial}{\partial x} + b)$ with real functions $a = a(x) \geq 0$ and $b = b(x)$ acting on functions $f = f(x)$. Upon transforming $f \to g$ with $f = \exp(-b/a)g$, $F \to F_1$ with $F_1 = -[\frac{\partial}{\partial x} - (b/2a)]a[\frac{\partial}{\partial x} + (b/2a)]$. The operator F_1 (acting on g) is non-negative. The preceding argument is closely related to the one to be employed later in Section 2.9 regarding (62). Going backwards, that enables us to give a meaning to $F^{-1/4}$, a_+, and a_-. A can be formally regarded as an operator-continued fraction. Even if a rigorous characterization for it is lacking at present, Equation (54) indicates that $D_n(0)$ factorizes into the n-independent operator $F^{-1/4}AF^{-1/4}$ times an explicit function of n. For large n, $\frac{\Gamma((n/2)+1/2)}{\Gamma((n/2)+1)}$ behaves as $2n^{-1/2}$ [44] and $D_n(0)$ tends to vanish proportionally to $n^{-1/2}$. This suggests that the operator $D_n(s)$ in Equation (51) also tends to vanish as $n^{-1/2}$ for large n. We shall employ this large n behavior in order to assess, for the general hierarchy (24), the convergence of $D_n(s)$ and $N_{n+1,n}(s)$ in Equations (48) and (49), respectively. That is, in these assessments for continued fractions and infinite series in the exact solutions, we estimate that the exact operator $D_n(s)$ behaves for large n and fixed and finite s as $n^{-1/2}$ (times an n-independent operator). The application of such estimates to the operators $N_{n+1,n}(s)$ and some judicious guess for the last factor in each operator contribution ($M_{n+3,n}$ in the first contribution in (49) and so on) leads us to infer that the operator $N_{n+1,n}$ is more negligible compared to $M_{n+1,n}$ the larger n is. Then, upon estimating D_{n+1} in the right-hand-side of (48) to be of order $n^{-1/2}$ and noticing that $M_{n+1,n}$ is of order n, it follows that D_n in the left-hand-side of (48) is of order $n^{-1/2}$ consistently. This suggests that D_1 in Equation (50) is finite. Even if full mathematical justifications are lacking, it is felt that the arguments above provide reasonable support towards the finiteness and well-posedness of the present developments based upon non-equilibrium moments and hierarchy.

2.7. Properties at Equilibrium: $\epsilon_{2,0}$

The equilibrium Wigner function in (4) and the definitions of the equilibrium moments using $Weq(x,q)$ as the generating function implies that all $W_n = W_{eq,n} = 0$ for $n = 1,2,3,\ldots$, while only the lowest one is non-vanishing:

$$W_0 = W_{eq,0} = \frac{1}{q_{eq}} \sum_j \exp[-\beta E_j]\varphi_j(x)\varphi_j^*(x) \tag{55}$$

In this case, $W_{\text{in},n} = 0$ for any $n = 0,1,2,\ldots$. On the other hand, one has the general equation:

$$M_{1,0}W_n = -\frac{q_{eq}}{m}\frac{\partial}{\partial x}((\epsilon_{2,0})W_n) + \frac{1}{q_{\text{eq}}}\frac{\partial V}{\partial x}W_n \tag{56}$$

Equation (24) for equilibrium and $n = 1$ imply

$$M_{1,0}W_{eq,0} = 0 \tag{57}$$

Equation (57) and the expression for $M_{1,0}$ from (56) with the condition that in the classical ($\hbar \to 0$) limit, $-\epsilon_{2,0} = 1/2$, can be recast as two alternative and exact representations, each having its own interest. First, Equation (57) implies:

$$-\epsilon_{2,0}(x)W_{eq,0}(x) = (1/2)\exp[-\frac{m}{q_{eq}^2}\int_x^{+\infty} dx' \frac{1}{\epsilon_{2,0}(x')}\frac{\partial V}{\partial x'}], \tag{58}$$

Since $W_{eq,0} \geq 0$, it follows that $-\epsilon_{2,0} \geq 0$ for any x.

Second, an explicit representation of $-\epsilon_{2,0}$ in terms of $W_{eq,0}$ and V is:

$$\epsilon_{2,0}(x) = -\frac{1}{2W_{eq,0}(x)} - \frac{m}{q_{eq}^2 W_{eq,0}(x)}\int_x^{+\infty} dx' W_{eq,0}(x')\frac{\partial V}{\partial x'}, \tag{59}$$

The formally infinite integration limits in Equations (58) and (59) are to be understood in the framework (very large L) and conditions on V stated in Section 2.1. In particular, since V is appreciable in a finite interval ($-a < x < a$), the integrals in Equations (58) and (59) converge.

2.8. One-Dimensional Non-Equilibrium Hierarchy: Small Thermal Wavelength

Let δx be a typical scale of variation of V: for instance, a fraction (say, 1/5 to 1/10) of one nanometer. We shall consider a quantum regime with a relatively small thermal wavelength with, say, $\lambda_{th}/\delta x < 1$: some features in it are treated as in the classical regime while others behave quantum mechanically. In such a regime, chemical reactions of the kind we are interested in occur. Such a regime is not the strict classical high-temperature limit. For assumptions and estimates characterizing such a regime, see Section 3.2 in [30]. A summary of those estimates is the following. Let the mass m be about one to two orders of magnitude larger than the neutron mass. The range of V may be about 0.1 to 0.5 nanometers. The value of k_BT may lie, for instance, between 300 K (room temperature) and, say, 1200 K. $\mid V \mid$ may vary between 1 and a few eV, and let V_0 be a positive constant energy having a value in such a range. Let $\frac{\partial^n V}{\partial x^n}$ be estimated as $\frac{\delta^n V}{(\delta x)^n}$. In turn, δV is estimated as one order of magnitude smaller than V_0, and $\delta^n V$ is estimated as one order of magnitude smaller than $\delta^{n-1} V$, $n = 2, 3, \ldots$.

In Equation (39), $\frac{\hbar^2}{q_{\rm eq}^3}\frac{\partial^3 V}{\partial x^3}$ appears to be smaller than $\frac{1}{q_{\rm eq}}\frac{\partial V}{\partial x}$ by a factor $(\frac{\lambda_{\rm th}}{\delta x})^2$ times a contribution smaller than unity. This suggests we can neglect the quantum correction containing $\frac{\hbar^2}{2^2 q_{\rm eq}^3}\frac{\partial^3 V}{\partial x^3}[-6+\frac{\epsilon_{4,2}}{\epsilon_{2,0}}]$ compared to the one containing $\frac{1}{q_{\rm eq}}\frac{\partial V}{\partial x}$ in Equation (39). Similar approximations can be carried out in the equation for $\partial W_5/\partial t$ (by neglecting the contribution containing $(\hbar^2/q_{\rm eq}^3)\partial^3 V/\partial x^3)$ and in the equation for $\partial W_6/\partial t$ (by neglecting $(\hbar^2/q_{\rm eq}^3)(\partial^3 V/\partial x^3)$ and $(\hbar^4/q_{\rm eq}^5)(\partial^5 V/\partial x^5)$) and so on. In general, we shall accept as a plausible approximation (applicable in applications for chemical reactions) that in Equation (24), one can neglect, on average, all contributions due to all $M_{n,n-n'}W_{n'}$ with $n' = 2, \ldots n-1$ compared to $M_{n,n-1}W_{n-1}$. Then, Equation (24) becomes the approximate (t-reversible) three-term hierarchy

$$\frac{\partial W_n}{\partial t} = -M_{n,n+1}W_{n+1} - M_{n,n-1}W_{n-1}, \tag{60}$$

but we still retain the full $M_n, n-1$ (including its own quantum corrections) in Equation (26). The latter makes Equation (60) differ from Equation (40), in which $M_n, n-1$ is given in Equation (43).

Alternatively, in D_1, the contribution $M_{2,3}D_3M_{3,4}D_4M_{4,1}$ (with a cyclic structure in the indices of M) is smaller than $M_{2,1}$ thanks formally to the $\hbar^2$ in the contribution $M_{4,1}$. In physical terms, we proceed to the quantum regime with a relatively small thermal wavelength as characterized above: then, on average, the magnitude of the values implied by the action of the operator $M_{2,3}D_3M_{3,4}D_4M_{4,1}$ on a generic function may be smaller (by about one to two orders of magnitude) than those by the operator $M_{2,1}$. Then, in D_1, it appears plausible to discard the contribution $M_{2,3}D_3M_{3,4}D_4M_{4,1}$ and, by a similar argument, the full $M_{2,3}D_3M_{3,4}D_4M_{4,5}D_5M_{5,6}D_6M_{6,1} + \ldots$ compared to $M_{2,1}$. It is stressed that the full $M_{2,1}$ is still kept (thereby still taking into account certain quantum effects). The actual counterpart to Equation (48) for the three-term hierarchy (60) is, for $n = 1, 2, 3\ldots$:

$$D_n = D_n(s) = \frac{1}{sI - M_{n,n+1}D_{n+1}M_{n+1,n}} \tag{61}$$

That is consistent with the operator-continued fractions that follow directly from Equation (60). We shall continue to assume the initial condition $W_{in,0} \neq 0$, $W_{in,0} \neq W_{eq,0}$, and $W_{in,n} = 0$ for $n \neq 0$ for simplicity. This amounts to discarding all contributions in the operators containing cyclic structures: that is, to discard $M_{2,3}D_3M_{3,4}D_4M_{4,5}D_5M_{5,6}D_6M_{6,1} + \ldots$ in D_1 and so on for D_n for $n = 2, 3, \ldots$.

The following remark can be regarded as a gratifying check of consistency. At very high temperatures, practically in the classical regime, and based upon [6], we shall approximate the one-dimensional equilibrium quantum distribution to a leading order by the classical distribution: $W_{\rm eq} \simeq c_0 f_{\rm eq}$ with $f_{\rm eq} = \exp[-\beta((q^2/2m) + V)]$, thereby neglecting the corrections computed in [6]. Then, the computations of all $\epsilon_{n,n-j}$ boil down to computing Gaussian integrals. From (15) and (16), one easily finds: $\epsilon_{2,0} = -1/2$ and $\epsilon_{4,2} = -3$. Then, under that approximation corresponding to the classical regime, one finds $-6 + \frac{\epsilon_{4,2}}{\epsilon_{2,0}} = 0$ in Equation (39) and, consequently, the hierarchy Equations (35)–(39) reduce to a three-term one. The same reduction of Equation (24) to a three-term hierarchy occurs for any n.

Notice as well the following behavior. Consider the Laplace transform of Equation (60), which implies $\tilde{W}_n(s) = -D_nM_{n,n-1}\tilde{W}_{n-1}(s)$ for $n \geq 1$. For fixed and finite s and large n, D_n behaves as $n^{-1/2}$. Then, $s\tilde{W}_n(s)$ is subdominant compared to $M_{n,n+1}\tilde{W}_{n+1}(s) + M_{n,n-1}\tilde{W}_{n-1}(s)$. In turn, that behavior will be consistent with the long-t approximation in the next subsection.

2.9. One-Dimensional Non-Equilibrium Hierarchy: Long-Time Approximation

We shall proceed to approximations adequate for a long-time non-equilibrium evolution based upon Equation (60), with chemical reactions in mind.

The operators $M_{n,n+1}$ and $M_{n,n-1}$ in (60) have dimensions (time)$^{-1}$.

The characteristic or effective evolution times associated with those operators, denoted as $(\tau^*)^{-1}$, have orders of magnitude that can be estimated easily. Thus, the τ^* associated with $M_{n,n+1}$ is about $\lambda_{\text{th}} m(\delta x)/\hbar)$ and so on for the various terms contributing to $M_{n,n-1}$ (the estimates of which depend on x).

We shall consider t about and larger than those effective evolution times. Then, as large t corresponds to small s, the simplest (long-time) approximation can be formally conjectured for each $n = 1, 2, \ldots$ as follows: We replace $D_n(s)$ in Equation (61) with the s-independent operator $D_n(s = \epsilon)$ (with fixed and small $s = \epsilon > 0$), and then, one has the approximation: $\tilde{W}_n(s) \simeq D_n(s = \epsilon)[-M_{n,n-1}]\tilde{W}_{n-1}$ (also regarded as a short-memory approximation). The system formed by the inverse Laplace transform of $\tilde{W}_1(s) \simeq D_1(s = \epsilon)[-M_{1,0}]\tilde{W}_0$ together with Equation (60) for $n = 0$ completes the long-time approximation scheme. This amounts to the argument that the t-dependence of $W_n(t)$, $n = 1, 2, \ldots$ would be approximately enslaved by that of $W_{n-1}(t)$. That immediately yields the following quantum equation:

$$\frac{\partial W_0}{\partial t} = \frac{q_{\text{eq}}}{m}\frac{\partial}{\partial x}[D_1(s = \epsilon)M_{1,0}W_0] \tag{62}$$

with the above initial condition $W_{in,0}$. Providing a suitable approximation method or ansatz yielding $D_1(s = \epsilon)$ is a difficult open problem: see [30]. The diffusion-like Equation (62) appears to be t-irreversible. However, at the present stage, we do not warrant that all eigenvalues of $-\frac{q_{\text{eq}}}{m}\frac{\partial}{\partial x}D_1(s = \epsilon)[-M_{1,0}]$ are nonnegative.

For the sake of a complementary understanding, we shall accept as a plausible approximation that the linear operator $D_1(s = \epsilon)$ be replaced by a non-negative function $D = D_1(s = \epsilon) \geq 0$ of x.

We introduce $f_0(x;t)$ through:

$$W_0 = \exp\left[\int_0^x dx' u(x')\right] f_0, \tag{63}$$

$$u(x) = -\frac{q_{\text{eq}}/m)(\partial(-\epsilon_{2,0})/\partial x) + (1/q_{\text{eq}})(\partial V/\partial x)}{(2q_{\text{eq}}/m)(-\epsilon_{2,0})}. \tag{64}$$

Then, Equation (62) becomes

$$\frac{\partial f_0}{\partial t} = \left[\frac{\partial}{\partial x} + u(x)\right]\frac{q_{\text{eq}}^2 D(-\epsilon_{2,0})}{m^2}\left[\frac{\partial}{\partial x} - u(x)\right] f_0. \tag{65}$$

Recall that $-\epsilon_{2,0}$ is nonnegative for any x (Section 2.7). Then, all eigenvalues of $[\frac{\partial}{\partial x} + u(x)]\frac{q_{\text{eq}}^2 D(-\epsilon_{2,0})}{m^2}[\frac{\partial}{\partial x} - u(x)]$ are nonnegative, and the solution of Equation (65) tends towards $W_{\text{eq},0}$ for $t \to +\infty$ for any $W_{\text{in},0}$ (irreversibility and thermalization).

Regarding stochastic equations, see [18,31,47–49].

The $1D$ developments in this section will be very useful at a certain stage (Sections 4.4 and 4.5) in the $3D$ model for polymerization explained in the following sections.

3. Towards a Model for Catalyzed Polymerization

Throughout this and the following section, we consider a fluid at rest in thermal equilibrium at absolute temperature T in an interval about room temperature and in three-dimensional (3D) space. The fluid plays the role of an HB. Then, we consider three ensembles immersed in the fluid: (1) an ensemble of widely separated independent unbound units (atoms and/or small molecules); (2) an ensemble of widely separated, independent, freely jointed (fj) chains as templates (*te*); and (3) an ensemble of widely separated independent catalysts.

3.1. Ensemble of Widely Separated Independent Unbound 3D Units

Let an ensemble of identical N_{euu} non-relativistic microscopic units (atoms and/or small molecules) with equal masses be widely separated from and interacting negligibly with one another. To fix the idea, we concentrate on one of them, denoted as "1", with mass M_1, position vector $\mathbf{R}_1$, momentum $\mathbf{\Pi}_1(=-i\hbar\nabla_1)$, and quantum kinetic energy $\mathbf{\Pi}_1^2/2M_1$, ∇_1 as the $3D$ gradient operator with respect to $\mathbf{R}_1$. Any wavefunction of the ensemble of N_{euu} units factorizes into the product of N_{euu} individual wavefunctions.

The 3D scalar product of two individual wavefunctions ψ_j, $j = 1,2$ for the same unit 1 that depends on the same $\mathbf{R}_1$ reads:

$$(\psi_1, \psi_2)_3 \equiv \int d^3\mathbf{R}_1 \psi_1^* \psi_2 \,, \tag{66}$$

with the integration being carried out over $\mathbf{R}_1$, and $*$ denoting the complex conjugate.

3.2. Ensemble of 3D Freely Jointed (fj) Chains as Templates (te)

We consider, also immersed in the fluid, a very dilute solution of identical *te* molecular chains described below that are adequately separated (and then independent) from one another. That is, the fluid contains an ensemble of such template chains. Any chain is supposed to be adequately long. Regarding molecular chains from various standpoints, see [21,23,31,32,50–55].

This subsection will remind the reader of a model for a single $3D$ *te* formed by $N_{te} - 1$ non-relativistic atoms or small molecules as an open, linear, freely jointed (fj) molecular chain. Let $\mathbf{R}_{te,i}$, $\P_{te,i}$, and $M_{te,i}$ be the position and momentum vectors and the mass, respectively, of the i-th atom in the *te* ($i = 2, \ldots, N_{te}$). The total mass of the *te* is: $M_{te} = \sum_{i=2}^{N_{te}} M_{te,i}$. Let $\mathbf{P}_{te,CM} = \sum_{i=2}^{N_{te}} \mathbf{P}_{te,i}$ be the total momentum operator of the *te*, and let $\mathbf{y}_i = \mathbf{R}_{te,i+1} - \mathbf{R}_{te,i}$, $i = 2, \ldots, N_{te} - 1$ be the relative position vectors (the bond vectors) along the *te*. The latter is treated in the framework of the Born–Oppenheimer approximation [56] so that the most rapidly varying electronic degrees of freedom have already been integrated out. $E_{el}(< 0)$ is the electronic energy (essentially, a constant), which will always be subtracted out from the outset.

In $3D$ spherical coordinates, the three-momentum operator associated with $\mathbf{y}_i$ reads:

$$-i\hbar\nabla_{\mathbf{y}_i} = -\frac{\mathbf{a}_{3,i}}{y_i} - i\hbar\mathbf{u}_i\frac{\partial}{\partial y_i} \,, \tag{67}$$

with

$$\mathbf{y}_i = y_i\mathbf{u}_i \,, \mathbf{a}_{3,i} = i\hbar\mathbf{u}_{\theta_i}\frac{\partial}{\partial\theta_i} + i\hbar\mathbf{u}_{\varphi_i}\frac{1}{\sin\theta_i}\frac{\partial}{\partial\varphi_i} \,, \tag{68}$$

with the three orthonormal vectors:

$$\mathbf{u}_i = (\cos\varphi_i \sin\theta_i, \sin\varphi_i \sin\theta_i, \cos\theta_i) \,, \tag{69}$$

$$\mathbf{u}_{\theta_i} = (\cos\varphi_i \cos\theta_i, \sin\varphi_i \cos\theta_i, -\sin\theta_i) \,, \tag{70}$$

$$\mathbf{u}_{\varphi_i} = (-\sin\varphi_i, \cos\varphi_i, 0) \tag{71}$$

Let:

$$\mathbf{e}_{3,l} \equiv i\hbar\mathbf{u}_l - \mathbf{a}_{3,l}, \tag{72}$$

$l = 2, \ldots, N_{te} - 1$ [53–55]. The total quantum kinetic energy of the *te* is:

$$\mathbf{P}_{te,CM}^2/2M_{te} + \sum_{i,j=2}^{N_{te}-1} A_{ij}[-i\hbar\nabla_{\mathbf{y}_i}][-i\hbar\nabla_{\mathbf{y}_j}] \tag{73}$$

The constants A_{ij} are given by: $M_i^{-1} + M_{i+1}^{-1}$ if $i = j$, $-M_i^{-1}$ if $j = i - 1$, $-M_j^{-1}$ if $j = i + 1$, and 0, otherwise. So: $A_{ij} = A_{ji}$.

Approximate models for a *te* as a $3D$ molecular chain with constrained distances (bond lengths) between successive pairs of neighbor atoms due to strong harmonic-oscillator-like vibrational potentials (covalent bonds that are not as strong as electrical degree-of-freedom interactions) have been constructed [33,53–55]. In short, the *te* is modeled as a freely jointed, or fj, molecular chain. For the sake of a short justification of the latter, as a dominant effective approximation of the covalent bonding (neglecting other weaker interactions), let nearest-neighbor atoms interact through harmonic-oscillator-like potentials with vibrational frequencies $\omega_{0,j}$ [53]. The vibrational energies $\hbar\omega_{0,j}$ (much smaller than E_{el}) are supposed to be larger than $K_B T$ [53–55]. One is also assuming that, on that energy scale, angular degrees of freedom are not constrained. In such a regime, y_j equals, approximately, the constant equilibrium distance d_j (bond length).

One can also entertain other molecular *te* with additional constrained distances between successive pairs of next-to-nearest-neighbor units due to (somewhat weaker) harmonic-oscillator-like vibrational potentials: freely rotating molecular chains. Freely rotating chains can be approximated by fj ones that include persistent lengths (namely, effective bond lengths, which amount to constraining approximately both the above y_j and also the angles between neighboring bond vectors) [21]. Such models do provide useful approximations for real single-polymer chains under various conditions [21]. In the present work, dealing with fj chains will suffice, with the understanding that the bond lengths can be interpreted as either fixed bond lengths in fj chains or as persistent lengths in freely rotating chains.

Let θ, φ denote, collectively, the two sets $\theta_2, \ldots, \theta_{N_{te}-1}$, $\varphi_2, \ldots, \varphi_{N_{te}-1}$, respectively. Different $3D$ variational computations [33,53–55] enable us to consistently derive the same $3D$ model for an fj molecular chain with fixed $y_j = d_j$ while allowing for purely angular motion of the bond vectors. From those coinciding results, by omitting the total zero-point energy E_{zp} of the vibrations and factoring out the overall center-of-mass motion, one arrives at the following $3D$ quantum purely angular Hamiltonian and at the scalar product for the fj molecular chain:

$$\tilde{H}_{3,fj} = \sum_{i,j=2}^{N_{te}-1} \frac{A_{ij}}{2d_i d_j} \mathbf{e}_{3,i} \mathbf{e}_{3,j} \tag{74}$$

$$(\psi_1, \psi_2)_{3,fj} \equiv \int [\mathbf{d\Omega}]_3 \psi_1(\theta,\varphi)^* \psi_2(\theta,\varphi) \,, \tag{75}$$

with $[\mathbf{d\Omega}]_3 = \Pi_{i=2}^{N_{te}-1} d\theta_i d\varphi_i$. The integration is carried out over the whole $N_{te} - 2$ set of solid angles. The angular motion of the *te* is described by wavefunctions $\psi = \psi(\theta, \varphi)$ and eigenfunctions of the stationary Schrodinger equation, with energy eigenvalues E:

$$\tilde{H}_{3,fj}\psi = E\psi \tag{76}$$

The purely angular motions described by $\tilde{H}_{3,fj}$ will be essential for the understanding of what follows.

The analysis in the remainder of this subsection is intended only as a methodological and essential step towards the study of polymerization in Section 4.

So, we suppose that $K_B T$ is adequately smaller than all $\hbar\omega_{0,i}$ and that all vibrational states of any fj chain are the ground-state ones, so that the relevant degrees of freedom of the fj chain are the rotational ones (say, in principle, those corresponding to $\tilde{H}_{3,fj}$).

The evolution of an individual fj chain in the fluid at $K_B T$ can be described in principle by a non-equilibrium Wigner distribution and equation. However, due to the influence of the fluid at such $K_B T$, it will be physically adequate to approximate the quantum description of the rotational motions of the template fj chain as provided by classical statistical mechanics and, hence, by classical Liouville distribution functions.

For units $2, \ldots, N_{te}$ forming the template fj chain, the approximate transition to classical mechanics reads: $\mathbf{e}_{3,i} \rightarrow -\mathbf{a}_{3,c,i}$, $-i\hbar\frac{\partial}{\partial\theta_i} \rightarrow \pi_{\theta_i,c}$, $-i\hbar\frac{\partial}{\partial\varphi_i} \rightarrow \pi_{\varphi_i,c}$. So, $-\mathbf{a}_{3,c,i} = \pi_{\theta_i,c}\mathbf{u}_{\theta_i} + \pi_{\varphi_i,c}\mathbf{u}_{\varphi_i}$. The small terms proportional to $i\hbar\mathbf{u}$ are taken as negligible and, so, are disregarded. The terms $\pi_{\theta_i,c}$, $\pi_{\varphi_i,c}$ are classical momenta and are canonically conjugate to θ_i, φ_i, respectively. The terms $\pi_{\theta,c}$, $\pi_{\varphi,c}$ denote the set of all $\pi_{\theta_i,c}$, $\pi_{\varphi_i,c}$, respectively.

Then, the quantum Hamiltonian $\tilde{H}_{3,fj}$ for the angular motion of the *te* chain is approximated by the following classical one (factoring out the center-of-mass motion):

$\tilde{H}_{3,fj,c} = \sum_{i,j=2}^{N_{te}-1}(A_{ij}/2d_i d_j)(-\mathbf{a}_{3,c,i})(-\mathbf{a}_{3,c,j})$. Let $f_{te,c} = f_{te,c}(\theta, \varphi, \pi_{\theta,c}, \pi_{\varphi,c}; t)$ be the non-equilibrium classical Liouville distribution function for the *te*. Accordingly, $[\tilde{H}_{3,fj,c}, f_{te,c}]_{Pb}$ denotes the standard classical Poisson bracket (*Pb*) of $\tilde{H}_{3,fj,c}$ and $f_{te,c}$ [57]. The non-equilibrium classical Liouville equation is:

$$\frac{\partial f_{te,c}}{\partial t} = [\tilde{H}_{3,fj,c}, f_{te,c}]_{Pb} \tag{77}$$

Let $[\mathbf{d\Omega}]_{3,L} = \Pi_{i=2}^{N_{te}-1} d\theta_i d\pi_{\theta_i,c} d\varphi_i d\pi_{\varphi_i,c}$. Total probability is consistently conserved: $\partial \int [\mathbf{d\Omega}]_{3,L} f_{te,c}/\partial t = 0$. These classical approximations for the *te* chain are instrumental and will be implemented directly in Sections 4.1 and 4.2.

3.3. Ensemble of Widely Separated Independent 3D Catalyst (cat) Molecules

Let an ensemble of identical non-relativistic individual catalyst units be widely separated from (with negligible interactions with) one another. Each catalyst unit (*cat* for short) is a medium-size molecule (formed by other small molecules, monomers...). The total mass of a *cat* unit is M_0.

The fluid is at rest and in thermal equilibrium at a $K_B T$ adequately smaller than all vibrational energies of a *cat* molecule. Then, by assumption, all relevant vibrational states in the *cat* are the ground ones, and the relevant degrees of freedom in it are the rotational ones and those associated with its corresponding center of mass.

With an enormous simplification, at the given T, the *cat* is modeled as a slow and free massive molecule of mass M_0 with position vector $\mathbf{R}_0$, which approximates the location of its center-of-mass and disregards the spatial extension of the catalyst. Alternatively, $\mathbf{R}_0$ can also be considered approximately by assumption as the location of a quite reduced domain of the catalyst (the "active" site or center), which will interact with the individual unit and the relevant part of the template and without taking into account the spatial extension of the *cat*.

To fix the ideas, in a quantum-mechanical setting, let $\mathbf{\Pi}_{3,0}$ be the quantum momentum canonically conjugate to $\mathbf{R}_0$. The quantum Hamiltonian of the free catalyst is $\tilde{H}_{3,enz} \simeq \mathbf{\Pi}_{3,0}^2/2M_0$. We advance that from Section 4.1 onward, at the given T, it will suffice to approximate the behavior of the *cat* by employing classical mechanics.

4. 3D Catalyzed Polymerization of One Single Atom by a Chain: Mixed Quantum–Classical Description

We shall consider a model for chain growth through polymerization: technically, for what is known as insertion polymerization. A (slightly simplified) example is: $\cdots - (CHR - CH_2)_n - [Ti] + (CHR - CH_2) \rightarrow \cdots - (CHR - CH_2)_n - (CH_2 - CHR) - [Ti]$. $(CHR - CH_2)$ plays the roles of the individual "small" molecule 1 and of one "small" molecule in the *te* chain (denoted, in turn, as $\cdots - (CHR - CH_2)_n$). $[Ti]$, symbolizing, in short, the complex $TiCl_4 - AlR_3$), is the catalyst. See [32].

We suppose that the number of unbound units N_{euu} is approximately equal both to the number of template chains and also to the number of individual *cat* molecules. Then, the fluid at equilibrium and at rest can also be regarded approximately as an ensemble formed by copies adequately separated from one another: with each copy, in turn, being a triplet formed by one unbound unit, one fj chain, and one *cat* molecule. We consider that each copy has a finite volume, albeit it is quite large on the microscopic scale. Those three entities are now regarded as not separated from one another, on average, so that interactions among them occur. Those interactions will be modeled in what follows.

The action of the catalyst enables an unbound unit to become attached to the fj *te* chain so as to give rise to another larger chain made up of N_{te} atoms. The individual unit is treated quantum-mechanically since its binding to the *te* is a chemical reaction.

4.1. Interactions of a Unit, an fj Chain, and a Catalyst

An individual unit (certainly influenced by its interaction with the catalyst and with the *te* fj chain) is dealt with quantum-mechanically, as this genuinely applies to its binding process to the template fj chain. In principle, the dynamics are accounted for by a quantum Wigner function [4,6,37].

We introduce the relative vector from that unit to the first atom in the template: $\mathbf{y}_1 = \mathbf{R}_{te,2} - \mathbf{R}_1$. We suppose a rotational-invariant interaction potential between the individual unit and the first atom in the *te* chain: $U_1(\mathbf{y}_1) = U_1(y_1)$ ($y_1 =| \mathbf{y}_1 |$). This is appreciable in $0 \le y_1 \le y_{1,2}$. Specifically, $U_1(y_1)$: (1) is repulsive (> 0) for short distances in $0 \le y_1 \le y_{1,0}$, (2) is attractive (< 0) in $y_{1,0} \le y_1 \le y_{1,1}$, (3) is repulsive in $y_{1,1} \le y_1 \le y_{1,2}$, and (4) vanishes very quickly for $y_1 > y_{1,2}$. The term y_1 varies inside the microscopically large but finite volume of each copy of the ensemble referred to at the beginning of this section. In short, $U_1(y_1)$ vanishes very quickly beyond a domain having a size of order of at most a few bond lengths in the *te* chain.

Another crucial interaction potential U_0, activation of the polymerization process, is supposed among the individual units, unit 2 in *te*, and the *cat*. The catalyst (also denoted here as unit 0) interacts simultaneously with the atomic unit (unit 1) in the ensemble and unit 2 in the *te* through the real, spherically symmetric potential $U_0(\mathbf{y}_0, \mathbf{y}_0 + \mathbf{y}_1)$, with $\mathbf{y}_0 = \mathbf{R}_1 - \mathbf{R}_0$. The properties of $U_0(\mathbf{y}_0, \mathbf{y}_0 + \mathbf{y}_1)$ will not be discussed at this stage. Its assumed effective behavior (specifically, that of the average of $U_0(\mathbf{y}_0, \mathbf{y}_0 + \mathbf{y}_1)$ over $\mathbf{y}_0$) will be considered a posteriori in Section 4.3.

Units 0 and 1 will be included in the overall CM and in the general description in Section 3.2, where they are enlarged with $\mathbf{y}_0$, $\mathbf{y}_1$. The model is being constructed with the following crucial numbering convention. The set of all material entities is numbered successively in the following sequence: 0 (*cat*), 1 (individual unit), and 2,3,... N_{te} (those in the *te*) consistently with the definitions and numbering of the $\mathbf{y}$. In turn, such a numbering convention and the above choice of potentials $U_1(y_1)$ and $U_0(\mathbf{y}_0, \mathbf{y}_0 + \mathbf{y}_1)$ will be consistent with the individual unit 1 to be bound to unit 2 in the *te*. Accordingly, we also introduce $A_{00} = M_0^{-1} + M_1^{-1}$, $A_{01} = A_{10} = -M_1^{-1}$ and so on for A_{11}, A_{12}, and A_{21} upon consistently extending the definitions of A_{ij} in Section 3.2.

The quantum purely kinetic Hamiltonian for the *cat*, the individual units, and the template is $\mathbf{\Pi}_{3,0}^2/2M_0 + \mathbf{\Pi}_1^2/2M_1 + \mathbf{P}_{te,CM}^2/2M_{te} + 2^{-1}\sum_{i,j=2}^{N_{te}-1} A_{ij}[-i\hbar\nabla_{\mathbf{y}_i}][-i\hbar\nabla_{\mathbf{y}_j}]$.

At this stage, one performs the following transformations: (i) one introduces and factors out the contribution of the total center-of-mass (for *te*, unit 1, and *cat*); (ii) one introduces all relative vectors $\mathbf{y}_0$, $\mathbf{y}_1$, $\mathbf{y}_2, \ldots, \mathbf{y}_{N_{te}-1}$ and makes use of Equation (74) for $\mathbf{y}_2, \ldots, \mathbf{y}_{N_{te}-1}$; and (iii) one adds U_0, U_1, and all vibrational potentials accounting for the structure of the fj chain and implements the transition giving rise to fixed bond lengths in the *te*. Then, one infers the following effective quantum Hamiltonian (omitting the overall center-of-mass for all material entities): $\sum_{i,j=0}^{1} 2^{-1}A_{ij}[-i\hbar\nabla_{\mathbf{y}_i}][-i\hbar\nabla_{\mathbf{y}_j}] + \frac{A_{12}+A_{21}}{2d_2}\mathbf{e}_{3,2}[-i\hbar\nabla_{\mathbf{y}_1}] + \sum_{i,j=2}^{N_{te}-1}\frac{A_{ij}}{2d_i d_j}\mathbf{e}_{3,i}\mathbf{e}_{3,j} + U_0 + U_1$.

It will be supposed that the *cat*, evolving in the fluid at the temperature T assumed, can be described approximately through classical statistical mechanics and that it has random motion during the effective duration of the process catalyzed by it. Let $\mathfrak{B}_{3,c,0}$ be the classical momentum, which is canonically conjugate to $\mathbf{y}_0$, for the classical *cat*.

As for the *cat*, the evolution of the fj *te* chain inside the fluid at thermal equilibrium at those temperatures is also described approximately through classical statistical mechanics: recall the analysis and the transition to classical variables in Section 3.2. Then, one can entertain the reasonableness of the following formal mixed quantum–classical Hamiltonian structure:

$$\frac{A_{00}\text{ß}_{3,c,0}^2}{2} + \frac{(A_{01}+A_{10})\text{ß}_{3,c,0}[-i\hbar\nabla_{\mathbf{y}_1}]}{2} + 2^{-1}A_{11}[-i\hbar\nabla_{\mathbf{y}_1}][-i\hbar\nabla_{\mathbf{y}_1}] + \frac{A_{12}+A_{21}}{2d_2}(-\mathbf{a}_{3,c,2})[-i\hbar\nabla_{\mathbf{y}_1}] + \sum_{i,j=2}^{N_{te}-1}\frac{A_{ij}}{2d_id_j}(-\mathbf{a}_{3,c,i})(-\mathbf{a}_{3,c,j}) + U_0 + U_1 \tag{78}$$

with the sole purpose of using it as a key guide to directly formulate a mixed quantum–classical Wigner–Liouville equation, as we shall do in Section 4.2.

4.2. Mixed Wigner–Liouville Equations for the Ensemble, te, and Catalyst

Let $\text{ß}_{3,c,1}$ be a momentum, which is canonically conjugate to $\mathbf{R}_1$, for the individual unit treated quantum-mechanically. Use will be made of the classical variables employed in Sections 3.2 and 4.1 for the *te* and catalyst, respectively.

The system formed by a unit, an fj chain, and a *cat* is described, by assumption, by a mixed (quantum–classical) distribution function in phase-space. The following quantum Wigner-like one for unit 1 and a classical Liouville-like one for the chain and the *cat* will be considered: $f_m = f_m(\mathbf{y}_1, \theta, \varphi, \mathbf{y}_0, \text{ß}_{3,c,1}, \pi_{\theta,c}, \pi_{\varphi,c}, \text{ß}_{3,c,0}; t)$. By assumption, f_m fulfills the time (t)-reversible Wigner–Liouville equation that follows naturally by starting from the mixed Hamiltonian structure in (78) by operating quantum-mechanically (via Wigner) with it for the variables of the individual unit and classically (via Liouville) for those of the *te* and the catalyst. One finds directly:

$$\begin{aligned}\frac{\partial f_m}{\partial t} &= -(A_{11}\text{ß}_{3,c,1} + A_{21}(-\mathbf{a}_{3,c,2}) + A_{01}\text{ß}_{3,c,0})(\nabla_{\mathbf{y}_1}f_m) + \int d^3\text{ß}_{3,c,1,0}f_m(\text{ß}_{3,c,1,0})\int\frac{id^3\mathbf{y}_{1,0}}{\hbar(\pi\hbar)^3}\times\\ &\exp\left(\frac{2i(\text{ß}_{3,c,1}-\text{ß}_{3,c,1,0})\mathbf{y}_{1,0}}{\hbar}\right)[U_1(\mathbf{y}_1+\mathbf{y}_{1,0}) - U_1(\mathbf{y}_1-\mathbf{y}_{1,0}) +\\ &U_0(\mathbf{y}_0, \mathbf{y}_0+\mathbf{y}_1+\mathbf{y}_{1,0}) - U_0(\mathbf{y}_0, \mathbf{y}_0+\mathbf{y}_1-\mathbf{y}_{1,0})] +\\ &+A_{12}\text{ß}_{3,c,1}\Big[\frac{\partial(-\mathbf{a}_{3,c,2})}{\partial\theta_2}\frac{\partial f_m}{\partial\pi_{\theta_2}} -\\ &\frac{\partial(-\mathbf{a}_{3,c,2})}{\partial\pi_{\theta_2}}\frac{\partial f_m}{\partial\theta_2} + \frac{\partial(-\mathbf{a}_{3,c,2})}{\partial\varphi_2}\frac{\partial f_m}{\partial\pi_{\varphi_2}} - \frac{\partial(-\mathbf{a}_{3,c,2})}{\partial\pi_{\varphi_2}}\frac{\partial f_m}{\partial\varphi_2}\Big] + [\tilde{H}_{3,fj,c}, f_m]_{Pb} -\\ &(A_{00}\text{ß}_{3,c,0} + A_{10}\text{ß}_{3,c,1})(\nabla_{\mathbf{y}_0}f_m) + (\nabla_{\mathbf{y}_0}U_0(\mathbf{y}_0, \mathbf{y}_0+\mathbf{y}_1))(\nabla_{\text{ß}_{3,c,0}}f_m),\end{aligned} \tag{79}$$

with $\tilde{H}_{3,fj,c} = \sum_{i,j=2}^{N-1}(A_{ij}/2d_id_j)(-\mathbf{a}_{3,c,i})(-\mathbf{a}_{3,c,j})$ (recall Section 3.2). $[\tilde{H}_{3,fj,c}, f_m]_{Pb}$ denotes the standard classical Poisson bracket [57] (recall Equation (77)). It approximates, in the classical regime, an integral contribution for the chain analogous to the one for the individual unit in Equation (79). We have interpreted the formulation of this mixed quantum–classical formulation by writing directly the pair $A_{01}\text{ß}_{3,c,0}(\nabla_{\mathbf{y}_1}f_m)$ and $A_{10}\text{ß}_{3,c,1}(\nabla_{\mathbf{y}_0}f_m)$ together with the pair $A_{21}(-\mathbf{a}_{3,c,2})(\nabla_{\mathbf{y}_1}f_m)$ and $A_{12}\text{ß}_{3,c,1}$ times (the Poisson bracket of $(-\mathbf{a}_{3,c,2})$ and f_m). Instead of the mixed quantum–classical Equation (79), a more basic treatment would have taken a fully quantum-mechanical Wigner equation for all (individual unit, *te*, and *cat*) entities, with Poisson brackets replaced by the corresponding integrals, as a starting point. However, in order not to encumber the analysis, it seemed more economical not to proceed like that but to start out from the mixed quantum–classical Equation (79). We advance that both *te* and *cat* will be supposed to be in classical states at approximate thermal equilibrium in the next subsection, which also supports such a shortened strategy. The term $f_m(\text{ß}_{3,c,1,0})$ inside the integral in Equation (79) is (omitting the writing of repeated variables) obtained just by replacing $\text{ß}_{3,c,1}$ with $\text{ß}_{3,c,1,0}$ in f_m.

Through direct partial integrations, total probability is shown to be conserved consistently: $\partial\int d^3\mathbf{y}_1\int d^3\text{ß}_{3,c,1}\int d^3\mathbf{y}_0\int d^3\text{ß}_{3,c,0}\int[\mathbf{d\Omega}]_{3,L}f_m/\partial t = 0$.

4.3. Template and Catalyst at Thermal Equilibrium: Integration over Their Degrees of Freedom and Non-Standard Effective Hamiltonian

The dynamics will be considered for a sufficiently long time (t). As the fluid is at thermal equilibrium at absolute temperature T, it seems physically reasonable that in each triplet, the *te* chain and *cat* are approximately at thermal equilibrium at the same T, and that under their influence, the individual atomic unit evolves off-equilibrium so as to give rise to polymerization. The resulting dynamics of the individual atomic unit are not expected to alter the statistical equilibrium states of the fj chain and of the *cat*. Actually, such an assumption underlies the very formulation of Equation (79).

We accept the approximate factorization of the non-equilibrium distribution: $f_m \simeq f_1 f_{eq,2} f_{eq,0}$ with an off-equilibrium $f_1 = f_1(\mathbf{y}_1, ß_{3,c,1}, t)$ for the individual atomic unit. The *cat* is described by the (t-independent) classical Boltzmann equilibrium distribution ($A_{10} = A_{01}$):

$f_{eq,0} = Z_{eq,0}^{-1} \exp[-(K_B T)^{-1}[(A_{00}/2)ß_{3,c,0}^2 + A_{10}ß_{3,c,0}ß_{3,c,1} + U_0]]$, which depends on $ß_{3,c,0}$, $ß_{3,c,1}$, $\mathbf{y}_0$, and $\mathbf{y}_1$ (with $Z_{eq,0} = \int d^3\mathbf{y}_0 d^3ß_{3,c,0} \exp[-(K_B T)^{-1}[(A_{00}/2)ß_{3,c,0}^2 + U_0)]]$).

We also accept that the *te* chain is described by the (t-independent) classical Boltzmann equilibrium distribution ($A_{12} = A_{21}$): $f_{eq,2} = Z_{eq,2}^{-1} \exp[-(K_B T)^{-1}[(A_{12}/d_2)ß_{3,c,1}(-\mathbf{a}_{3,c,2}) + \tilde{H}_{3,fj,c,2}]]$, which depends on all variables of the *te* and $ß_{3,c,1}$ (with $Z_{eq,2} = [\mathbf{d\Omega}]_{3,L} \exp[-(K_B T)^{-1}\tilde{H}_{3,fj,c,2}])$, which includes $ß_{3,c,1}(-\mathbf{a}_{3,c,2})$.

Notice that $(-\mathbf{a}_{3,c,2})$ is coupled to $\nabla_{\mathbf{y}_1}$ in Equation (79).

We integrate Equation (79) with $\int [\mathbf{d\Omega}]_{3,L} \int d^3\mathbf{y}_0 d^3ß_{3,c,0}$ and perform the approximate replacement $f_m \simeq f_1 f_{eq,2} f_{eq,0}$

The contribution of $A_{12}ß_{3,c,1}$ times the Poisson bracket involving $f_1 f_{eq,2} f_{eq,0}$ and multiplying it plus the contribution of the Poisson bracket ($[\tilde{H}_{3,fj,c,2}, f_1 f_{eq,2} f_{eq,0}]_{Pb}$) gives a vanishing result.

Then: $\int [\mathbf{d\Omega}]_{3,L} f_m \simeq f_1 f_{eq,0} \int [\mathbf{d\Omega}]_{3,L} f_{eq,2}$, and $\int [\mathbf{d\Omega}]_{3,L} f_m(-\mathbf{a}_{3,c,2}) \simeq f_1 f_{eq,0} \int [\mathbf{d\Omega}]_{3,L} f_{eq,2}(-\mathbf{a}_{3,c,2}) = f_1 f_{eq,0}[-((K_B T d_2)/A_{12})(\nabla_{ß_{3,c,1}} \int [\mathbf{d\Omega}]_{3,L} f_{eq,2})]$.

The function $\int [\mathbf{d\Omega}]_{3,L} f_{eq,2} \equiv f_2 = f_2(ß_{3,c,1})$ is studied in the Appendix A.

Consequently: $\int d^3\mathbf{y}_0 d^3ß_{3,c,0} \int [\mathbf{d\Omega}]_{3,L} f_m \simeq f_1 f_2 f_0$, with

$f_0 = \int d^3\mathbf{y}_0 d^3ß_{3,c,0} f_{eq,0} = \exp[(2K_B T)^{-1}(A_{10}^2/A_{00})ß_{3,c,1}^2]$.

Equation (79) becomes, in terms of $W = W(\mathbf{y}_1, ß_{3,c,1}, t) = f_1 f_2 f_0$:

$$\frac{\partial W}{\partial t} = -[A_{11}ß_{3,c,1} - (K_B T)(\nabla_{ß_{3,c,1}} \ln f_2) - (A_{10}^2/A_{00})ß_{3,c,1}](\nabla_{\mathbf{y}_1} W) + \int d^3ß_{3,c,1,0} \times W(\mathbf{y}_1, ß_{3,c,1,0}, t)] \int \frac{i d^3\mathbf{y}_{1,0}}{\hbar(\pi\hbar)^3} \exp\left(\frac{2i(ß_{3,c,1} - ß_{3,c,1,0})\mathbf{y}_{1,0}}{\hbar}\right)[U_{1,eff}(|\,\mathbf{y}_1 + \mathbf{y}_{1,0}\,|) - U_{1,eff}(|\,\mathbf{y}_1 - \mathbf{y}_{1,0}\,|)] \tag{80}$$

$$U_{1,eff}(|\,\mathbf{y}_1\,|) = U_1(|\,\mathbf{y}_1\,|) + U_{0,eff}(|\,\mathbf{y}_1\,|) \tag{81}$$

$$U_{0,eff}(|\,\mathbf{y}_1\,|) = \frac{\int d^3\mathbf{y}_0 U_0(\mathbf{y}_0, \mathbf{y}_0 + \mathbf{y}_1) \exp[-(K_B T)^{-1} U_0)]}{\int d^3\mathbf{y}_0 \exp[-(K_B T)^{-1} U_0]} \tag{82}$$

We shall assume the following properties of $U_{0,eff}(|\,\mathbf{y}_1\,|) = U_{0,eff}(y_1)$ ($y_1 = |\,\mathbf{y}_1\,|$): (a_1) it is repulsive for $0 \leq y_1 \leq y_{1,3}$, where $y_{1,0} \leq y_{1,3} < y_{1,1}$; ($a_2$) it is attractive for $y_{1,3} \leq y_1 \leq y_{1,2}$; and ($a_3$) it tends to vanish for adequately large values of $y_1 (> y_{1,2})$. We allow, at this stage, for $U_0(\mathbf{y}_0, \mathbf{y}_0 + \mathbf{y}_1)$ to give rise to bound states of the catalyst for the system formed by the individual unit and the *te*. Such a possibility can be entertained at the level of Equation (79) but lies outside the scope of Equation (80) and its consequences, which concentrate on the individual atom 1. See the comments in Section 5.

Upon recalling the properties assumed for U_1, it follows that $U_{1,eff}(|\,\mathbf{y}_1\,|) = U_{1,eff}(y_1)$: (1) is repulsive in $0 \leq y_1 \leq y_{1,0}$; (2) is attractive in $y_{1,0} \leq y_1 \leq y_{1,1}$; (3) is repulsive in $y_{1,1} \leq y_1 \leq y_{1,2}$; and (4) tends to vanish for adequately large values of $y_1 (> y_{1,2})$. Two important additional points are: (5) $U_{1,eff}(y_1)$ continues to be attractive in $y_{1,0} \leq y_1 \leq y_{1,1}$

in spite of the possibility that $U_{0,eff}(y_1)$ can be repulsive in $y_{1,0} \leq y_1 \leq y_{1,3}$; and (6) $U_{1,eff}(y_1)$ is considerably less repulsive than $U_1(y_1)$ in $y_{1,1} \leq y_1 \leq y_{1,2}$. The basic effect due to $U_{0,eff}(y_1)$ (and, hence, due to the catalyst) is to offset and make lower the positive values of $U_1(y_1)$ in $y_{1,1} \leq y_1 \leq y_{1,2}$. We shall assume later (Section 4.5) that $U_{1,eff}(y_1)$ does give rise to bound states (specifically, to one bound state).

Equation (80) depends only on the degrees of freedom of the individual unit. Equation (80) with (81) is the standard Wigner equation for the non-standard quantum Hamiltonian: $\tilde{H}_{n-s,1} = -(\hbar^2/2)((A_{11} - (A_{10}^2/A_{00})))\nabla^2_{\mathbf{y}_1} + [\ln f_2](\mathfrak{B}_{3,c,1} \to -i\hbar\nabla_{\mathbf{y}_1}) + U_{1,eff}(|\ \mathbf{y}_1\ |)$. In so doing, we are correcting some misprint in the non-standard quantum Hamiltonian $\tilde{H}_{n-s,1}$ in Section 5.1 in [33]: such a misprint is inconsequential regarding the developments in [33].

Equation (80) directly yields the probability flux conservation:

$$\frac{\partial}{\partial t}\int d^3\mathfrak{B}_{3,c,1}W = -\nabla_{\mathbf{y}_1}\int d^3\mathfrak{B}_{3,c,1}[A_{11}\mathfrak{B}_{3,c,1} - (K_BT)(\nabla_{\mathfrak{B}_{3,c,1}}\ln f_2) - $$
$$-(A_{10}^2/A_{00})\mathfrak{B}_{3,c,1}]W \tag{83}$$
$$\frac{\partial}{\partial t}\int d^3\mathbf{y}_1\int d^3\mathfrak{B}_{3,c,1}W = 0 \tag{84}$$

4.4. Standard Approximate Effective Quantum Hamiltonian for the Individual Unit

As it is difficult to handle $\tilde{H}_{n-s,1}$, it will be approximated by the new effective Hamiltonian $\tilde{H}_{eff,1}$ below. Accordingly, one approximates: $-(K_BT)(\nabla_{\mathfrak{B}_{3,c,1}}\ln f_2) \simeq A_{12}\sigma\mathfrak{B}_{3,c,1}$: see Appendix A. The constant $\sigma(>0$ and dimensionless) accounts for the influence of the classical fj chain on the dynamics of unit 1. After this approximation, Equation (80) becomes the standard Wigner equation for the effective standard quantum Hamiltonian for unit 1: $\tilde{H}_{eff,1} = -(\hbar^2/2)A_{11,eff}\nabla^2_{\mathbf{y}_1} + U_{1,eff}(|\ \mathbf{y}_1\ |)$, where $A_{11,eff} = (A_{11} - (A_{10}^2/A_{00})) + A_{12}\sigma$, which yields a (quasi-)continuous spectrum of eigenvalues and one bound state associated with (quasi-)unbound motion and polymerization of the individual unit, respectively.

Let $W = W(\mathbf{y}_1, \mathfrak{B}_{3,c,1}, t)$ be the (effective) non-equilibrium Wigner function for a quantum particle with mass $A_{11,eff}^{-1}$ and subject to the potential $U_{1,eff}(|\ \mathbf{y}_1\ |)$. The corresponding non-equilibrium Wigner equation reads:

$$\frac{\partial W}{\partial t} = -[A_{11,eff}\mathfrak{B}_{3,c,1}](\nabla_{\mathbf{y}_1}W) + \int d^3\mathfrak{B}_{3,c,1,0}W(\mathbf{y}_1, \mathfrak{B}_{3,c,1,0}, t)] \times$$
$$\int \frac{id^3\mathbf{y}_{1,0}}{\hbar(\pi\hbar)^3}\exp(\frac{2i(\mathfrak{B}_{3,c,1} - \mathfrak{B}_{3,c,1,0})\mathbf{y}_{1,0}}{\hbar})[U_{1,eff}(|\ \mathbf{y}_1 + \mathbf{y}_{1,0}\ |) - U_{1,eff}(|\ \mathbf{y}_1 - \mathbf{y}_{1,0}\ |)] \tag{85}$$

The equilibrium distribution determined by Equation (85) is W_{eq}.

4.5. Extension of Sections 2.8 and 2.9 to Equation (85): Small Thermal Wavelength and Long-Time Approximations

The one-dimensional analysis in Section 2 can now be directly extended to the $D = 3$ Equation (85): namely, equilibrium distribution W_{eq}, the family of orthogonal polynomials generated by the former, non-equilibrium moments and hierarchy, and small thermal wavelength (corresponding to the absolute temperature T assumed for the fluid) and long-time approximations. Details will be omitted. Then, the $1D$ long-time approximations in Section 2.9, extended directly to $3D$ by following [30], lead to the irreversible $D = 3$ Smoluchowski equation for the lowest moment $W_{[0]} = W_{[0]}(\mathbf{y}_1, t)(= \int d^3\mathfrak{B}_{3,c,1})W(\mathbf{y}_1, \mathfrak{B}_{3,c,1}, t))$ for the individual atomic unit ($\mathbf{y}_1 = (y_{1,1}, y_{1,2}, y_{1,3})$):

$$\frac{\partial W_{[0]}}{\partial t} = Dq_{eq}A_{11,eff}\sum_{\alpha=1}^{3}\frac{\partial}{\partial y_{1,\alpha}}M_{[1_\alpha],[0]}W_{[0]} \tag{86}$$

$$M_{[1_\alpha],[0]}W_{[0]} = -q_{eq}A_{11,eff}\frac{\partial}{\partial y_{1,\alpha}}(\epsilon_{[2],[0]}W_{[0]}) + +\frac{1}{q_{eq}}\frac{\partial U_{1,eff}}{\partial y_{1,\alpha}}W_{[0]} \tag{87}$$

D (assumed to be $\mathbf{y}_1$-independent) and $\epsilon_{[2],[0]}$ are the natural $3D$ counterparts of the $1D$ ones in Section 2.9. The equilibrium distribution for (87) is:

$W_{[0],eq} = \sum_j \exp(-E_j/(K_BT))\phi_j(\mathbf{y}_1)^*\phi_j(\mathbf{y}_1)$.

The terms $\phi_j(\mathbf{y}_1)$ and E_j are, for all possible values of the set of subindices j, the (almost) continuum and bound-state eigenfunctions and energies of $\tilde{H}_{eff,1}$ ($\tilde{H}_{eff,1}\phi_j(\mathbf{y}_1) = E_j\phi_j(\mathbf{y}_1)$). Since $U_{1,eff} = U_{1,eff}(|\ \mathbf{y}_1\ |) = U_{1,eff}(y_1)$, W_{eq} is seen to depend on $\mathbf{y}_1^2$, $\text{ß}_{3,c,1}^2$, and $(\mathbf{y}_1\text{ß}_{3,c,1})^2$. Then: $\epsilon_{[2],[0]} = -\int d^3\text{ß}_{3,c,1}W_{eq}(\text{ß}_{3,c,1})_\alpha^2/(q_{eq}^2\int d^3\text{ß}_{3,c,1}W_{eq})(<0)$ is seen to be independent on $\alpha = 1,2,3$ and to depend only on y_1. Consistently, $W_{[0],eq}$ fulfills: $M_{[1_\alpha],[0]}W_{[0],eq} = 0$, which is a set of three partial differential equations that are compatible and explicitly solvable for $W_{[0],eq}$, since $U_{1,eff}$ and $\epsilon_{[2],[0]}$ depend on y_1. We shall assume that $U_{1,eff}$ gives rise to just one bound state.

The physically interesting solution is, naturally, spherically symmetric $W_{[0]} = W_{[0]}(y_1,t)$ so that Equation (86) becomes:

$$\frac{\partial W_{[0]}}{\partial t} = Dq_{eq}A_{11,eff}(\frac{\partial}{\partial y_1} + \frac{2}{y_1})(-q_{eq}A_{11,eff}\frac{\partial}{\partial y_1}(\epsilon_{[2],[0]}W_{[0]}) + \frac{1}{q_{eq}}\frac{\partial U_{1,eff}}{\partial y_1}W_{[0]}) \quad (88)$$

now with $q_{eq} = (2K_BT/A_{11,eff})^{1/2}$. It is convenient to replace $W_{[0]}$ with another distribution $f = f(y_1,t) = y_1^{-2}W_{[0]}(y_1,t)$. Equation (88) becomes:

$$\frac{\partial f}{\partial t} = Dq_{eq}A_{11,eff}\frac{\partial}{\partial y_1}(-q_{eq}A_{11,eff}\frac{\partial}{\partial y_1}(\epsilon_{[2],[0]}f) + \frac{1}{q_{eq}}\frac{\partial U_{1,eff}}{\partial y_1}f + \frac{2}{y_1}q_{eq}\epsilon_{[2],[0]}A_{11,eff}f \quad (89)$$

4.6. Mean First Passage Time

It is important to compute approximately the (average) time required for unit 1 to become attached to the chain as a next-neighbor of unit 2 in the *te* chain in the presence of and under the influence of the *cat*. For that purpose, the mean first passage time (MFPT) formalism, which provides an estimate of the latter time, is very useful. For references about the MFPT formalism, see [18,47,58]. In the present application of the MFPT, we shall extend [30,33,59]. The MFPT $t(y_1)$ function is the solution of the following so-called adjoint equation associated with Equation (89):

$$Dq_{eq}^2A_{11,eff}^2(-\epsilon_{[2],[0]})\frac{\partial^2 t(y_1)}{\partial y_1^2} - Dq_{eq}A_{11,eff}(\frac{1}{q_{eq}}\frac{\partial U_{1,eff}}{\partial y_1} + \frac{2q_{eq}\epsilon_{[2],[0]}A_{11,eff}}{y_1})\frac{\partial t(y_1)}{\partial y_1} = -1 \quad (90)$$

provided that suitable boundary conditions are added. The properties of $U_{1,eff}(y_1)$ have been explained in Section 4.3. Accordingly, the boundary conditions adequate for polymerization are the following: $t(y_{abs}) = 0$ (absorption) and $[\partial t(y_1)/\partial y_1]_{y_1=y_{ref}} = 0$ (reflection). The radial distance y_{ref} is supposed to fulfill: $y_{1,0} \leq y_{ref} \leq y_{1,1}$. The radial distance y_{abs} is supposed to be larger than (but close to) $y_{1,2}$. The term $t(y_1)$ is interpreted here as an estimate of the time required for the individual atom 1—near y_{abs} and, thus, far from unit 2 of the *te*—to reach y_1 under the action of the *cat*. If $y_{1,0} \leq y_1 \leq y_{1,1}$, then $t(y_1)$ estimates the time required for unit 1 to become bound to unit 2 of the *te*.

Then, by direct integration, the solution of Equation (90) with those boundary conditions is:

$$t(y_1) = \int_{y_1}^{y_{abs}} \frac{ds_1}{s_1^2D(q_{eq}A_{11,eff})^2}J(s_1) \quad (91)$$

$$J(s_1) = -\int_{y_{ref}}^{s_1} \frac{ds_2s_2^2}{\epsilon_{[2],[0]}(s_2)}\exp[-\frac{1}{2K_BT}\int_{s_2}^{s_1}\frac{ds_3(\partial U_{1,eff}/\partial s_3)}{\epsilon_{[2],[0]}(s_3)}] \quad (92)$$

For definiteness, we have chosen $y_{1,2} = y_{abs}$ in Equation (91).

Reference [30] studied an MFPT for a simpler chemical reaction between two atoms with neither a *te* molecular chain nor a *cat* (and not having required to start out from any sort of mixed Liouville–Wigner equation, as has been the case here). In spite of that relative simplicity, the resulting MFPT is rather similar to that described by Equations (91) and (92). The detailed analysis of the MFPT in [30] can be extended rather directly to Equations (91) and (92) and will be omitted. The resulting MFPT in [30] displays a temperature dependence consistent with the Arrhenius formula for rate constants in chemical reactions, and the same is true for Equations (91) and (92). For brevity, we shall limit ourselves to a direct estimate based upon the properties of $U_{1,eff}$: yielding the Arrhenius formula here.

We suppose that y_1 fulfills $y_{ref} \leq y_1 \leq y_{1,1}$. We choose suitable estimates for $-\epsilon_{[2],[0]}$ to let them to be displaced without large errors outside $\int_{s_2}^{s_1} ds_3$, and we subsequently perform the resulting integration $\int_{s_2}^{s_1} ds_3(\partial U_{1,eff}/\partial s_3)$. Then, we argue that an estimate of the dominant contribution to the exponential inside $J(s_1)$ is: $\exp[\frac{1}{2K_BT}[\frac{U_{1,eff}}{(-\epsilon_{[2],[0]})}(y_+) - \frac{U_{1,eff}}{(-\epsilon_{[2],[0]})}(y_-)]]$. The radial distance y_+ is larger than $y_{1,1}$ and is not far from the y_1 at which $\frac{U_{1,eff}}{(-\epsilon_{[2],[0]})}$ is positive (with repulsive $U_{1,eff}$) and takes on its largest values. The radial distance y_- is smaller than $y_{1,1}$ and is not far from the y_1 at which $\frac{U_{1,eff}}{(-\epsilon_{[2],[0]})}$ is negative (with attractive $U_{1,eff}$) and takes on its minimum values. We remind that $-\epsilon_{[2],[0]}$ is positive (Section 2.7). It is plausible that $-\epsilon_{[2],[0]}(y_+)$ $(-\epsilon_{[2],[0]}(y_-))$ is dominated by the almost continuous spectrum (the bound state). Then, an estimate of the the MFPT is:

$$t(y_1) \simeq \int_{y_1}^{y_{abs}} \frac{ds_1}{s_1^2 D(q_{eq} A_{11,eff})^2} \int_{y_{ref}}^{s_1} \frac{ds_2 s_2^2}{(-\epsilon_{[2],[0]}(s_2))} \times \\ \exp[\frac{1}{2K_BT}[\frac{U_{1,eff}}{(-\epsilon_{[2],[0]})}(y_+) - \frac{U_{1,eff}}{(-\epsilon_{[2],[0]})}(y_-)]] \tag{93}$$

Other contributions to $J(s_1)$ are regarded as subdominant and, hence, discarded. The term $t(y_1)^{-1}$ for $y_{ref} \leq y_1 \leq y_{1,1}$ provides an approximate estimate of a rate constant for polymerization activated by the *cat*. Then, notice that factor $\exp[-\frac{1}{2K_BT}[\frac{U_{1,eff}}{-\epsilon_{[2],[0]}}(y_+) - \frac{U_{1,eff}}{-\epsilon_{[2],[0]}}(y_-)]]$ is responsible for and characteristic of an Arrhenius behavior. It also explains why the inclusion of the *cat* activates polymerization: in fact, the latter (due to $U_{0,eff} < 0$) makes $U_{1,eff}(> 0)$ be smaller than $U_1(> 0)$ in $y_{1,1} \leq y_1 \leq y_{1,2}$.

One can obtain directly another solution of Equation (90) with other boundary conditions: namely, with reflection for $y_1 \geq y_{1,2}$ and reflection for $y_1 \leq y_{1,1}$. However, such a solution has been discarded as it implies properties that disagree with the physically expected ones (implied correctly by Equations (91) and (92)).

5. Conclusions and Discussion

The first part (Section 2) of this work dealt with the basic quantum Wigner function and non-equilibrium equation for a microscopic particle subject to a potential V and to a heat bath (HB) at thermal equilibrium. Previous analyses were extended non-trivially. For simplicity, only $1D$ was considered, with the extension to $3D$ being direct. The case in which V has one bound state (plus an infinite number of discrete states that approximate the standard continuum or scattering states) was considered. The equilibrium Wigner distribution generates an infinite number of orthogonal polynomials in momentum. The latter enabled us to define an infinite family of non-equilibrium moments. Commonly, Wigner functions are employed to evaluate expectation values for suitable operators in phase space (see, for instance [7,8]). On the other hand, the present approach is not specifically concerned with phase space; rather, the former focuses on the information encoded in non-equilibrium moments (depending on the spatial positions of the particles as time evolves and becomes long): specifically, in the lowest moment, which can give probabilistic information. The non-equilibrium Wigner equation yields a general n-term hierarchy for the corresponding moments. A new non-trivial solution of the non-equilibrium hierarchy that combines operator-continued fractions and the infinite series thereof is obtained and analyzed; arguments are given to support its finiteness. In a short thermal wavelength regime (retaining quantum features and adequate for chemical reactions), the non-equilibrium hierarchy is approximated by a three-term one. In a long-time approximation, the approximate three-term hierarchy is, in turn, approximated by a Smoluchovski equation. Among other open issues left open by the present study, we quote here the following: (i) further improvements regarding the theory of the generalized orthogonal polynomials in [41], which can benefit the construction of the $H_{Q,n}(y)$; (ii) further analysis for infinite n-term hierarchies ($n > 3$); and (iii) mathematically rigorous analysis of continued fractions of operators. In fact, in previous works (see [29,30] and references therein) and in the present one, we have handled continued fractions of operators and the infinite series thereof in a formal way, with effort to provide arguments to justify consistency, convergence, and an approximate approach to equilibrium for long t. Recall, for instance, the operator A in Section 2.6. However, regarding (iii) in particular, we have been recognizedly unable to obtain mathematically rigorous results that, to the best of our knowledge, appear to be lacking.

In the second part (Sections 3 and 4) of this work, a new model of the growth (polymerization) of a molecular chain (template, *te*) by binding an individual atom activated by a catalyst is developed in $3D$. The atom, *te*, and *cat* move randomly as solutions in a fluid at rest (playing the role of an HB) in thermal equilibrium. Classical statistical mechanics describe *te* and *cat* approximately. The individual atom is treated quantum-mechanically. Mixed non-equilibrium quantum–classical Wigner–Liouville functions and dynamical equations for the individual atom and for the *te* and *cat*, respectively, are employed. By assuming the latter two to be at thermal equilibrium, integrating over their degrees of freedom, and through a further approximation regarding the degrees of freedom of the *te*, a standard $3D$ effective non-equilibrium Wigner equation is obtained for the individual atom. Upon extending to the latter Wigner equation the moment methods together with the short thermal wavelength and long-time approximations in Sections 2.8 and 2.9, respectively, an approximate $3D$ Smoluchowski equation is obtained for the individual atom. The mean first passage time (MFPT) for the individual atom to become bound to the *te*, facilitated by the *cat*, is considered. The resulting MFPT displays a temperature dependence consistent with the Arrhenius formula for rate constants in chemical reactions.

The following properties of typical insertion polymerizations occurring in practice should be mentioned (see Section 3.6.1 in [32]). Firstly, the *cat* does not emerge unchanged as a consequence of the polymerization process: that is, its final state is different from the initial one. Secondly, it continues to reside, in the form of deteriorated fragments, at the relevant end of the augmented *te* chain.

Motivated by those facts, two comments regarding the model presented here seem in order. Our treatment allows for the catalyst to remain weakly bound to the enlarged chain formed by atom 1 bound to the initial *te* chain: that can, in principle, occur for suitable $U_0(\mathbf{y}_0, \mathbf{y}_0 + \mathbf{y}_1)$. The model, with the chosen numbering of material units and dynamical variables, has been formulated precisely in order to account for that: the individual unit is numbered atom 1 and will become bound to atom 2 (one end atom in the *te*). Then, at the end of the process, the *cat* (numbered as 0) may remain bound to atom 1 as a possible additional extension of the chain (see Figure 1). On the other hand, our treatment does not seem to account for the deterioration of the catalyst: a generalization (outside our scope here) would be required for that.

Figure 1. Catalyst (circle), individual unit atom (box), and freely jointed template formed by 11 units (small boxes). Dashed lines represent binding (bonds). The successive numbering is: catalyst (unit 0), individual atom 1, and, in the template, atom 2, atom 3,. . . , atom 11. After polymerization, atom 1 becomes bound to atom 2, and the catalyst is bound (weakly) to atoms 1 and 2. The term $\mathbf{y}_1$ is the vector (not displayed) along the dashed line from atom 1 (box) to atom 2 (small box).

Funding: This work and the author are related to the following project of Ministerio de Ciencia, Innovacion y Universidades (Spain), Agencia Estatal de Investigacion (AEI, Spain, 10.13039/501100011033) and European Regional Development Fund (ERDF, A Way of making Europe): grant PID2022-136374NB-C21.

Data Availability Statement: Data are contained within the article.

Acknowledgments: The author is grateful to Department of Theoretical Physics, Complutense University, Madrid for its hospitality. The author is grateful to David Fernandez for kind informatics help regarding the present work. The author is an Associate Member of the Instituto de Biocomputación y Física de los Sistemas Complejos, Zaragoza University, Zaragoza, Spain. The author acknowledges the Guest Editors of the Special Issue "180th Anniversary of Ludwig Boltzmann" for their kind invitation to contribute to it.

Conflicts of Interest: The author declares no conflicts of interest.

Appendix A. 3D Single-Unit Polymerization by Classical fj Chain at Equilibrium: Computations

The formula $f_2 = f_2(ß_{3,c,1}) = \int [\mathbf{d\Omega}]_{3,L} f_{eq,2}$ will be studied here by extending [33]. First, Gaussian integrations over $\Pi_{i=2}^{N_{te}-1} d\varphi_i d\pi_{\varphi,c,i}$ in $[\mathbf{d\Omega}]_{3,L}$ are performed by generalizing directly the rotational invariant methods in [60]. The result is:

$$f_2 = \exp[\frac{A_{12}^2(A_2^{-1})_{22}ß_{3,c,1}^2}{2K_BT}]([\int(\frac{[\mathbf{d\Omega}]_{3,2}}{det\Delta_2})^{1/2}]^{-1}\int(\frac{[\mathbf{d\Omega}]_{3,2}}{det\Delta_2})^{1/2}\times$$
$$\exp[-\frac{A_{12}}{2K_BT}\sum_{i,j=2}^{N_{te}-1}(A_{12}(A_2^{-1})_{2i})(ß_{3,c,1}\mathbf{u}_i)((\Delta_{2,1})^{-1})_{ij}(A_{12}(A_2^{-1})_{j2})(ß_{3,c,1}\mathbf{u}_j)] \quad (A1)$$

The $(N_{te}-2)\times(N_{te}-2)$ matrix A_2 with non-vanishing elements $A_{ij}, i,j = 2,\ldots,N_{te}-1$ is symmetric, tridiagonal, and has positive eigenvalues. A_2^{-1} and $detA_2$ are the inverse and the determinant, respectively, of A_2. The $(N_{te}-2)\times(N_{te}-2)$ matrices Δ_2 and $\Delta_{2,1}$ have elements: $(\Delta_2)_{ij} = (A_2^{-1})_{ij}\mathbf{u}_i\mathbf{u}_j$ and $(\Delta_{2,1})_{ij} = A_{12}(A_2^{-1})_{ij}\mathbf{u}_i\mathbf{u}_j$, respectively. The integral in Equation (A1) with $(\Delta_{2,1})^{-1} = 0$ has been studied in [53,60]: it was found that the dominant contributions are equal to one another and come from all tiny domains with $(\mathbf{u}_i\mathbf{u}_j)^2$ close to $+1$. One finds:

$$f_2 \simeq \exp[\frac{A_{12}^2(A_2^{-1})_{22}ß_{3,c,1}^2}{2K_BT}][\int_0^{2\pi}d\varphi_2\int_0^{\pi}d\theta_2\sin\theta_2]^{-1}\int_0^{2\pi}d\varphi_2\int_0^{\pi}d\theta_2\sin\theta_2\times$$
$$\exp[-\frac{A_{12}^2(A_2^{-1})_{22}(\mathbf{u}_2ß_{3,c,1})^2}{2K_BT}] \quad (A2)$$

A_{12} and $(A_2^{-1})_{22}$ account, respectively, for the influences of unit 2 and of all units $2,\ldots,N_{te}-1$ on the dynamics of unit 1. At this stage, one approximates: $-(K_BT)(\nabla_{ß_{3,c,1}}\ln f_2) \simeq A_{12}\sigma ß_{3,c,1}$, $\sigma = -A_{12}(A_2^{-1})_{22}\sigma_1$ with constant σ_1, $0<\sigma_1<1$.

References

1. Balescu, R. *Equilibrium and Nonequilibrium Statistical Mechanics*; Wiley: New York, NY, USA, 1975.
2. Zubarev, D.; Morozov, V.G.; Röpke, G. *Statistical Mechanics of Nonequilibrium Processes*; Akademie Verlag: Berlin, Germany, 1996; Volume 1.
3. Kreuzer, H.J. *Nonequilibrium Thermodynamics and Its Statistical Foundations*; Clartendon Press: Oxford, UK, 1981.
4. Liboff, R.L. *Kinetic Theory: Classical, Quantum and Relativistic Descriptions*, 3rd ed.; Springer: New York, NY, USA, 2003.
5. Weiss, U. *Quantum Dissipative Systems*, 4th ed.; World Scientific: Singapore, 2012.
6. Wigner, E.P. On the quantum correction to thermodynamical equilibrium. *Phys. Rev.* **1932**, *40*, 749–759. [CrossRef]
7. Zakos, C.K.; Fairlie, D.B.; Curtwright, T. *Quantum Mechanics in Phase Space. An Overview with Selected Papers*; World Scientific: Singapore, 2005.
8. Schleich, W.P. *Quantum Optics in Phase Space*; Wiley VCH: Berlin, Germany, 2001.
9. Tasaki, H. From quantum dynamics to the canonical distribution: General picture and a rigorous example. *Phys. Rev. Lett.* **1998**, *80*, 1373–1376. [CrossRef]
10. Golstein, S.; Lebowitz, J.; Tumulka, R.; Zanghi, N. Canonical typicality. *Phys. Rev. Lett.* **2006**, *96*, 050403. [CrossRef] [PubMed]
11. Linden, N.; Popescu, S.; Short, A.J.; Winter, A. Quantum mechanical evolution towards thermal equilibrium. *Phys. Rev. E* **2009**, *79*, 061103. [CrossRef]
12. Reiman, P. Canonical thermalization. *New J. Phys.* **2010**, *12*, 055027. [CrossRef]
13. Short, A.J. Equilibration of quantum systems and subsystems. *New J. Phys.* **2011**, *13*, 053009. [CrossRef]
14. Reimann, P. Foundation of statistical mechanics under experimentally realistic conditions. *Phys. Rev. Lett.* **2008**, *101*, 190403. [CrossRef]
15. Short, A.J.; Farrelly, T.C. Quantum equilibration in finite time. *New J. Phys.* **2012**, *14*, 013063. [CrossRef]
16. Goold, J.; Huber, M.; Riera, A.; del Rio, L.; Skrzypczyk, P. The role of quantum information in thermodynamics—A topical review. *J. Phys. A Math. Theor.* **2016**, *49*, 143001. [CrossRef]
17. Kosloff, R. Quantum Thermodynamics. *Entropy* **2013**, *15*, 2100–2128. [CrossRef]
18. Van Kampen, N.G. *Stochastic Processes in Physics and Chemistry*; Elsevier: Amsterdam, The Netherlands, 2001.
19. Santillan, M. *Chemical Kinetics, Stochastic Processes and Irreversible Thermodynamics*; Springer: Berlin/Heidelberg, Germany, 2014.
20. Mayer, J.E.; Mayer, M.G. *Statistical Mechanics*; John Wiley and Sons: New York, NY, USA, 1977.
21. McQuarrie, D.A. *Statistical Thermodynamics*; Harper and Row: New York, NY, USA, 1964.
22. Overby, J.; Chang, R. *Chemistry*, 4th ed.; McGraw-Hill: New York, NY, USA, 2022.
23. Volkenshtein, M.V. *Biophysics*; Mir Publishers: Moscow, Russia, 1983.
24. Gyftopoulos, E.P.; Beretta, G.P. *Thermodynamics. Foundations and Applications*; Dover Pub. Inc.: New York, NY, USA, 2005.
25. Grandy, W.T., Jr. *Foundations of Statistical Mechasnics Volume II: Nonequilibrium Phenomena*; Reidel: Dordrecht, The Netherlands, 1988.
26. Wilde, R.E.; Singh, S. *Statistical Mechanics. Fundamentals and Modern Applications*; John Wiley and Sons: New York, NY, USA, 1998.

27. Ottinger, H.C. *Beyond Equilibrium Thermodynamics*; John Wiley and Sons, Inc.: Hoboken, NJ, USA, 2005.
28. Lebon, G.; Jou, D.; Casas-Vazquez, J. *Understanding Non-Equilibrium Thermodynamics*; Oxford University Press: Oxford, UK, 2008.
29. Alvarez-Estrada, R.F. Approach to Equilibrium of Statistical Systems: Classical Particles and Quantum Fields Off-Equilibrium. *Dynamics* **2023**, *3*, 345–378. [CrossRef]
30. Alvarez-Estrada, R.F.; Calvo, G.F. Chemical reactions using a non-equilibrium Wigner function approach. *Entropy* **2016**, *18*, 369. [CrossRef]
31. Doi, M.; Edwards, S.F. *The Theory of Polymer Dynamics*; Oxford University Press: Oxford, UK, 1988.
32. Elias, H.-G. *An Introduction to Polymer Science*; VCH: New York, NY, USA, 1997.
33. Alvarez-Estrada, R.F. Quantized Constrained Molecular Chains: Vibrations, Internal Rotations and Polymerization. *Contemp. Math.* **2022**, *3*, 353–385. [CrossRef]
34. Messiah, A. *Quantum Mechanics*; North-Holland: Amsterdam, The Netherlands, 1961; Volume 1.
35. Landau, L.D.; Lifshitz, E.M. *Quantum Mechanics: Non-Relativistic Theory*; Pergamon: London, UK, 1977.
36. Huang, K. *Statistical Mechanics*, 2nd ed.; John Wiley and Sons: New York, NY, USA, 1987.
37. Hillery, M.; O'Connell, R.F.; Scully, M.O.; Wigner, E.P. Distributions functions in Physics: Fundamentals. *Phys. Rep.* **1984**, *106*, 121–167. [CrossRef]
38. Hochstrasser, U.W. Orthogonal Polynomials. In *Handbook of Mathematical Functions*; Abramowitz, M., Stegun, I.A., Eds.; Dover: New York, NY, USA, 1965.
39. Ballantine, L.E. *Quantum Mechanics: A Modern Development*; World Scientific: Singapore, 2000.
40. Hudson, R.L. When is the Wigner quasi-probability density non-negative? *Rep. Math. Phys.* **1974**, *6*, 249–252. [CrossRef]
41. Chihara, T.S. *An Introduction to Orthogonal Polynomials*; Gordon and Breach: New York, NY, USA, 1978.
42. Brinkman, H.C. Brownian motion in a field of force and the diffusion theory of chemical reactions. *Physica* **1956**, *22*, 29–34. [CrossRef]
43. Garcia-Palacios, J.L.; Zueco, D. The Caldeira-Leggett quantum master equation in Wigner phase space: Continued-fraction solutions and applications to Brownian motion in periodic potentials. *J. Phys. A Math. Gen.* **2004**, *37*, 10735–10770. [CrossRef]
44. Jones, W.H.; Thron, W.J. *Continued Fractions, Analytic Theory and Applications*; Addison-Wesley Pub. Co.: Reading, MA, USA, 1980.
45. Antippa, A.F.; Phares, A.J. General formalism solving linear recurrence relations. *J. Math. Phys.* **1977**, *18*, 173–181. [CrossRef]
46. Gautschi, W. Error Function and Fresnel Integrals. In *Handbook of Mathematical Functions*; Abramowitz, M., Stegun, I.A., Eds.; Dover: New York, NY, USA, 2008.
47. Risken, H. *The Fokker-Planck Equation*, 2nd ed.; Springer: Berlin, Germany, 1989.
48. Coffey, W.T.; Kalmykov, Y.P. *The Langevin Equation*, 3rd ed.; World Scientific: Singapore, 2012.
49. Coffey, W.T.; Kalmykov, Y.P.; Titov, S.V.; Mulligan, B.P. Wigner function approach to the quantum Brownian motion of a particle in a potential. *Phys. Chem. Chem. Phys.* **2007**, *9*, 3361–3382. [CrossRef]
50. des Cloiseaux, J.; Jannink, J.F. *Polymers in Solution. Modelling and Structure*; Clarendon Press: Oxford, UK, 1990.
51. Grosberg, A.Y.; Khokhlov, A.R. *Statistical Physics of Macromolecules*; American Institute of Physics Press: New York, NY, USA, 1994.
52. Kleinert, H. *Path Integrals in Quantum Mechanics, Statistics, Polymer Physics and Financial Marketing*, 4th ed.; World Scientific: Singapore, 2006.
53. Alvarez-Estrada, R.F.; Calvo, G.F. Models for polymers and biopolymers based on Quantum Mechanics. *Mol. Phys.* **2002**, *100*, 2957–2970. [CrossRef]
54. Alvarez-Estrada, R.F.; Calvo, G.F. Constrained macromolecular chains at thermal equilibrium. *Eur. Phys. J. Spec. Top.* **2011**, *200*, 225–258. [CrossRef]
55. Calvo, G.F.; Alvarez-Estrada, R.F. Three-dimensional models for homogeneous DNA near denaturation. *J. Phys. C Condens. Matter* **2005**, *17*, 7755–7781. [CrossRef]
56. Messiah, A. *Quantum Mechanics*; North Holland: Amsterdam, The Netherlands, 1962; Volume 2.
57. Landau, L.D.; Lifchitz, E.M. *Mechanics. Volume 1. Course of Theoretical Physics*, 3rd ed.; Elsevier: Amsterdam, The Netherlands, 1976.
58. Haengi, P.; Talkner, P.; Borkovec, M. Reaction-rate theory: Fifty years after Kramers. *Rev. Mod. Phys.* **1990**, *62*, 251–342. [CrossRef]
59. Calvo, G.F.; Alvarez-Estrada, R.F. The time duration for DNA thermal denaturation. *J. Phys. C Condens. Matter* **2008**, *20*, 035111. [CrossRef]
60. Alvarez-Estrada, R.F. Models of macromolecular chains based on Classical and Quantum Mechanics: Comparison with Gaussian models. *Macromol. Theory Simul.* **2000**, *9*, 83–114. [CrossRef]

entropy

MDPI

Article

Probability Turns Material: The Boltzmann Equation

Lamberto Rondoni [1,2,*] and Vincenzo Di Florio [1,2,3,*]

1 Dipartimento di Scienze Matematiche, Politecnico di Torino, Corso Duca Degli Abruzzi 24, 10129 Turin, Italy
2 INFN, Sezione di Torino, Via Pietro Giuria 1, 10125 Turin, Italy
3 Istituto Italiano di Tecnologia (IIT), Via Melen 83, B Block, 16163 Genoa, Italy
* Correspondence: lamberto.rondoni@polito.it (L.R.); vincenzo.diflorio@polito.it (V.D.F.)

Abstract: We review, under a modern light, the conditions that render the Boltzmann equation applicable. These are conditions that permit probability to behave like mass, thereby possessing clear and concrete content, whereas generally, this is not the case. Because science and technology are increasingly interested in small systems that violate the conditions of the Boltzmann equation, probability appears to be the only mathematical tool suitable for treating them. Therefore, Boltzmann's teachings remain relevant, and the present analysis provides a critical perspective useful for accurately interpreting the results of current applications of statistical mechanics.

Keywords: dynamical systems; probability; observables

1. Introduction

Statistical mechanics arose to interpret thermodynamic results from an atomistic perspective. That is an objectively difficult task because thermodynamics is deterministic. It views matter as a continuum characterized by a reduced number of variables, and it is irreversible in time. Partial differential equations are the suitable mathematical tool for such a description of physical objects. On the contrary, the microscopic dynamics of atoms are considered deterministic when dealing with fluids in non-extreme conditions. However, they are time-reversal invariant, with a few notable exceptions, such as Prigogine [1]. Being discrete rather than continuous, these dynamics are represented by systems of many ordinary differential equations. The difficulty, then, does not lie in doubts about the atomistic hypothesis, which have been dispelled long ago [2], when the Brownian motion was explained [3]. The difficulty concerns the above very different mathematical descriptions that have been developed for macroscopic and microscopic dynamics. Indeed, they are so different that they could even look incompatible.

With statistical physics, the gap has been filled adding statistical notions to the purely mechanical ones, and profiting from the mind-boggling number of degrees of freedom that must be integrated out to go from atoms to continuum [4]. This second aspect was initially overlooked by Boltzmann's contemporaries, who criticized him on the grounds of the laws of mechanics [5]. As a matter of fact, certain sentences in Boltzmann's fundamental paper [6] prompted criticism because they lack reference to probabilistic arguments and seem to imply a direct derivation of irreversible relaxation to equilibrium from the purely reversible dynamics of atoms:

> *"It has thus been rigorously proved that, whatever may be the initial distribution of kinetic energy, in the course of a very long time it must always necessarily approach the one found by Maxwell. The procedure used so far is of course nothing more than a mathematical artifice employed in order to give a rigorous proof of a theorem whose exact proof has not previously been found. It gains meaning by its applicability to the theory of polyatomic gas molecules".*

Whether this interpretation of Boltzmann's intention is correct or not, Loschmidt and Zermelo legitimately, and correctly, pointed out that mechanics by itself does not justify the result. Boltzmann's reply, then, set things straight once and for all. Although, to date, the argument can be considered quite subtle. In a sense, Boltzmann agreed with the criticism, without admitting any error in his original point of view, but simply elaborating on it. He invoked a probabilistic argument, which does not belong to mechanics, but is powerfully unavoidable: given the number of molecules that make a macroscopic object, it is unreasonable not to take seriously the probabilistic predictions. In his own words, this is Boltzmann's answer to Zermelo:

> *"The applicability of probability theory to a particular case cannot of course be proved rigorously. If, out of 100,000 objects of a certain kind, about 100 are annually destroyed by fire, then we cannot be sure that this will happen next year. On the contrary, if the same conditions could be maintained for* $10^{10^{10}}$ *years, then during this time it would often happen that all 100,000 objects would burn up on the same day; and likewise there will be entire years during which not a single object is damaged. Despite this, every insurance company relies on probability theory. It is even more valid, on account of the huge number of molecules in a cubic millimetre, to adopt the assumption (which cannot be proved mathematically for any particular case) that when two gases of different kinds or at different temperatures are brought in contact, each molecule will have all the possible different states corresponding to the laws of probability and determined by the average values at the place in question, during a long period of time. These probability arguments cannot replace a direct analysis of the motion of each molecule; yet if one starts with a variety of initial conditions, all corresponding to the same average values (and therefore equivalent from the viewpoint of observation), one is entitled to expect that the results of both methods will agree, aside from some individual exceptions which will be even rarer than in the above example of 100,000 objects all burning on the same day. The assumption that these rare cases are not observed in nature is not strictly provable (nor is the entire mechanical picture itself) but in view of what has been said it is so natural and obvious, and so much in agreement with all experience with probabilities, from the method of least squares to the dice game, that any doubt on this point certainly cannot put in question the validity of the theory when it is otherwise so useful. It is completely incomprehensible to me how anyone can see a refutation of the applicability of probability theory in the fact that some other argument shows that exceptions must occur now and then over a period of eons of time; for probability theory itself teaches just the same thing".*

(We would like to stress one aspect of this reply that is also quite important, although not directly related to the debate between Boltzmann and Zermelo, hence usually not considered. Boltzmann clearly indicates that a mathematical explanation of a physical phenomenon cannot be taken as applicable without restrictions. Indeed, any mathematical model is just that: it cannot replace reality in all its aspects. It suffices to keep this in mind to realize paradoxes may simply lie in the incorrect application of a theory to situations exiting its range of validity. The problem may be not lack of mathematical rigour, but of physical relevance.) One should note that, strictly speaking, Boltzmann reasonings apply to a special class of macroscopic objects, the rarefied gases. Nevertheless, over the years, the Boltzmann equation has been extended to an incredibly wide range of systems and phenomena. First, within the realm of physics, it has been extended to classical systems subjected to external forces, as well as to quantum and relativistic mechanics, with applications, e.g., in solid state physics and astrophysics, cf., e.g., [7,8]. Successively, applications have been produced in fields such as social sciences, economics, and traffic studies [9–13], just to mention a few examples. In mathematics, the study of the Boltzmann equation has also produced fundamental results, of general interest. For instance, the proof of the existence of solutions [14]. More recently, Cercignani's conjecture [15], that has been influential in understanding dissipation from a microscopic point of view, was proven to be substantially correct [16].

Even through all that, one comes to realize that probability, the primary tool of statistical physics, is crucial for accurately handling quantities related to discrete lumps of matter composed of numerous microscopic constituents, especially on the scale where they seem to exhibit continuous behavior. But, thanks to this tool, statistical physics may in principle be used even beyond of the realm of thermodynamics if is it true, as it is, that even insurance companies rely on it. In fact, statistical physics has turned its attention more and more to the study of small systems, which are not guaranteed to be in Local Thermodynamic Equilibrium (LTE), and require thermodynamic quantities to be interpreted in a different or wider sense than in thermodynamics. Indeed, LTE requires, at least, that systems should possess a microscopic, mesoscopic and macroscopic scale in space and time w.r.t the observer. This condition cannot be ensured for all systems to which statistical physics is applied. This can lead, for instance, to finite size effects and fluctuations entailing blurred boundaries between different states of aggregation matter. This gives probabilities an even wider role in the description of small systems than in the statistical mechanics of macroscopic objects. Often, probability is the only sensible thing to compute. The trouble is that probability is immaterial.

In this paper, we analyze various notions of probability, highlighting this fact, and then deomnstrate how it may be seen turning material within the framework of the Boltzmann equation. This may be useful in correctly interpreting a large fraction of present day statistical mechanics results.

2. Probability: Obscure and Practical

How can probability be useful? Karl Popper maintained that probabilistic results cannot be falsified, thus contradicting his concept of science: *For although probability statements play such a vitally important role in empirical science, they turn out to be in principle impervious to strict falsification"* [17]. Further:

> *"In whatever way we may define the concept of probability, or whatever axiomatic formulations we choose: so long as the binomial formula is derivable within the system, probability statements will not be falsifiable. Probability hypotheses do not rule out anything observable; probability estimates cannot contradict, or be contradicted by, a basic statement; nor can they be contradicted by a conjunction of any finite number of basic statements; and accordingly not by any finite number of observations either"* [17].

Of course, Popper's view of science can be criticized, but there is a fact: suppose meteorologists have computed that the probability of rain tomorrow is 0.999. We go out carrying an umbrella, and we find the Sun is shining. Have we refuted the meteorologists' prediction? Of course not. Consequently, if it rains it also does not prove anything about the mentioned calculations. So, why is probability interesting?

In the mathematical sciences, probability is a rather recent subject. In the XVI century, it was conceived to improve chances of winning in gambling. However, it rapidly came to constitute the basis of statistical and quantum mechanics, as well as of chaos theory and of the sciences of "complexity". It is dominant in finance, in the interpretation of medical tests, in epidemiology, weather forecasts, etc. On the other hand, can we grasp the meaning of a statement like:"there is a probability of 10^{-40} that a nuclear power plant blows up"? When and how will we experience a number like that, so that we develop an opinion about its significance? What do scientists say?

Joh Archibald Wheeler, who would constantly make use of probabilities in his scientific activity, stated that *"Probability, like time, is a concept invented by humans, and humans have to bear the responsibility for the obscurities that attend it. Obscurities there are whether we consider probability defined as frequency or defined a la Bayes"* [18].

The great probabilist, Bruno de Finetti, in the preface of his treatise on probability theory said,

> *My thesis, paradoxically, and a little provocatively, but nonetheless genuinely, is simply this:*
>
> *PROBABILITY DOES NOT EXIST*
>
> *The abandonment of superstitious beliefs about the existence of Phlogiston, the Cosmic Ether, Absolute Space and Time, ..., or Fairies and Witches, was an essential step along the road to scientific thinking. Probability, too, if regarded as something endowed with*

some kind of objective existence, is no less a misleading misconception, an illusory attempt to exteriorize or materialize our true probabilistic beliefs [19].

Clearly, profound familiarity with the subject does not eliminate obscurities, and probability appears *immaterial*. Simple considerations may help us understand these statements.

The classical notion of probability gives an objective rule to attribute a probability to some expected event E. For **finitely many** possible events, N say, the classical probability is objectively defined by counting as

$$\text{Prob}(E) = \frac{N(E)}{N} \tag{1}$$

where $N(E)$ is number of possible outcomes yielding E. This means that the single events are declared equally probable. Coin tossing is the usual example of this kind: the single events are *heads* and *tails*; therefore, each of them is attributed a probability of $1/2$. It is a simple and appealing notion of probability, that looks particularly useful to deal with situations that totally defeat our understanding. However, it is obviously not suitable e.g., for tossing loaded coins or dice; or for YES or NO answers to "Is it going to rain?" Moreover, paradoxes arise if the set of possible events cannot be counted, particularly for continuous sets of events. Think, for instance, of the Bertrand Paradox: there are three different, apparently equally legitimate, probabilities for the event L, which represents the probability that a chord of a circle of unit radius is longer than $\sqrt{3}$, the length of the side of the inscribed equilateral triangle:

- given two points on the circumference, consider the chord joining them, and take one of the two points as the vertex of one inscribed equilateral triangle. If the chord lies outside the triangle, it is shorter than $\sqrt{3}$; if it intersects the triangle, it is longer than or equal to $\sqrt{3}$. Now, the vertices of the triangle delimit three arcs of equal length, and only one of them corresponds to a chord longer than $\sqrt{3}$. Therefore, the probability of L is $1/3$.
- consider a radius of the circle, and choose a point on it. There is a chord perpendicular to this radius, and its length is larger than $\sqrt{3}$ if the point is closer to the center of the circle than to the circumference. Therefore, the probability of L is $1/2$.
- take a point anywhere within the circle and construct a chord with this point as its midpoint. The chord is longer than $\sqrt{3}$ if the point lies within a concentric disk of radius $1/2$ the radius of the larger circle. The area of the smaller circle is $1/4$ of the full disk; therefore, the probability of L is $1/4$.

Despite the numerous *solutions* that have been devised to solve this conundrum, see, e.g., Ref. [20], it remains that they all refer to some kind of superior principle that can only be subjective. Hence, they cannot be taken as objectively valid, at least not in the sense of the mass of a stone. One may then adopt a subjective notion of probability at the start. This amounts to saying that probability should be assigned by the one interested in it, defining

$$\text{Prob}(E) = \text{maximum someone consistently puts at stake to win 1 in an unbiased game}$$

There is an immediate advantage with respect to the classical notion of probability: it is always possible to unambiguously associate a probability with an event, even in the question of whether or not I am a robot. But there are difficulties as well. First of all, different people may have different views on the event E, and that may make the scientific activity cumbersome.

In some cases, one may think that probabilities are objectively imposed by the sequence of events. One then may rely on a different kinds of counting events, thinking that at each time only a finite number of events has occurred, but assuming that extrapolation to infinitely many events is possible. This constitutes the frequentist approach. Let N_t be the number of observations, or experiments repeated in time up to time t. Then,

$$\text{Prob}(E) = \lim_{t\to\infty} \frac{N(E;t)}{N_t}$$

where $N(E;t)$ is the number of outcomes regarding E in the time t. This notion of probability is very suggestive e.g., in dealing with situations that are often repeated in time or space, such as weather forecasts or car accidents. However, many events happen only once or a few times in a lifetime, and can be dramatically positive or negative, like nuclear power plants accidents or becoming Captain Regent of the San Marino Republic. Even harder problems are posed by events that happen frequently, but not enough for us to extract a reliable predictive rule, e.g., earthquakes or pandemics.

From these considerations, one might ask why bother at all to use probabilities. The fact is that the above has not stopped people from referring to probabilities. This is understandable if one realizes that probability, like any other mathematical tool, yields a model, one description of one aspect of reality, and does not intend to represent the whole reality. However, as such, it can change its meaning depending on the problem at hand; and it can do so without causing any inconvenience, enabling us to better understand the problems at hand. Indeed, James Clerk Maxwell argued that

> *They say that Understanding ought to work by the rules of right reason. These rules are, or ought to be, contained in Logic; but the actual science of Logic is conversant at present only with things either certain, impossible, or entirely doubtful, none of which (fortunately) we have to reason on. Therefore the true Logic for this world is the Calculus of Probabilities, which takes account of the magnitude of the probability (which is, or which ought to be in a reasonable man's mind). This branch of Math., which is generally thought to favour gambling, dicing, and wagering, and therefore highly immoral, is the only 'Mathematics for Practical Men', as we ought to be* [21].

How can one be practical? The approach that has proven successful is refraining from adopting a specific interpretation of probability, in general, and to develop a mathematical theory within which unambiguous conclusions can be reached, by means of suitable calculations. In doing so, one must also bear in mind that results will refer to a rational framework, not to reality, and connection with reality will have to be assessed case by case. The mathematical framework suitable to this task is that of axiomatic probability, briefly summarized as follows:

- There is a set Ω called sample space, which contains all the "elementary" events ω;
- there is a collection of subsets of Ω, $\mathcal{F}(\Omega)$ that has the structure of a $\sigma-$algebra of subsets $E \subset \Omega$, representing all (elementary and combined) events to which a probability is to be assigned;
- for all $E \in \mathcal{F}$, $P(E) \geq 0$, $P(\emptyset) = 0$, and $P(\Omega) = 1$, where $\emptyset$ is the empty set;
- for $E_1 \cap E_2 = \emptyset$ one has $P(E_1 \cup E_2) = P(E_1) + P(E_2)$;
- given an infnite sequence of disjoint events, $\{E_i\}_{i=1}^{\infty}$ with $E_i \cap E_j = \emptyset$ if $i \neq j$, one has

$$P\left(\bigcup_{i=1}^{\infty} E_i\right) = \sum_{i=1}^{\infty} P(E_i) \tag{2}$$

Given these axioms, the rest follows, provided one has a suitable method of assigning probabilities to the elementary events. However, this method does not necessarily need to be a general rule endowed with a specific interpretation.

When do we need probabilities? When events (apparently) follow no rule, like the outcomes of coin tossing, or the trajectories of pollen grains in water, that are called random phenomena. Singularly, such phenomena follow no rule, but it may happen that their collective large-scale behaviour is described by (possibly evolving) probabilities that follow precise rules, which are called stochastic processes. Deterministic processes, on the other hand, are singularly subjected to strict rules, such as those of classical mechanics, in which the future is totally determined by the initial conditions. However, because such evolutions require exact knowledge of the initial conditions, something that is never achievable in practice, probabilities play a fundamental role also in the case of deterministic evolutions, especially those that are known as chaotic. As a matter of fact, determinism and randomness are but points of view, more or less convenient, depending on many factors. One cannot empirically decide the "true" nature of a given phenomenon. One can only try to make our descriptions effective: i.e., satisfactory for our purposes, and we are satisfied when we can make reliable predictions.

Let us investigate how such an elusive and immaterial notion as probability became concrete and, in fact, "material" thanks to Boltzmann's work and the Boltzmann equation. This may then serve as a guide to the use of probabilities in realms in which they often are the only suitable mathematical tool, but still bear the consequences of being "obscure" and "immaterial".

3. Mass vs. Probability

Consider N particles, each with d degrees of freedom, which obey the equations of motion $\dot{x} = G(x)$, on their phase space $\mathcal{M}$, which is n-dimensional, with $n = 2dN$. Let Φ^t represent the operator, which yields the solution of the equation of motion at time t, $\Phi^t x$, if x is the initial condition. The functions of phase, $\mathcal{O} : \mathcal{M} \longrightarrow \mathbb{R}$, are considered microscopic properties of such particles. Then, time averages

$$\overline{\mathcal{O}}(x) = \lim_{t\to\infty} \frac{1}{t} \int_0^t \mathcal{O}(\Phi^s x) ds \tag{3}$$

are postulated to represent macroscopically observable quantities. The idea is that a measurement tool and the macroscopic object of interest interact for a macroscopic time, which is very long compared to microscopic times, and the result is represented by the time average of the quantity of interest. Assuming that one is treating objects amenable to thermodynamic analysis, which are described by a handful of observables, the $t \to \infty$ limit results in a very good mathematical idealization of the measurement time, since that may last 10^{12} or more microscopic units. Moreover, the dependence on the initial condition x of a trajectory, $\gamma(x) = \{\Phi^t x\}_{t\in\mathbb{R}}$, representing the microscopic history of a given (single) object of interest, is irrelevant. (This is not necessarily true in small systems, in which local thermodynamic equilibrium is not verified; hence, thermodynamic relations may be inapplicable.) This suggests that the time average (3) may be obtained by replacing the impossible-to-compute phase space trajectory with a phase space probability distribution μ that weighs the different regions of $\mathcal{M}$ proportionally to the frequency with which $\gamma(x)$ visits them:

$$\overline{\mathcal{O}}(x) = \int_{\mathcal{M}} \mathcal{O} d\mu = \mathbb{E}_\mu[\mathcal{O}] . \tag{4}$$

Here, the right hand side is called the phase space average. Because the left hand side of Equation (4) depends on x, while the right hand side does not, one must specify that the equality is not expected to hold for all $x \in \mathcal{M}$. At most, one can ask that it holds for *μ-almost every $x \in \mathcal{M}$*, i.e., as long as its possible violations are confined within a set $E \in \mathcal{M}$ of vanishing probability:

$$\mu(E) = \int_E d\mu = 0,$$

which by definition does not affect the integral in Equation (4).

The investigation of the validity of Equation (4) has led to the development of ergodic theory, concerning the properties of dynamical systems and their invariant measures. In its standard formulation, this mathematical theory states that there are four equivalent properties called *ergodicity*, including the validity of Equation (4). Letting $(\mathcal{M}, \mu, \Phi)$ be a dynamical system with μ, an invariant measure under the dynamics Φ defined on the phase space $\mathcal{M}$, the following properties are equivalent:

(E1) For every integrable function $\mathcal{O} : \mathcal{M} \mapsto \mathbb{R}$, time average and phase space average coincide for μ-almost every x, which means for all $x \in \mathcal{M}$ apart from a set E of vanishing measure, i.e., such that $\mu(E) = 0$:

$$\overline{\mathcal{O}}(x) = \mathbb{E}_\mu(\mathcal{O}) , \quad \mu - \text{a.e. } x \in \mathcal{M} \tag{5}$$

(E2) For every measurable $A \subset \mathcal{M}$ and for μ-a.e. $x \in \mathcal{M}$, the fraction of time a trajectory spends in A is given by:

$$\tau_A(x) = \mu(A) , \quad \mu - \text{a.e. } x \in \mathcal{M} \tag{6}$$

where the fraction of time is defined by

$$\tau_A(x) = \lim_{t\to\infty} \frac{1}{t} \int_0^t \chi_A(\Phi^s x) ds \tag{7}$$

(E3) There are no non-trivial integrals of motion. In other words, let $\mathcal{O}$ be a function of phase, such that $\mathcal{O}(\Phi^t x) = \mathcal{O}(x)$ for all t and for μ-a.e. $x \in \mathcal{M}$, and let $\mathcal{O}$ be integrable. Then,

$$\mathcal{O}(x) = \text{constant} , \quad \mu - \text{a.e.} x \in \mathcal{M} \tag{8}$$

(E4) The dynamical system is metrically indecomposable. In other words, let A be an invariant measurable set, i.e., $\Phi^t A = A$ for all t. Then, either $\mu(A) = 1$ or $\mu(A) = 0$. When that is the case, the expression $\mathcal{M} = A \cup (\mathcal{M} \backslash A)$ is called metrically trivial decomposition of $\mathcal{M}$.

It is easy to see that ergodicity may be realized in simple dynamical systems whose phase space trajectories $\gamma(x)$ move parallel to each other and with the same phase space "velocity", i.e., same $G(x)$. Take for instance a two-dimensional torus as $\mathcal{M}$, and let trajectories be oriented so that they do not close on a periodic orbit. (Cutting the torus to obtain a square, with periodic boundary conditions, it suffices that the components of G be not rational with respect to each other.) This is but an example of an ergodic system whose trajectories remain close to each other in time, while they explore the whole phase space, thus satisfying E2. Now, imagine that probability belongs to phase space sets like mass belongs to chunks of matter moving in space. In other words, let us mimic the conservation of mass in real space with the conservation of probability in phase space, although the difference between the two is huge, as is shown even by the dimensions of real space, which is 3, and of $\mathcal{M}$, which is $O(2 \times 3 \times 10^{23})$ for spherical atoms. That means that probabilities may evolve, obeying

$$\mu_t(E) = \mu_0(\Phi^{-t} E) \tag{9}$$

and, if the evolving probability has a density, it satisfies the (generalized) Liouville equation

$$\frac{\partial f}{\partial t} = -\text{div}_\Gamma(fG) = -G \cdot \nabla_\Gamma f - f \, \text{div}_\Gamma G \tag{10}$$

If the equilibrium distribution is the microcanonical ensemble, which, for an ideal gas, is the uniform distribution of microstates with specific energy, and if the initial distribution is not uniform, the ergodic property does not lead in time to the equilibrium distribution. In our example of parallel trajectories, a lump of probability continues to go round and round in $\mathcal{M}$, never spreading to become uniform. This will be revealed by the phase space averages of observables that will correspondingly fluctuate in time, as the support of this initial probability moves within $\mathcal{M}$. If one pushes further the analogy between probability and mass, taking the spread of probability in phase space as representing the spread of the molecules of a gas in their container, ergodicity does not suffice to describe the *irreversible* relaxation of macroscopic states to their equilibrium. Therefore, a stronger property, that implies ergodicity, has been considered to describe the most common irreversible phenomenon, see, e.g., Ref. [22]. This condition is called *mixing*, and is defined by two equivalent properties:

(M1) For every pair of measurable sets in the phase space, $A, B \subset \mathcal{M}$, the following holds:

$$\lim_{t\to\infty} \mu(\Phi^{-t}A \cap B) = \mu(A)\mu(B), \quad \text{or} \quad \lim_{t\to\infty} \mu(A \cap \Phi^{t}B) = \mu(A)\mu(B) \tag{11}$$

(M2) for every pair of observables, e.g., $\mathcal{O}, \mathcal{P} \in L_2(\mathcal{M}, \mu)$, the following holds:

$$\lim_{t\to\infty} \int_{\mathcal{M}} (\mathcal{O} \circ \Phi^t)\mathcal{P} d\mu = \int_M \mathcal{O} d\mu \int_M \mathcal{P} d\mu, \tag{12}$$

i.e., $\lim_{t\to\infty} \mathbb{E}_\mu\left[(\mathcal{O} \circ \Phi^t)\mathcal{P}\right] = \mathbb{E}_\mu(\mathcal{O})\mathbb{E}_\mu(\mathcal{P})$

The meaning of $M1$ is that, in time Φ^t, B will occupy a fraction of A equal to its measure in the full $\mathcal{M}$. As A can be taken in any region of phase space, in particular within the support of an invariant measure under consideration, this means that an initial ensemble supported in B will reach in time and will stay in any corner of the support of μ. When the microcanonical ensemble is the relevant invariant measure, this means that the probability "contained" in B will eventually be uniformly distributed over the whole $\mathcal{M}$. This can be seen observing that the invariance of μ implies $\mu(A \cap \Phi^t B) = \mu(\Phi^{-t}(A \cap \Phi^t B)) = \mu(\Phi^{-t}A \cap B)$, and $\mu(A)\mu(\Phi^t B) = \mu(A)\mu(B)$. Then, for sufficiently long times, one may approximately write

$$\frac{\mu(A \cap \Phi^t B)}{\mu(A)} \approx \mu(B), \quad \text{provided} \quad \mu(A) \neq 0 \tag{13}$$

which becomes exact in the $t \to \infty$ limit.

Property M2, in particular, expresses the decay of correlations of microscopic events in phase space. When the correlation function of the functions $\mathcal{O}$ and $\mathcal{P}$, defined by

$$G(t) = \mathbb{E}_\mu\left[(\mathcal{O} \circ \Phi^t)\mathcal{P}\right] - \mathbb{E}_\mu(\mathcal{O})\mathbb{E}_\mu(\mathcal{P}) \tag{14}$$

does not vanish, the measurements of $\mathcal{O}$ and $\mathcal{P}$ are, in a sense, not independent. The decay in time of such correlation indicates a sort of irreversible loss of memory about the initial state. Indeed, a simple calculation appears to confirm that relaxation to an equilibrium state, for a Hamiltonian particle system, is achieved under mixing [23]:

$$\begin{aligned} \mathbb{E}_{\mu_t}(\mathcal{O}) &= \int_{\mathcal{M}} \mathcal{O}\rho_t \mathrm{d}x = \int_{\mathcal{M}} \mathcal{O}(\rho_0 \circ \Phi^{-t})\mathrm{d}x \\ &= \int_{\mathcal{M}} (\mathcal{O} \circ \Phi^t)\rho_0 \mathrm{d}x \overset{t\to\infty}{\longrightarrow} \int_{\mathcal{M}} \mathcal{O}\mathrm{d}x \int_{\mathcal{M}} \rho_0 \mathrm{d}x = \mathbb{E}_{\mathrm{d}x}(\mathcal{O}). \end{aligned} \tag{15}$$

Here, the dynamics is assumed to be phase space volume-preserving, so that the Jacobian determinant denoted by J^{-t} equals 1. Then, $\rho_0 =$ constant is the invariant probability density (its integral equals 1), and mixing has been applied, considering the probability density as a single observable. The result is that the initial value of the observable irreversibly converges to its microcanonical value. A mixing counterpart to the uniform translation on the two-dimensional torus is the so-called Arnold's CAT map, defined by

$$\begin{pmatrix} x' \\ y' \end{pmatrix} = \begin{pmatrix} 1 & 1 \\ 1 & 2 \end{pmatrix} \begin{pmatrix} x \\ y \end{pmatrix} \mod 1\,. \tag{16}$$

Despite the apparent formal consistency, numerous difficulties arise with these phase space calculations. Although the same rules applying to mass have been applied to probability, the second obviously remains an *immaterial* abstract concept. For instance, mixing also holds for systems made of a single particle confined within convex walls [24], which surely cannot expand to fill its container, while the probability does. One conceptual difficulty with Equation (15) is that the probability density ρ_t is treated like a microscopic physical property of a given object, which it is not. It represents a probability, that is a completely different object. One microscopic quantity, the kinetic energy say, is an objective quantity defined for a single concrete object of interest, which is represented by a single geometric point in the phase space. On the other hand, a probability may either refer to collections of systems, or sequences of experiments, each represented by a single phase in the phase space, or it can be subjectively introduced. In all cases, a distribution of phases in phase space is obtained, but the degree of crowding of phase points in different regions has no bearing on the evolution of a given single system. For example, beyond the cases of systems with a single particle, numerous large systems (high probability) could exist within a small region of phase space, each corresponding to uniform mass density. In such instances, the probability would evolve while the mass of the systems of interest remains unchanged. The evolution of phase space probabilities is not directly related to what pertains to thermodynamic quantities. That is solely determined by the equations of motion and the corresponding initial condition. Being associated with a higher or lower amount of companion phases does not modify its fate.

The question arises: under which conditions does probability turn *material*? Indeed, acknowledging, as Maxwell implied, that we have no better tool than probability does not imply that it can be used uncritically. We get a glimpse of that in Poincaré's considerations on the laws that govern the large number of molecules constituting a gas:

> *"It seems at first that the orderless collisions of this innumerable dust can only engender an inextricable chaos before which the analyst must retire. But the law of great numbers, that supreme law of chance, comes to our assistance. In face of a semi-disorder we should be forced to despair, but in extreme disorder this statistical law re-establishes a kind of average or mean order in which the mind can find itself again"* [25].

Understanding these facts definitely helps in uncharted territories. Great numbers and extreme disorder seem to be essential, but often they are not present.

4. Hamiltonian Particle Systems and Ensembles

Because a priori it is not possible to decide which microscopic dynamics are better suited to the task of describing macroscopic systems, one almost universally begins with classical Hamiltonian mechanics, proven extremely successful in describing an incredible variety of macroscopic phenomena. This was Laplace's opinion:

> *"We may regard the present state of the universe as the effect of its past and the cause of its future. An intellect which at a certain moment would know all forces that set nature in motion, and all positions of all items of which nature is composed, if this intellect were also vast enough to submit these data to analysis, it would embrace in a single formula the movements of the greatest bodies of the universe and those of the tiniest atom; for such an intellect nothing would be uncertain and the future just like the past would be present before its eyes".*

(Note, however, that Laplace made that statement in an essay on probabilities. Indeed, he considered such an intellect out of our reach. This was not the manifesto of reductionism, as it is often portrayed, but the vindication of probabilities [26]). This was a very bold statement, considering the fact that experience was limited to large, observable objects, which typically come in small numbers, like planets and comets. In principle, there is no relation with the exceedingly large assemblies of invisibly tiny molecules that constitute a macroscopic body. This may be a cause of concern; hence, it must be subjected to testing whenever new phenomena are considered. The outcome is that statistical mechanics has greatly advanced by relying on Hamiltonian dynamics.

Consider a system of N particles, each having d degrees of freedom, characterized by a Hamiltonian $H = H(\Gamma)$, where $\Gamma = (\mathbf{q}, \mathbf{p}) \in \mathcal{M}$, and by the following equations of motion:

$$\dot{\mathbf{q}} = \frac{\partial H}{\partial \mathbf{p}}\,, \quad \dot{\mathbf{p}} = -\frac{\partial H}{\partial \mathbf{q}} \tag{17}$$

As already mentioned, the connection with physical measurements begins from the observation that they take a positive time, and from the assumption that they yield the time average of one phase function. If the measurement takes a time τ, and it starts when the microstate is Γ, the result of the measurement is postulated to be

$$\overline{\mathcal{O}}_{0,\tau}(\Gamma) = \frac{1}{\tau}\mathcal{O}_{0,\tau} = \frac{1}{\tau}\int_0^\tau \mathcal{O}(\Phi^t\Gamma)\,dt \tag{18}$$

where by the symbol $\mathcal{O}_{s,t}$ we mean the time integral of $\mathcal{O}$ from time s to time t, and the bar denotes time averaging.

If one accepts this picture, it follows that measurements depend on τ and are therefore subjective because different observers may choose different initial times and different τ's. They also depend on Γ, hence their result is a random variable because Γ is unknown. However, this contradicts Thermodynamics, which is an objective and deterministic theory, universally confirmed by experimental tests. One possibility to accord Equation (18) with experience is to assume that τ is very large, virtually infinitely larger than the microscopic scales concerning the evolution of Γ, and that over such a long time, $\mathcal{O}$ has explored many times—hence with proper frequencies—its range of values so that the initial condition Γ is irrelevant. The same would hold if, in the time τ, generic phase space trajectories, with negligible exceptions, thoroughly explore the phase space, which is property E2 of Section 3. However, while E2 is sufficient for the result, it is not necessary, and typically impossible. One way or another, we obtain

$$\frac{1}{\tau}\int_0^\tau \mathcal{O}(\Phi^t\Gamma)\,dt \approx \overline{\mathcal{O}}(\Gamma) = \lim_{\tau\to\infty}\frac{1}{\tau}\int_0^\tau \mathcal{O}(\Phi^t\Gamma)\,dt \approx o \in \boldsymbol{R} \tag{19}$$

where mathematically the first approximate equality is due to the fact that τ should be very large, although it cannot be infinite, and the second to the fact that the range of $\mathcal{O}$ may not be perfectly explored. As long as the approximations fall below the scale of thermodynamic interest, equality can thus be used, and o represents the result of a measurement. This picture is justified by Fermi as follows:

> *"Studying the thermodynamical state of a homogeneous fluid of given volume at given temperature [...] we observe that there is an infinite number of states of molecular motion that correspond to it. With increasing time, the system exists successively in all the dynamical states that correspond to the given thermodynamical state. From this point of view we may say that a thermodynamical state is the ensemble of all the dynamical states through which, as a result of the molecular motion, the system is rapidly passing"* [27].

Callen adds that, if the system passes sufficiently rapidly through representative atomic states, a measurement effectively averages over all them, and the system is in equilibrium [28]. If, on the other hand, the transitions from atomic state to atomic state is ineffective (e.g., the system is trapped in a small subset of states, or the transition rate is too slow) a macroscopic measurement does not yield a proper average over all possible atomic states, and the system is not in equilibrium. A different description must be adopted in such situations, which are in fact well known; think, e.g., of metal alloys subjected to thermal treatments. However, when the microscopic values of $\mathcal{O}$ are indeed sufficiently rapidly explored, compared to macroscopic observation times, a single system of interest reveals itself through $\mathcal{O}$ as the average over the *ensemble* of such possibilities. To the observer, the system revealed by $\mathcal{O}$ *appears* like the average over the ensemble of its microscopic phases, and the result of the measurement equals an average with respect to a phase space probability distribution, called *ensemble*. However, this implicitly requires the macroscopic state to be stationary; if it shifts during the measurement, the observable cannot explore its range with the frequencies corresponding to that state, and different initial microstates may lead to different observable values. In fact, this is the case of, e.g., ageing systems.

Given a dynamical system like Equation (17), it is not obvious that Equation (19) holds, and even if it does, the statement is so complex that it may be impossible to prove. Therefore, the validity of Equation (19) is commonly postulated together with the assumption that there is an invariant probability distribution μ on $\mathcal{M}$, such that a measurement yields Equation (4) for μ-almost every $x \in \mathcal{M}$, and for every phase function $\mathcal{O}$. What is obvious is that dense exploration of the hugely dimensional phase space cannot be the reasons why experience taught us that Equation (4) can be used, at least in the case of equilibrium systems, with the classical ensembles as probability distributions.

This fact can be explained without invoking the ergodic properties E1–E4. Given a single variable $\mathcal{O}$, its time average may appear like a full phase space average, even if its phase space trajectory $\Phi^t \Gamma$ explores a limited region of $\mathcal{M}$. It suffices that the range of values of $\mathcal{O}$ be visited with proper frequencies; then the result is the same. Does this ever happen? Khinchin proved that this does indeed happen for the relevant observables of rarefied gases [29]. He noted that the subtleties of microscopic dynamics count little compared to the following:

(a) macroscopic systems are made of very many particles: $N \ggg 1$;
(b) only several and special phase functions are physically relevant;
(c) it does not matter if ensemble averages disagree with time averages on limited sets of trajectories.

For rarefied gases, the relevant phase functions are sums of molecular contributions, $\mathcal{O}(\Gamma) = \sum_{n=1}^{N} o(\mathbf{q}_n, \mathbf{p}_n)$, where $(\mathbf{q}_n, \mathbf{p}_n)$ is the vector of configurations and momentum of the n-th particle. These functions are appropriate for the pressure, temperature and density of rarefied gases, whose energy can also be expressed as $H = \sum_{n=1}^{N} h(\mathbf{q}_n, \mathbf{p}_n)$, because interactions among particles are energetically negligible. (Interactions are, however, essential for the condition of LTE, hence for the existence of thermodynamic properties, to be established.) Then, in the microcanonical case, which here means a uniform probability distribution in $\mathcal{M}$, Khinchin proved the validity of the following relation:

$$\text{Prob}\left(\frac{|\overline{\mathcal{O}} - \mathbb{E}_{\text{mc}}(\mathcal{O})|}{|\mathbb{E}_{\text{mc}}(\mathcal{O})|} \geq K_1 N^{-1/4} \right) \leq K_2 N^{-1/4} , \qquad (20)$$

where K_1 and K_2 are positive constants, $\overline{\mathcal{O}}$ is the time average of $\mathcal{O}$, and $\mathbb{E}_{\text{mc}}(\mathcal{O})$ its phase space average in the microcanonical ensemble. In other words, the probability (in the microcanonical sense) that time averages differ by a small amount from the phase space averages is small if N is large: the larger N the smaller the probability of even smaller differences.

In this framework, the physically relevant ergodicity follows by and large from the $N \ggg 1$ condition, combined with the validity of the law of large numbers. That makes the range of sum variables of interest quite narrow, close to a single constant, when the single molecular contributions o can be considered independent, identically distributed, random variables. Then, exploration of the range of all the observables of interest does not take a long time. (See, e.g., Ref. [30] for an interesting analysis of the observables that relax to an equilibrium state, while others do not relax, because the underlying Toda dynamics are integrable, hence, not ergodic.) The details of the microscopic dynamics become irrelevant, while ensembles, i.e., probabilities in phase space, turn considerably useful. Even though probability, per se, is an "immaterial" and abstract mathematical notion, it becomes quite useful in the description of equilibrium states, under the above conditions. With the derivation of the Boltzmann equation a further step is taken forward, allowing time evolution.

5. Boltzmann Equation from the Probabilistic BBGKY Hierarchy

Let us specify the above general treatment of probabilities in phase space to the case of N identical spherical hard particles of mass 1. These particles are subjected to no external forces, and only interact elastically when they collide. We may think of their interaction potential as being zero when the particle centers are at a distance larger than their diameter, and infinitely high when this distance equals their diameter, so that particles are neither deformed, nor do they penetrate each other, when they collide. Denote a probability density in the phase space $\mathcal{M}$ by

$$f_N^{(N)}(\Gamma;t) = f_N^{(N)}(\mathbf{q}_1,\mathbf{p}_1;\ldots;\mathbf{q}_N,\mathbf{p}_N;t) \tag{21}$$

with notation indicating the joint probability of N particles out of N. In absence of external forces, there are no accelerations between collisions; hence, $\dot{\mathbf{p}}_i = 0$ for every particle, and $\dot{\Gamma} = (\dot{\mathbf{q}}_1, 0;\ldots;\dot{\mathbf{q}}_N, 0)$. The Liouville equation can be written as

$$\frac{\partial f_N^{(N)}}{\partial t} + \left(\mathrm{div}_\Gamma \dot{\Gamma} + \dot{\Gamma}\cdot\nabla_\Gamma\right) f_N^{(N)} = \frac{\partial f_N^{(N)}}{\partial t} + \sum_{i=1}^{N} \mathbf{p}_i \cdot \nabla_{\mathbf{q}_i} f_N^{(N)} = 0 \tag{22}$$

where we have used the fact that the dynamics is Hamiltonian; hence, the first term in brackets is null. If we introduce the s-particle distribution function

$$f_N^{(s)}(\mathbf{q}_1,\mathbf{p}_1;\ldots;\mathbf{q}_s,\mathbf{p}_s;t) = \int d\mathbf{q}_{s+1} d\mathbf{p}_{s+1} \ldots d\mathbf{q}_N d\mathbf{p}_N\, f_N^{(N)}(\mathbf{q}_1,\mathbf{p}_1;\ldots;\mathbf{q}_N,\mathbf{p}_N;t) \tag{23}$$

and we integrate the Liouville equation over the variables $\mathbf{q}_{s+1}, \mathbf{p}_{s+1} \cdots \mathbf{q}_N, \mathbf{p}_N$, we obtain [31]:

$$\frac{\partial f_N^{(s)}}{\partial t} + \sum_{i=1}^{s} \mathbf{p}_i \cdot \nabla_{\mathbf{q}_i} f_N^{(N)} = F^{(s)}\left(f_N^{(s+1)}, f_N^{(s+1)'}\right) \tag{24}$$

where $F^{(s)}$ is a function of the joint $(s+1)$ particles distribution before collision $f_N^{(s+1)'}$, and of the corresponding distribution after collision $f_N^{(s+1)}$. Therefore, Equation (24) is not closed: computing $f_N^{(s)}$ requires knowledge of $f_N^{(s+1)}$. The set of distributions $f_N^{(s)}$ and their evolution equations, Equation (24), constitute the *BBGKY hierarchy*. In particular, taking $s = 1$ and making $F^{(1)}$ explicit by solving the elastic collision dynamics, one obtains

$$\frac{\partial f_N^{(1)}}{\partial t} + \mathbf{p}\cdot\nabla_{\mathbf{q}} f_N^{(1)} = (N-1)\sigma^2 \int_{\mathbb{R}} \int_{S_-} [f_N^{(2)'} - f_N^{(2)}] |(\mathbf{p}-\mathbf{p}_*)\cdot\mathbf{n}| d\mathbf{p}_* d\mathbf{n} \tag{25}$$

for the one-particle probability distribution function, obtained integrating out the coordinates and momenta of all particles but one. Here, σ is the diameter of one particle; hence, σ^2 is the collision cross section; $\mathbf{n}$ is the unit vector in the direction going from the center of the test particle of position $\mathbf{q}$ and momentum $\mathbf{p}$, labelled by 1, to that of another particle of momentum $\mathbf{p}_*$; the range of integration S_- is the hemisphere within which $(\mathbf{p}-\mathbf{p}_*)\cdot\mathbf{n} < 0$, i.e., particles directed toward each other before collision; and $(N-1)$ is the number of particles with which the test particle, can collide, assuming they all have the same probability of doing it, having integrated over all their momenta $\mathbf{p}_*$. Moreover, having denoted by $f_N^{(2)}$ the two-particle distribution function, whose expression immediately before and after the collision respectively takes the form

$$f_N^{(2)'} = f_N^{(2)}(\mathbf{q},\mathbf{p}';\mathbf{q}+\sigma\mathbf{n},\mathbf{p}'_*;t) \quad \text{and} \quad f_N^{(2)} = f_N^{(2)}(\mathbf{q},\mathbf{p};\mathbf{q}+\sigma\mathbf{n},\mathbf{p}_*;t) \tag{26}$$

where $\mathbf{q}_* = \mathbf{q} + \sigma\mathbf{n}$ gives the position of the second particle at the moment of collision, and

$$\mathbf{p}' = \mathbf{p} - [\mathbf{n}\cdot(\mathbf{p}-\mathbf{p}_*)]\mathbf{n} \quad \text{and} \quad \mathbf{p}'_* = \mathbf{p}_* + [\mathbf{n}\cdot(\mathbf{p}-\mathbf{p}_*)]\mathbf{n} \tag{27}$$

where $\mathbf{p}'$ and $\mathbf{p}'_*$ are, respectively, the momenta of the test particle and of the other colliding particle before collision, while $\mathbf{p}$ and $\mathbf{p}_*$ denote the corresponding momenta after collision.

To solve this equation, one needs a closure assumption on $f_N^{(2)}$. The equation

$$f_N^{(2)}(\mathbf{q},\mathbf{p};\mathbf{q}_*,\mathbf{p}_*;t) = f_N^{(1)}(\mathbf{q},\mathbf{p};t) f_N^{(1)}(\mathbf{q}_*,\mathbf{p}_*;t) \tag{28}$$

is known in the kinetic theory of gases as the *stosszahlansatz*, or hypothesis of molecular chaos. It amounts to assuming that when two particles collide, they are independent. As the dynamics of particles is deterministic, this statement may only have a statistical meaning, justified by some kind of dynamical *randomness*, that can be legitimately called *chaos*. (In the theory of dynamical systems, the term chaos is often used to indicate systems that have at least one positive Lyapunov exponent. This notion was not available to Boltzmann and, indeed, he did not need this kind of chaos.) Then, writing f in place of $f_N^{(1)}$ to simplify the notation, and recalling that $\mathbf{v} = \mathbf{p}$ because we assumed unit masses, Equation (25) takes the form

$$\frac{\partial f}{\partial t} + \mathbf{v} \cdot \nabla_{\mathbf{q}} f = (N-1)\sigma^2 \int_{\mathbb{R}} \int_{S_-} [f' f'_* - f f_*] |(\mathbf{v} - \mathbf{v}_*) \cdot \mathbf{n}| d\mathbf{v}_* d\mathbf{n} \tag{29}$$

where primes mean before collision and $*$ second particle, and small particles, small σ, imply $\mathbf{q}_* \approx \mathbf{q}$. This is the celebrated *Boltzmann equation*. Given its probabilistic derivation, which begins with the distribution $f_N^{(N)}$ in phase space, $f(\mathbf{q}, \mathbf{p}, t)d\mathbf{q}d\mathbf{p}$ represents the probability that particle 1 be found in the volume element $d\mathbf{q}$ around $\mathbf{q}$, with momentum within the set $d\mathbf{p}$ around $\mathbf{p}$, at time t. The classical interpretation of probability we have so far adopted, then states that, given a very large collection of independent macroscopically equal systems, each made of N hard spheres, the fraction of them having particle 1 within $d\mathbf{q}d\mathbf{p}$ equals $f d\mathbf{q}d\mathbf{p}$. If one assumes, in addition, that particles are indistinguishable, a simple combinatorial factor turns $f d\mathbf{q}d\mathbf{p}$ into the fraction of ensemble members having one particle, whichever it is, in $d\mathbf{q}d\mathbf{p}$ at time t. It remains to connect this probability to material properties, like the density of the gas in $d\mathbf{q}d\mathbf{p}$.

6. Boltzmann Equation from Mass Balance in Real Space

The probability distribution f appearing in Equation (29) is obtained projecting the one for N particles defined on the corresponding $2dN$ dimensional phase space, down to the $2d$ dimensional one-particle space. Given a single system of interest and the ensemble interpretation of the probability in phase space, the condition $f_N^{(1)}(\mathbf{q}, p; t) > 0$ does not imply that particle 1 actually lies within the volume $d\mathbf{q}d\mathbf{p}$, since only a fraction of the systems of the ensemble has particle 1 in $d\mathbf{q}d\mathbf{p}$: particle 1 of the complementary fraction of systems does not lie in $d\mathbf{q}d\mathbf{p}$. (If particles are indistinguishable and the 1-particle probability distribution has been correspondingly normalized, the same can be stated for any single particle of the systems of the ensemble.) In particular, $d\mathbf{q}d\mathbf{p}$ could be empty because the BBGKY hierarchy can be developed for any $N \geq 1$, and a few particles only occupy a limited region of space. Therefore, one could conclude that even the Boltzmann equation is *immaterial*, and there is no compelling reason to take it as a description of a given material system.

However, a physical derivation of the Boltzmann equation is also available. It does not start from a probability distribution on a phase space $\mathcal{M}$, which is a continuum of geometric points, each of which represents a whole N-particles system. This derivation refers to a single system made of a very large number N of small, but finite size, particles of mass 1, with positions and velocities in a given volume $\Delta\mathbf{q}\Delta\mathbf{p}$. The particles in such a volume constitute a discrete rather than a continuous set, and can be represented by a continuous density of mass only when many of them populate $\Delta\mathbf{q}\Delta\mathbf{p}$. Therefore, given the finite size of the particles, such a volume has to be small, to appear like a point on the macroscopic scale. However, unlike the volumes $d\mathbf{q}d\mathbf{p}$ of the previous section, which can be arbitrarily small, it cannot be infinitesimal. Indeed, it has to be much larger than the volume of a single particle, so that many of them can be lodged in it [32,33]. This highlights another essential difference between probability and mass densities. Let us denote by C_{ij} of size $\Delta\mathbf{q}\Delta\mathbf{p}$ one cell centered around a discrete set of lattice points $(\mathbf{q}_i, \mathbf{p}_j)$. The mass density within C_{ij} is defined as the mass of one particle times the number density $\rho(\mathbf{q}_i, \mathbf{p}_j; t)$, the number of particles per unit volume around $(\mathbf{q}_i, \mathbf{p}_j)$. Letting n_{ij} be the number of particles in cell C_{ij}, at time t, one may thus write

$$n_{ij}(t) = \rho(\mathbf{q}_i, \mathbf{p}_j; t)\Delta\mathbf{q}\Delta\mathbf{p}\,; \quad \sum_{i,j} n_{ij}(t) = N \tag{30}$$

and may try to approximate this set of discrete mass densities using a continuous function ρ, so that

$$N = \int_{V_\mu} \rho(\mathbf{q}, \mathbf{p}; t)\mathrm{d}\mathbf{q}\mathrm{d}\mathbf{p} \tag{31}$$

where V_μ is the set of positions and momenta available to particles of the system. This makes no sense unless n_{ij} is very large in each cell C_{ij}; hence, such cells must be relatively large. Furthermore, particles have to collide: their motion has to be random, so that a uniform distribution can be rapidly achieved within each C_{ij}. Large n_{ij}, together with dynamical randomness, makes the particles entering or leaving a cell just a negligible number, compared to those staying inside. At the same time the cells must be sufficiently small compared to the observation scale, that they appear like a point. A wide separation of scales, possible only if N is very large, is required.

Let us now follow the motion of one set of particles lying in one cell $\Delta\mathbf{q}\Delta\mathbf{p}$ at time t, as they stroll in V_μ. Let us assume that ρ accurately describes the mass density as a continuum. In the absence of external forces, collisions are the only mechanism that allows velocities to change. If, in a time $\mathrm{d}t$, the particles in $\Delta\mathbf{q}\Delta\mathbf{p}$ undergo no collisions, the position $\mathbf{q}$ of one of them turns $\mathbf{q} + \mathbf{v}\mathrm{d}t$, while its momentum $\mathbf{p}$ is unchanged, and all the particles in the cell $\Delta\mathbf{q}\Delta\mathbf{p}$, centered about $(\mathbf{q}, \mathbf{p})$, will be found in the cell $\Delta\widehat{\mathbf{q}}\Delta\widehat{\mathbf{p}}$ centered about $(\mathbf{q} + \mathbf{v}\mathrm{d}t, \mathbf{p})$. This is achieved by evolving in time all points of the original cell with the same dynamics of the particles in it. Therefore, in the absence of collisions, one has

$$\rho(\mathbf{q} + \mathbf{v}\mathrm{d}t, \mathbf{p}, t + \mathrm{d}t)\Delta\widehat{\mathbf{q}}\Delta\widehat{\mathbf{p}} = \rho(\mathbf{q}, \mathbf{p}, t)\Delta\mathbf{q}\Delta\mathbf{p} \tag{32}$$

As $\mathbf{q}$ and $\mathbf{p}$ are canonically conjugate variables, and between two consecutive collisions each particle can be considered by itself as a whole Hamiltonian system, we obtain $\Delta\widehat{\mathbf{q}}\Delta\widehat{\mathbf{p}} = \Delta\mathbf{q}\Delta\mathbf{p}$, and

$$\rho(\mathbf{q} + \mathbf{v}\mathrm{d}t, \mathbf{p}, t + \mathrm{d}t) = \rho(\mathbf{q}, \mathbf{p}, t) \tag{33}$$

On the other hand, if particles collide, one additional term arises:

$$\rho(\mathbf{q} + \mathbf{v}\mathrm{d}t, \mathbf{p}, t + \mathrm{d}t) = \rho(\mathbf{q}, \mathbf{p}, t) + \left(\frac{\mathrm{d}\rho}{\mathrm{d}t}\right)_{coll} \mathrm{d}t \tag{34}$$

which defines the collision term. Finally, expanding the left-hand side of Equation (34) to first order in $\mathrm{d}t$, one obtains

$$\rho(\mathbf{q}, \mathbf{p}, t) + \frac{\partial\rho}{\partial t}\mathrm{d}t + \mathbf{v} \cdot \nabla_{\mathbf{q}}\, \rho\mathrm{d}t = \rho(\mathbf{q}, \mathbf{p}, t) + \left(\frac{\mathrm{d}\rho}{\mathrm{d}t}\right)_{coll} \mathrm{d}t \tag{35}$$

because ρ depends on t both explicitly and implicitly through the position coordinates. Dividing by $\mathrm{d}t$ yields:

$$\frac{\partial\rho}{\partial t} + \mathbf{v} \cdot \nabla_{\mathbf{q}}\rho = \left(\frac{\mathrm{d}\rho}{\mathrm{d}t}\right)_{coll} \tag{36}$$

where $\mathbf{v} \cdot \nabla_{\mathbf{q}}\rho$ represents streaming in position space. There is no streaming in momentum space, because no field acts on the particles, but particles may enter or exit $\Delta\mathbf{p}$ thanks to collisions.

Relation (36) expresses the conservation of mass, and does not require any interpretation of a probabilistic nature: it is based on the objective counting of particles (or measurement of mass density) and on their deterministic Hamiltonian dynamics. Assuming that instantaneous collisions of three or more particles are negligible in a gas of hard spheres, the collision term should be expressed as a function of the number of pairs of particles located inside $\Delta\mathbf{q}$ that may either enter (gain term) or leave (loss term) $\Delta\mathbf{p}$ because of collisions. Let $\rho^{(2)}$ be the density of such pairs. As the initial condition of the N particles is not known, going beyond Equation (36) requires some assumption on the collision term, analogously to Equation (25). One may then rely on the large value of N, on the smallness of the particles, and on the disorder produced in configuration and velocity space by the collisions, which make sensible the factorization of $\rho^{(2)}$:

$$\rho^{(2)}(\mathbf{q},\mathbf{p};\mathbf{q}+\sigma\mathbf{n},\mathbf{p}_*;t) = \rho^{(1)}(\mathbf{q},\mathbf{p};t)\rho^{(1)}(\mathbf{q}+\sigma\mathbf{n},\mathbf{p}_*;t) \quad (37)$$

That, indeed, means that each of the $n_{\Delta\mathbf{q}}$ particles in an elementary volume $\Delta\mathbf{q}$ may collide with any of the remaining $n_{\Delta\mathbf{q}} - 1 \approx n_{\Delta\mathbf{q}}$ either for a gain or a loss of particles in the elementary volume $\Delta\mathbf{p}$. This is a rarefied gas condition, because it means that no particle can hide behind another one. Then, integrating over all possible velocities yields an expression like Equation (29), which makes sense only if particles are many and small. The equation for the mass distribution turns identical to Equation (29), introducing $\rho = \rho^{(1)}/N$. Therefore, the Boltzmann equation is obtained again. Its solution ρ enjoys the same properties that f enjoys.

At the same time, the conceptual difference between f of Equation (25) and ρ of Equation (36) could not be greater. In the first case one considers the fraction of identical abstract systems that enjoy certain properties out of the total of an ideal hypothetical continuum of identical systems; in the second case one considers the number of particles of a single concrete, experimentally observable system. Probabilistically, all calculations can be performed for a large collection of identical systems made of any number $N \geq 1$ of particles; physically, a single systems is considered and N must be huge. At this level of description, there is no stringent reason for one quite sophisticated abstract object like f to behave similarly to a material tangible object, such as ρ. Nevertheless, this is precisely what happens if probability is required to evolve like a fluid made of geometric points obeying the equations of motion of the system that each of them separately represents, and the *stosszahlansatz* holds. That may well be mathematically too hard to prove within a general Hamiltonian framework, also because it is not verified in a series of systems, hard to fully classify, cf. integrable systems. However, experience has demonstrated that it does hold in very many situations; situations in which we may legitimately state that *probability* turns *material*: probability can be formally identified with mass.

What is the cost of such an identification? When can the stosszahlansatz be justified? In the first place, one would think that large N makes particles independent before colliding, because they may have hardly met before, but this is not enough. Particles are not independent after having collided with each other. Moreover, the factorization condition

$$f_N^{(s)}(q_1,p_1,\ldots,q_s,p_s;t) = \prod_{j=1}^{s} f_N^{(1)}(q_j,p_j;t) \quad (38)$$

is hard to obtain because particles do not overlap, and s of them occupy a volume of order $O(s\sigma^3)$. Thus, large s increases the region to be excluded from the free motion of particles, resulting in correlations among them that may persist over time. This suggests that $s = 2$ is the best candidate for the postulated independence. Also, large N helps in making rare the encounters between the same pair of particles only if the density is small; otherwise packed particles may collide forever with the same small group of companions. Then, σ should be small. On the other hand, collisions of hard spheres are defocussing, and have a randomizing effect on positions and velocities, which is beneficial for the decay of correlations. Therefore, N and σ should be tuned so that the total cross section is not negligible. Grad identified the scaling regime in which all the above is realized, now known as Boltzmann–Grad limit [34]. It consists in keeping $N\sigma^2$ positive and finite, while N grows:

$$N\sigma^2 = \text{constant} > 0 \quad \text{which implies} \quad \sigma^2 \sim \frac{1}{N}, \ N\sigma^3 \sim N^{-\frac{1}{2}} \tag{39}$$

In this limit, the volume occupied by particles is negligible, while the total cross section is finite. More precisely, let λ denote the mean free path and τ the mean free time as, respectively, the typical microscopic length scale of the system, which is the distance traveled on average by a particle between collisions, and the corresponding mean time. The first is defined by the the ratio between the volume and the total cross section of the particles, the second by the ratio between this distance and the typical speed. Then, denoting by $|\Omega|$ the available volume, and letting $\beta = 1/k_B T$, so that the typical speed for particles of mass 1 is $\sqrt{2/\beta}$, we have

$$\lambda = \frac{|\Omega|}{N\pi\sigma^2}, \quad \tau = \sqrt{\frac{\beta}{2}}\frac{|\Omega|}{N\pi\sigma^2} \tag{40}$$

which remain finite in the Boltzmann–Grad limit. This implies a net effect produced by particle collisions, which is necessary to randomize their motion. Together with the vanishing of the excluded volume, the limit of Equation (40) is also necessary to make correlations rapidly decay, when N is large. Physically, this picture corresponds to a rarefied gas, in which particles collide, but the time average of their interaction energy is negligible compared to the time average of their kinetic energy. Whether the assumptions apply or not, only experience can tell. An important result implied by the Boltzmann–Grad limit is that $f_N^{(2)}$ remains factorized in time, if the solution of the s-th equation of the BBGKY hierarchy, $f_N^{(s)}$, exists is unique, and starts factorized as s factors $f_N^{(1)}$ for all s [31].

Arguably, the most relevant conceptual result stemming from the Boltzmann equation is the H-theorem. It can be concisely illustrated as follows: let the H-functional be defined by

$$H(f) = \int_{V\times\mathbb{R}^3} f \log(f)\, d\mathbf{p} d\mathbf{q} \tag{41}$$

where $V \subset \mathbb{R}^3$ is the volume in which a gas is contained. For f a solution of the Boltzmann equation within the volume V, the H-theorem states

$$\frac{d}{dt}H(f) \le 0 \tag{42}$$

where the equality holds if and only if f is constant for $\mathbf{q} \in V$ and the $\mathbf{p}$ are distributed as a Maxwellian probability distribution.

A point that may at times be overlooked, is that the solution of the Boltzmann equation requires a specification of initial and boundary conditions, like any other integro-differential equation. The form of the boundary of V is crucial, particularly for the validity of the H-theorem and, in general, for the overall solution. For instance, reflecting or periodic boundaries are commonly used, and the boundary is typically assumed to be piece-wise smooth, although works on irregular boundaries have been performed, cf. [35]. Moreover, V must be compact for physical properties such as pressure to have a definite equilibrium positive value.

The H-theorem shows that the Boltzmann equation exhibits irreversible behaviour, which appears in contrast to the time-reversal invariant laws of classical mechanics governing the motion of particles. In reality, various approximations and hypotheses, which are extraneous to the dynamics of an N particle system, have been introduced to pass from the evolution of probabilities in phase space to the Boltzmann equation. The N particle system has been approximated by a system consisting of infinitely many particles, verifying the molecular chaos hypothesis, through the Boltzmann–Grad limit, thus breaking time reversibility. The interesting fact is that such approximations and hypotheses perfectly represent the physics of a rarefied gas.

We conclude this section by noting that there exists quantum mechanical and relativistic generalizations of the Boltzmann equation, cf., e.g., Refs. [7,36]. These generalizations have proven successful in an exceedingly wide range of applications, including transport of electrons in solid conductors and in astrophysical plasmas. The most direct extensions are the classical ones, specifically concerning particles that are accelerated by external fields, such as electric fields. In this case, streaming in velocity space is introduced in addition to the streaming in position space. Denoting by $\mathbf{F}$ the external force and by m the mass of the particles, its effect on $\mathbf{v}$ appears as the acceleration $\mathbf{F}/m$. This acceleration smoothly guides a particle inside or outside an element of velocity space, analogous to how velocity guides a particle inside or outside an element of position space. Consequently, the Boltzmann equation may be re-written as

$$\frac{\partial \rho}{\partial t} + \mathbf{v} \cdot \nabla_{\mathbf{q}} \rho + \frac{\mathbf{F}}{m} \cdot \nabla_{\mathbf{v}} \rho = \left(\frac{\mathrm{d}\rho}{\mathrm{d}t} \right)_{coll} \tag{43}$$

In sufficiently hot plasmas made of charged particles of charge c, collisions can be neglected. Thus, denoting the electric and magnetic fields acting on these particles as $\mathbf{E}$ and $\mathbf{B}$, respectively, Equation (43) can be re-written as

$$\frac{\partial \rho}{\partial t} + \mathbf{v} \cdot \nabla_{\mathbf{q}} \rho + \frac{c}{m} (\mathbf{E} + \mathbf{v} \times \mathbf{B}) \cdot \nabla_{\mathbf{v}} \rho = 0 \,, \tag{44}$$

which is known as the Vlasov equation. In reality, variations of the Boltzmann equation are numerous. The kinetic theory of wave turbulence [37] finds applications from quantum to astrophysical phenomena, including anomalous transport. There are Boltzmann equations for thermostatted rather than Hamiltonian dynamics, and for a wide spectrum of variously interacting systems, such as the components of blood, or the individuals in a social network etc. It is impossible to list all of them. Of course, there is a difference between these variations and the original Boltzmann equation: the interaction rules are often rather loosely known, which makes the fundamental hypothesis of molecular chaos hard to justify. Nevertheless, these variations provide valuable insights into non-standard and even non-physical phenomena. The interested readers may easily find the relevant literature, which is not within our scope.

7. Concluding Remarks

In this paper we have reviewed some of the reasons that, in principle, could make probability an abstract physically inapplicable mathematical tool, while it is clear that it is arguably the most effective tool physicists can use. The point is that such effectiveness does not arise for free; it is the result of informed, repeatedly tested, and verified choices, as revealed by the derivation of the Boltzmann equation. Uncritical use of probabilities can easily lead to errors. Keeping in mind the teachings of statistical mechanics, which have unfolded since Boltzmann's glorious work of 180 years ago, is thus beneficial for future developments, which will venture in largely uncharted territory, often at variance with our established knowledge of macroscopic objects. Indeed, the microscopic fluctuations are not observable in the behaviour of macroscopic objects in thermodynamic conditions, although present day technology allows us to measure them in microscopic observations [38,39]. On the contrary, fluctuations may even dominate the behaviour of small systems, and require suitable approaches, whose results must be properly interpreted, especially if one insists on using the thermodynamic language. For instance, when dealing with situations in which there is no dissipation, one may find that the derivations of thermodynamic relations are only formal, and that the relevant quantities only have a statistical meaning. However, even more problematic may be the extension to nonequilibrium (dissipative or non-dissipative) states, where even the formal thermodynamic analogies may fail [40]. At present, one must proceed by examples, as a comprehensive theory of small and nonequilibrium systems is still missing, although notable works are being performed, see, e.g., Ref. [41] for one review.

Author Contributions: Writing—original draft preparation, L.R.; Writing—review and editing, L.R. and V.D.F. All authors have read and agreed to the published version of the manuscript.

Funding: Research funded by the Italian Ministry of University and Research (MUR) through the grant PRIN2022-PNRR project (No. P2022Z7ZAJ) "A Unitary Mathematical Framework for Modelling Muscular Dystrophies" (CUP: E53D23018070001). This work has been performed under the auspices of Italian National Group of Mathematical Physics (GNFM) of the Istituto Nazionale di Alta Matematica (INdAM).

Institutional Review Board Statement: Not applicable.

Data Availability Statement: Data are contained within the article.

Acknowledgments: L.R. gratefully acknowledges the generous hospitality of the Department of Mathematics, School of Advanced Sciences, Vellore Institute of Technology, India, where part of this work has been performed.

Conflicts of Interest: The authors declare no conflict of interest.

References

1. Prigogine, I. *From Being to Becoming: Time and Complexity in the Physical Sciences*; Freeman & Co.: Dallas, TX, USA, 1980.
2. Feynman, R.P.; Leighton, R.B.; Sand, M. *The Feynman Lectures on Physics*; Addison Wesley: Reading, MA, USA, 1963; Volume 1.
3. Einstein, A. *Investigations on the Theory of the Brownian Movement*; Dover Publications: New York, NY, USA, 1956.
4. Chibbaro, S.; Rondoni, L.; Vulpiani, A. *Reductionism, Emergence and Levels of Reality. The Importance of Being Borderline*; Springer: Berlin/Heidelberg, Germany, 2014.
5. Cercignani, C. *Ludwig Boltzmann: The Man Who Trusted Atoms*; Oxford University Press: Oxford, UK, 1998.
6. Boltzmann, L. *Weitere Studien Liber das Wärmegleichgewicht unter Gasmolekiilen*; Sitzungsberichte der Akademie der Wissenschaften; k. k. Hof- und Staatsdruckerei: Vienna, Austria, 1972; Volume 66, pp. 275–370; English translation in: S.G. Brush *Kinetic Theory*; Irreversible Processes; Pergamon Press: Oxford, UK, 1966; Volume 2, pp. 88–175.
7. Cercignani, C.; Kremer, G.M. *The Relativistic Boltzmann Equation: Theory and Applications*; Birkhäuser: Basel, Switzerland, 2002.
8. Haug, H.; Jauho, A.-P. *Quantum Kinetics in Transport and Optics of Semiconductors*; Springer: Berlin/Heidelberg, Germany, 2007.
9. Preziosi, L.; Rondoni, L. Conservative energy discretization of Boltzmann collision operator. *Q. Appl. Math.* **1999**, *57*, 699–721. [CrossRef]
10. Pareschi, L.; Toscani, G. *Interacting Multiagent Systems: Kinetic Equations and Monte Carlo Methods*; Oxford University Press: Oxford, UK, 2013.
11. Tosin, A.; Zanella, M. Boltzmann-type description with cutoff of Follow-the-Leader traffic models. In *Trails in Kinetic Theory: Foundational Aspects and Numerical Methods*; Springer: Cham, Switzerland, 2021; pp. 227–251.
12. Toscani, G.; Brugna, C. Wealth redistribution in Boltzmann-like models of conservative economies. In *Econophysics and Economics of Games, Social Choices and Quantitative Techniques*; Springer: Milan, Italy, 2010; pp. 71–82.
13. Albi, G.; Cristiani, E.; Pareschi, L.; Peri, D. Mathematical models and methods for crowd dynamics control. In *Crowd Dynamics, Volume 2: Theory, Models, and Applications*; Birkhäuser: Cham, Switzerland, 2020; pp. 159–197.
14. DiPerna, R.J.; Lions, P.L. On the Cauchy Problem for Boltzmann Equations: Global Existence and Weak Stability. *Ann. Math.* **1989**, *130*, 321–366. [CrossRef]
15. Cercignani, C. H-theorem and trend to equilibrium in the kinetic theory of gases. *Arch. Mech. Stosow.* **1982**, *34*, 231–241.
16. Villani, C. Cercignani's conjecture is sometimes true and always almost true. *Commun. Math. Phys.* **2003**, *234*, 455–490. [CrossRef]
17. Popper, K. *The Logic of Scientific Discovery*; Routledge: London, UK, 1959.
18. Wheeler, J.A. Information, Physics, Quantum: The Search for Links. In Proceedings of the 3rd Internation Symposium on Foundations of Quantum Mechanics, Tokyo, Japan, 28–31 August 1989; pp. 54–368.
19. de Finetti, B. *Theory of Probability*; John Wiley & Sons: New York, NY, USA, 1990; Volume 1.
20. Jaynes, E.T. The Well-Posed Problem. *Found. Phys.* **1973**, *3*, 477. [CrossRef]
21. Maxwell, J.C. Chapter V—Letter to L Campbell Esq. In *The Life of James Clerk Maxwell*; Campbell, L., Garnett, W., Eds.; Mac Millan and Co.: London, UK, 1882.
22. Reichl, L.E. *A Modern Course in Statistical Physics*; Wiley-VCH: Weinheim, Germany, 1998.
23. Sinai, Y.G. *Introduction to Ergodic Theory*; Princeton University Press: Princeton, NJ, USA, 1976.
24. Bunimovich, L.A.; Sinai, Y.G.; Chernov, N.I. Statistical Properties of Two-Dimensional HYPERBOLIC billiards. *Russ. Math. Surv.* **1991**, *46*, 47. [CrossRef]
25. Poincaré, H. *Science and Method*; Thomas Nelson and Sons: London, UK, 1914.
26. Laplace, P. *Essai Philosophique sur les Probabilities*, 5th ed.; Courcier: Paris, France, 1825; English translation in 1952 as *Philosophical Essay on Probabilities*; Dover Publications: New York, NY, USA, 1814.
27. Fermi, E. *Thermodynamics*; Dover Publications: New York, NY, USA, 1956.
28. Callen, H.B. *Thermodynamics and an Introduction to Thermostatistics*; Wiley: Hoboken, NJ, USA, 1985.

29. Khinchin, A.I. *Mathematical Foundations of Statistical Mechanics*; Dover Publications: New York, NY, USA, 2013.
30. Baldovin, M.; Vulpiani, A.; Gradenigo, G. Statistical Mechanics of an Integrable System. *J. Stat. Phys.* **2021**, *183*, 41. [CrossRef]
31. Cercignani, C.; Illner, R.; Pulvirenti, M. *The Mathematical Theory of Dilute Gases*; Springer: Berlin/Heidelberg, Germany, 1994.
32. Kreuzer, H.J. *Nonequilibrium Thermodynamics and Its Statistical Foundations*; Claredon Press: Oxford, UK, 1981.
33. Falcioni, M.; Palatella, L.; Pigolotti, S.; Rondoni, L.; Vulpiani, A. Initial growth of Boltzmann entropy and chaos in a large assembly of weakly interacting systems. *Phys. A* **2007**, *385*, 170. [CrossRef]
34. Grad, H. On the kinetic theory of rarefied gases. *Commun. Pure Appl. Math.* **1949**, *2*, 331. [CrossRef]
35. Arkeryd, L.; Heintz, A. On the solvability and asymptotics of the Boltzmann equation in irregular domains. *Commun. Partial Differ. Equ.* **1997**, *22*, 2129. [CrossRef]
36. Ashcroft, N.W.; Mermin, N.D. *Solid State Physics*; Harcourt College Publishers: Orlando, FL, USA, 1976.
37. Nazarenko, S. *Wave Turbulence*; Springer: Berlin/Heidelberg, Germany, 2011.
38. Bonaldi, M.; Conti, L.; Gregorio, P.D.; Rondoni, L.; Vedovato, G.; Vinante, A.; Bignotto, M.; Cerdonio, M.; Falferi, P.; Liguori, N.; et al. Nonequilibrium Steady-State Fluctuations in Actively Cooled Resonators. *Phys. Rev. Lett.* **2009**, *103*, 010601. [CrossRef]
39. Conti, L.; Gregorio, P.D.; Karapetyan, G.; Lazzaro, C.; Pegoraro, M.; Bonaldi, M.; Rondoni, L. Effects of breaking vibrational energy equipartition on measurements of temperature in macroscopic oscillators subject to heat flux. *J. Stat. Mech.* **2013**, *2013*, P12003. [CrossRef]
40. Cohen, E.G.D.; Rondoni, L. Particles, Maps and Irreversible Thermodynamics I. *Phys. A* **2002**, *306*, 117. [CrossRef]
41. Rondoni, L. Introduction to Nonequilibrium Statistical Physics and Its Foundations. In *Frontiers and Progress of Current Soft Matter Research*; Liu, X.-Y., Ed.; Soft and Biological Matter Series; Springer: Singapore, 2021.

entropy MDPI

Opinion

Walking with the Atoms in a Chemical Bond: A Perspective Using Quantum Phase Transition

Sabre Kais

Department of Chemistry and Elmore Family School of Electrical and Computer Engineering, Purdue Quantum Science and Engineering Institute, Purdue University, West Lafayette, IN 47907, USA; kais@purdue.edu

Abstract: Phase transitions happen at critical values of the controlling parameters, such as the critical temperature in classical phase transitions, and system critical parameters in the quantum case. However, true criticality happens only at the thermodynamic limit, when the number of particles goes to infinity with constant density. To perform the calculations for the critical parameters, a finite-size scaling approach was developed to extrapolate information from a finite system to the thermodynamic limit. With the advancement in the experimental and theoretical work in the field of ultra-cold systems, particularly trapping and controlling single atomic and molecular systems, one can ask: do finite systems exhibit quantum phase transition? To address this question, finite-size scaling for finite systems was developed to calculate the quantum critical parameters. The recent observation of a quantum phase transition in a single trapped $^{171}Yb^{+}$ ion indicates the possibility of quantum phase transitions in finite systems. This perspective focuses on examining chemical processes at ultra-cold temperatures, as quantum phase transitions—particularly the formation and dissociation of chemical bonds—are the basic processes for understanding the whole of chemistry.

Keywords: quantum phase transitions; finite-size scaling; formation of chemical bond

1. Formation of the Chemical Bond

Classical phase transitions, like solid–liquid–gas or order–disorder spin magnetic phases, are all driven by thermal energy fluctuations by varying the temperature. On the other hand, quantum phase transitions happen at absolute zero temperature, with quantum fluctuations causing the ground state energy to show abrupt changes as one varies the system parameters like electron density, pressure, disorder, or external magnetic and electric fields [1]. As a consequence of Heisenberg's uncertainty principle, the position and the velocity of a quantum object cannot both be measured precisely, causing the ground state energy to show abrupt changes as one varies the system parameters. For example, one can look at the melting of a Wigner lattice (see Figure 1), as suggested theoretically by the physicist Eugene Wigner, a metal that typically conducts electricity could become an insulator when the density of electrons is reduced to ultra-cold temperatures. What has recently been observed experimentally by Park and Demler [2] is the melting of the crystal state into liquid because of quantum fluctuations near absolute zero temperature. These results highlight the beautiful fundamental fact of wave–particle duality in quantum mechanics, wherein electrons behave as "particle-like" in the solid phase, but as "wave-like" in the melted liquid phase. In quantum spin systems, continuous and discontinuous quantum phase transitions have been studied as a function of varying external magnetic fields, pressure, and disorder. In an analogy with the critical point at the classical liquid–gas phase transition in water, Jimenez et al. [3] provided experimental evidence of a quantum critical point in the geometrically frustrated quantum antiferromagnet $SrCu_2(BO_3)_2$ by controlling both the pressure and the applied magnetic field at low temperatures. These phase transitions are associated with singularities of the free energy that occur only in the thermodynamic limit, where the system size approaches infinity. M.E. Fisher and others developed the finite-size scaling, a systematic theoretical approach that allows one to extrapolate information on the criticality of the finite system to the thermodynamic limit.

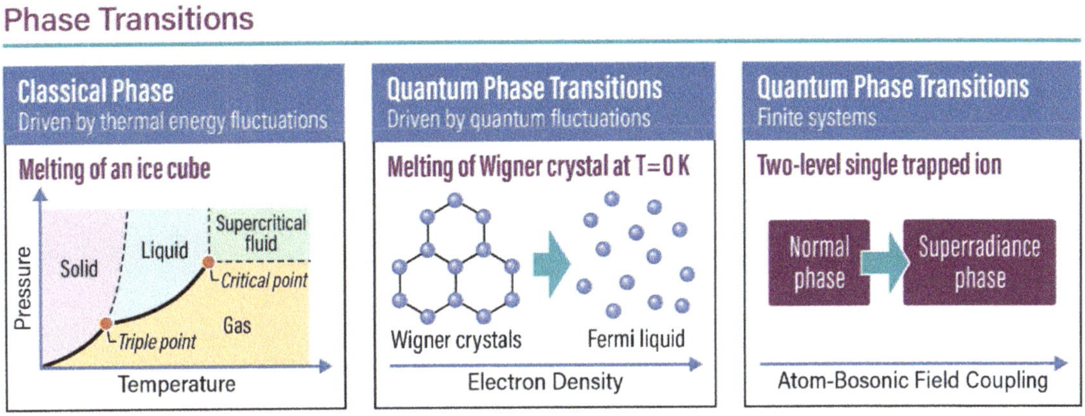

Figure 1. (**Left**) Classical phase transition in a macroscopic system, like water, as characterized in the (T,P) plane. The regimes wherein liquid water, solid ice, and water vapor are stable are represented in three different colors, whereas the black curves indicate phase boundaries wherein two phases can co-exist in equilibrium. (**Middle**) Melting of a Wigner crystal to a Fermi liquid due to increasing electron density. Being a quantum phase transition, such transformations can occur even at absolute zero due to quantum fluctuations. (**Right**) Quantum phase transition (QPT) in a finite system like a single two-level system coupled with an external Bosonic reservoir. Unlike in classical phase transitions wherein macroscopic variables like pressure, temperature, etc., are usually involved, the control variable for a QPT like this is simply the energy separation of the system relative to that of the bath.

However, theoretical and experimental results indicate the existence of quantum phase transitions of finite systems. Plenio and coworkers [4] show that the Quantum Rabi Model (QRM), a two-level spin interacting with a single-mode bosonic field, exhibits a continuous quantum phase transition as one varies the spin-field coupling constant and energies. They prove that in the limit where the spin energy over the field energy tends to infinity, the system has two phases, as follows: the normal phase where the boson field is in the ground state and the superradiant phase where the boson field is excited (see Figure 1). Recently, Cai et al. [5] experimentally observed this continuous quantum phase transition in a system composed of only the two hyperfine atomic levels interacting with the bosonic motional states of an $^{171}\ Yb^{+}$ ion in a Paul trap. They measured the spin-up populations in the two hyperfine states in the ground state manifold $^{2}S_{1/2}$ and the rescaled phonon number associated with the spatial motion of the ion along one of its principal axes. The motional degree of freedom can be well described as a quantum harmonic oscillator and thus serve as the Bosonic mode in the quantum Rabi model. This exciting new experimental observation of quantum phase transition in finite systems opens up a new field of research understanding in the quantum critical phenomena of finite systems at ultra-cold temperatures.

Theoretically, the quantum critical phenomena and symmetry breaking of electronic structure configurations of finite systems have been studied extensively for decades. Herschbach and coworkers [6] pioneered the field of dimensional scaling for electronic structure calculations and have discussed the symmetry breaking of electronic structure configurations of both single atoms and simple molecular systems [6]. This approach generalizes the Schrodinger equation to the large-dimensional space and solves the resulting simple equation at the limit $D \to \infty$. At this limit, the electrons take a fixed position in the large-D effective potential, allowing one to study the criticality and symmetry breaking as one varies the parameters controlling the system, such as the nuclear charge, interatomic distance in diatomic molecules, and external electric and magnetic fields. For example, symmetry breaking at the large-D limit of the electronic structure configurations of atoms and simple molecular systems has one-to-one mapping to the criticality of mean-field theory, in analogy to the mean-field criticality of fluid and magnetic systems [7]. To carry out the calculations to the physical space, D = 3, we have shown that one can describe such quantum phase transitions of finite systems in Hilbert space with the analogy between the thermodynamic limit and the size of the Hilbert space [8]. To obtain the exact quantum critical parameters, a finite-size scaling can be formulated in the Hilbert space by replacing the number of particles with the number of basis functions used to obtain the exact ground-state wave function (see Figure 2). This approach can be used to describe the symmetry breaking of electronic structure configurations and the quantum criticality of atomic and molecular systems as one varies parameters in the Hamiltonian [9]. Moreover, another approach to quantum phase transition in describing atom–diatomic phase transitions is the use of the inverse participation ratio and the Rényi entropy as other good markers for the critical point [10,11].

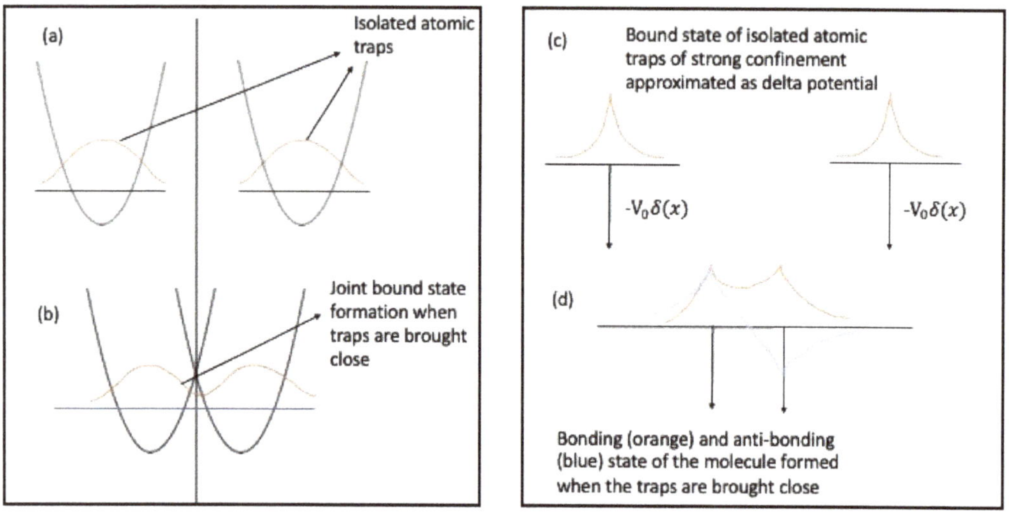

Figure 2. (**a**) Individual atomic sites trapped by a harmonic potential field generated from an optical tweezer. (**b**) Formation of a molecular bound state when the individual traps are adiabatically brought close to one another through the tunnelling of electronic density. (**c**) If the traps are strongly bound one can simplify the process using the model of delta function potential wells. (**d**) Formation of the bonding (orange) and anti-bonding (blue) electronic states when the wells are brought closer to a critical distance.

Quantum phase transitions are also used as a standard tool in experiments on ultra-cold atoms, particularly to demonstrate the BEC (Bose–Einstein Condensation) and BCS (Bardeen–Cooper–Schrieffer) cross-over [12]. Here, "Feshbach resonance" is used to change the effective interactions of atoms in a controlled way. At "criticality" the two-atom scattering length diverges which gives rise to many interesting phenomena for the interaction of a cloud of atoms. This occurs when the scattering state of an inter-atomic interaction potential is nearly resonant with the bound state of the potential above it. In this case, an atom pair with the resonant scattering energy have a large probability of forming the bound state. The scattering length diverges in this case if the scattering state energy is exactly resonant with that of the bound state. In an ultra-cold gas, the scattering length can be tuned through the resonance by applying an external magnetic field. Thus, the basic characteristics of Feshbach molecule formation are a divergence of the atom–atom scattering length when approaching the transition from the dissociated side and a divergence of the size of the molecule when approaching the transition from the bound-state side [13,14].

Next, we would like to argue that at ultracold temperatures, using optical tweezers that use a highly focused laser beam to hold and adiabatically move the two separate atoms, allows one to watch the formation of the chemical bond. This might describe the formation/dissociation of chemical bonds as a quantum phase transition between a free atomic state phase and a molecularly bonded state phase. In Figure 2 we describe this possibility wherein we have separate tweezers trapping two individual atoms (Figure 2a) when brought close to one another, leading to the formation of a joint bound state through tunnelling (Figure 2b). To gain insight into the process we envision the event in Figure 2c,d as two sharply peaked delta function wells signifying each strongly trapped atomic site. When the effective distance between the wells is below a certain limit we see a new-bonding (orange) and an anti-bonding (blue) molecular state develop. The bonding state has an enhanced electronic density in the region between the two sites which acts as a cohesive bond for molecule formation. By opening such new ways of examining the formation/dissociation of chemical bonds as quantum phase transitions, one might envision a universal classification of chemical systems. To illustrate this approach, we examine the bond formation of a simple generic homonuclear dimer A_2 formed from two atomic constituents of A. In the bonded configuration, the total ground state electronic energy of the system A_2 without correction due to vibrational degrees of freedom can be computed from either a simple Hartree-Fock treatment under the Born–Oppenheimer approximation and subsequently refined with advanced post-Hartree-Fock methods to include static and dynamic correlation. Just like in the QRM model, one might construct a fictitious two-level system in this problem wherein one of the 'levels' corresponds to the aforesaid bound state energy of A_2 at the equilibrium bond length and the other being the ground state energy computed similarly for the free atoms of A. The energy separation between the two forms the excitation energy in the two-level system, just like in the QRM. The interaction of this two-level system with the vibrational stretch mode of A_2 can now be envisioned to include non-Born–Oppenheimer effects. The said stretch mode with its characteristic frequency under harmonic approximation will form the Bosonic reservoir in the QRM, and the coupling parameter g in the QRM will be replaced by electronic–vibrational coupling strength. After mapping the details of the problem to the QRM model, one can find the critical point and see how it is related to the equilibrium distance ($R = R_{eq}$) in the formation of the molecular state of A_2 (see Figure 3). So far, we have not discussed the electron and nuclear spin interaction in the case of the formation of A_2, but one can add the spin degrees of freedom to the analysis and examine the hyperfine splitting changes from free atoms to the formation of molecules.

Adding the spin and the rotational degrees of freedom might give an exotic quantum molecular phase. The analysis of the chemical bond formation and breaking as a quantum phase transition at absolute zero temperature is general for any two-level model system interacting with an environment. This, we believe, would shed unforeseen insight into the very process of forming a chemical bond, which is at the heart of chemistry and molecular physics. Further experimentation is required to investigate the idea, using the optically controlled atomic beams we shall discuss in the next paragraph. One can similarly describe the formation of chemical bonds in any dimeric system, resonance in a benzene molecule, two base states for the molecule of the dye magenta, and nitrogen tunneling in ammonia, to name just a few systems. For example, in the symmetry-breaking transition in the ammonia molecule (NH_3) as a function of the distance of the atoms, the molecule will probably develop from a planar configuration with all atoms in a plane to the "buckled" configuration where the N atom is either on top of or below the plane defined by the 3 H atoms. This symmetry-breaking transition is driven by a vibrational stretch mode and can be viewed as an approximate quantum critical point.

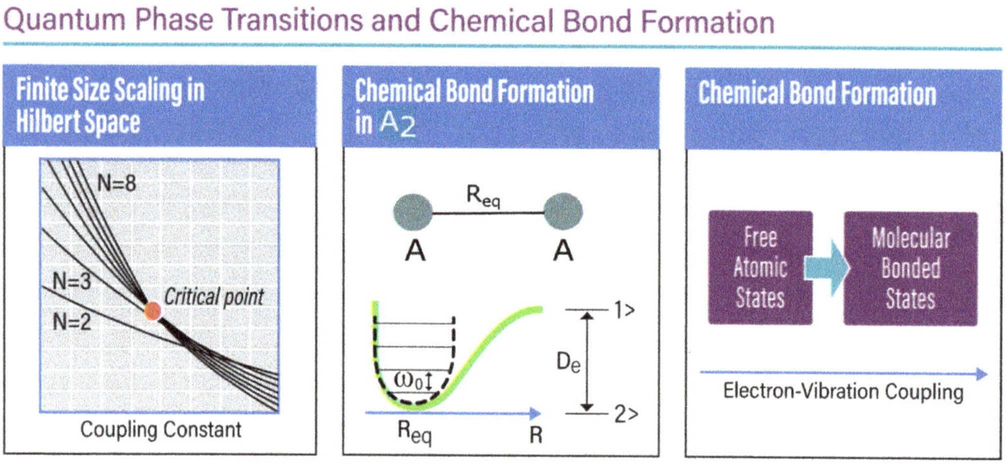

Figure 3. (**Left**) Detection of critical parameters in a quantum phase transition from a finite-size scaling approach. This can be reliably carried out without approaching the limit of infinitely many degrees of freedom, but by successive enhancement of the dimension of the Hilbert space. (**Middle**) The geometric depiction of a generic A_2 dimer and a single phonon reservoir to which it can be coupled. (**Right**) The two-level description of the chemical bond formation process in A_2, wherein one of the accompanying states is the free (A,A) units and the other is the bonded system.

For many decades, the scientific community has witnessed many Nobel prizes being granted for the basic inventions and discoveries in the area of low-temperature physics, the development of methods to cool and trap atoms with laser light, Bose–Einstein condensation in dilute gases of alkali atoms, and recently for groundbreaking inventions in the field of laser physics and the application of optical tweezers. The field of cold chemistry is an exciting field of research, particularly for quantum information and computing science where quantum mechanical wave-like behavior plays a central role, such as superposition, inference, and entanglement [15,16]. With the advancement of this field, recently Liu et al. [17] presented an experimental study where they combined exactly two atoms in a single, controlled reaction. The experimental apparatus traps two individual laser-cooled atoms, sodium and one cesium, in separate optical tweezers and then merges them into one optical dipole trap, forming one molecule. One must emphasize here that such a bond formation from two individual atoms would need the participation of a third body, the photon, which can act as a scavenger for the lost energy and drive the free atom states to an excited molecular state [18].

In the reverse direction, once a bond is formed, excitation from the lowest vibrational level of the ground electronic potential energy surface (PES) into the continuum (corresponding to loosely bound atoms) in the same ground electronic PES can be initiated through a detuned three-level excitation involving vibrational states of excited electronic PESs. Thus, methods of trapping and controlling single atoms to form molecules might open many exciting avenues for research, particularly quantum phase transitions and the formation of chemical bonds and the possibility of creating exotic molecular quantum phases [19–24].

In summary, the very process of chemical bond formation, cornerstone to the foundation of chemistry, can be viewed using the perspective of a quantum phase transition, an idea which thereby unifies the two disciplines and paves the road to further research to consolidate this newly emergent intuition.

Funding: We acknowledge funding from the U.S. Department of Energy (Office of Basic Energy Sciences) under Award No. DE-SC0019215, the National Science Foundation under Award No. 1955907, the NSF grant # 2124511, CCI Phase I: NSF Center for Quantum Dynamics on Modular Quantum Devices (CQD-MQD), and the U.S. Department of Energy, Office of Science, National Quantum Information Science Research Centers, Quantum Science Center.

Acknowledgments: I would like to thank the referees for their many useful comments and also Jonathan Hood and Manas Sajjan for reading and commenting on the draft.

Conflicts of Interest: The author declares no conflict of interest.

References

1. Sachdev, S. *Quantum Phase Transitions*; Cambridge University Press: Cambridge, UK, 2011.
2. Zhou, Y.; Sung, J.; Brutschea, E.; Esterlis, I.; Wang, Y.; Scuri, G.; Gelly, R.J.; Heo, H.; Taniguchi, T.; Watanabe, K.; et al. Bilayer Wigner crystals in a transition metal dichalcogenide heterostructure. *Nature* **2021**, *595*, 48–52. [CrossRef] [PubMed]
3. Jiménez, J.L.; Crone, S.P.G.; Fogh, E.; Zayed, M.E.; Lortz, R.; Pomjakushina, E.; Conder, K.; Läuchli, A.M.; Weber, L.; Wessel, S.; et al. A quantum magnetic analogue to the critical point of water. *Nature* **2021**, *592*, 370–375. [CrossRef]
4. Hwang, M.-J.; Puebla, R.; Plenio, M.B. Quantum Phase Transition and Universal Dynamics in the Rabi Model. *Phys. Rev. Lett.* **2015**, *115*, 180404. [CrossRef]
5. Cai, M.-L.; Liu, Z.-D.; Zhao, W.-D.; Wu, Y.-K.; Mei, Q.-X.; Jiang, Y.; He, L.; Zhang, X.; Zhou, Z.-C.; Duan, L.-M. Observation of a quantum phase transition in the quantum Rabi model with a single trapped ion. *Nat. Commun.* **2021**, *12*, 1126. [CrossRef] [PubMed]
6. Hershbach, D.R.; Avery, J.; Goscinsky, O. *Dimensional Scaling in Chemical Physics*; Kluwer: Dordercht, The Netherlands, 1993.
7. Serra, P.; Kais, S. Critical Phenomena for Electronic Structure at the Large-Dimension. *Phys. Rev. Lett.* **1996**, *77*, 466–469. [CrossRef] [PubMed]
8. Kais, S.; Serra, P. Finite Size Scaling for Atomic and Molecular Systems. *Adv. Chem. Phys.* **2003**, *125*, 1–99.
9. Kais, S.; Serra, P. Quantum critical phenomena and stability of atomic and molecular ions. *Int. Rev. Phys. Chem.* **2000**, *19*, 97–121. [CrossRef]
10. Baena, I.; Pérez-Fernández, P.; Rodríguez-Gallardo, M.; Arias, J.M. Entropies and IPR as Markers for a Phase Transition in a Two-Level Model for Atom–Diatomic Molecule Coexistence. *Entropy* **2022**, *24*, 113. [CrossRef]
11. Graefe, E.-M.; Graney, M.; Rush, A. Semiclassical quantization for a bosonic atom-molecule conversion system. *Phys. Rev. A* **2015**, *92*, 012121. [CrossRef]
12. Greiner, M.; Regal, C.A.; Jin, D.S. Emergence of a molecular Bose–Einstein condensate from a Fermi gas. *Nature* **2003**, *426*, 537. [CrossRef]
13. Chin, C.; Grimm, R.; Julienne, P.; Tiesinga, E. Feshbach resonances in ultracold gases. *Rev. Mod. Phys.* **2010**, *82*, 1225. [CrossRef]
14. Nikolić, P.; Sachdev, S. Renormalization-group fixed points, universal phase diagram, and $1/N$ expansion for quantum liquids with interactions near the unitarity limit. *Phys. Rev. A* **2007**, *75*, 033608. [CrossRef]
15. Wei, Q.; Cao, Y.; Kais, S.; Friedrich, B.; Herschbach, D. Quantum Computation using Arrays of N Polar Molecules in Pendular States. *Chemphyschem* **2016**, *17*, 3714–3722. [CrossRef] [PubMed]
16. Kais, S. *Quantum Information and Computation for Chemistry*; Wiley and Sons: Hoboken, NJ, USA, 2014; Volume 154.
17. Liu, L.R.; Hood, J.D.; Yu, Y.; Zhang, J.T.; Hutzler, N.R.; Rosenband, T.; Ni, K.-K. Building one molecule from a reservoir of two atoms. *Science* **2018**, *360*, 900–903. [CrossRef] [PubMed]
18. Cairncross, W.B.; Zhang, J.T.; Picard, L.R.B.; Yu, Y.; Wang, K.; Ni, K.-K. Assembly of a Rovibrational Ground State Molecule in an Optical Tweezer. *Phys. Rev. Lett.* **2021**, *126*, 123402. [CrossRef] [PubMed]
19. Kaufman, A.M.; Ni, K.-K. Quantum science with optical tweezer arrays of ultracold atoms and molecules. *Nat. Phys.* **2021**, *17*, 1324–1333. [CrossRef]
20. Son, H.; Park, J.J.; Lu, Y.-K.; Jamison, A.O.; Karman, T.; Ketterle, W. Control of reactive collisions by quantum interference. *Science* **2022**, *375*, 1006–1010. [CrossRef]
21. Bao, Y.; Yu, S.S.; Anderegg, L.; Chae, E.; Ketterle, W.; Ni, K.-K.; Doyle, J.M. Dipolar spin-exchange and entanglement between molecules in an optical tweezer array. *Science* **2023**, *382*, 1138–1143. [CrossRef] [PubMed]
22. Bayha, L.; Holten, M.; Klemt, R.; Subramanian, K.; Bjerlin, J.; Reimann, S.M.; Bruun, G.M.; Preiss, P.M.; Jochim, S. Observing the emergence of a quantum phase transition shell by shell. *Nature* **2020**, *587*, 583–587. [CrossRef]
23. Blodgett, K.N.; Peana, D.; Phatak, S.S.; Terry, L.M.; Montes, M.P.; Hood, J.D. Imaging a 6 Li Atom in an Optical Tweezer 2000 Times with Λ-Enhanced Gray Molasses. *Phys. Rev. Lett.* **2023**, *131*, 083001. [CrossRef]
24. Rivy, H.M.; Aljunid, S.A.; Lassalle, E.; Zheludev, N.I.; Wilkowski, D. Single atom in a superoscillatory optical trap. *Commun. Phys.* **2023**, *6*, 155. [CrossRef]

entropy

MDPI

Article

It Ain't Necessarily So: Ludwig Boltzmann's Darwinian Notion of Entropy

Steven Gimbel

Department of Philosophy, Gettysburg College, Gettysburg, PA 17325, USA; sgimbel@gettysburg.edu

Abstract: Ludwig Boltzmann's move in his seminal paper of 1877, introducing a statistical understanding of entropy, was a watershed moment in the history of physics. The work not only introduced quantization and provided a new understanding of entropy, it challenged the understanding of what a law of nature could be. Traditionally, nomological necessity, that is, specifying the way in which a system must develop, was considered an essential element of proposed physical laws. Yet, here was a new understanding of the Second Law of Thermodynamics that no longer possessed this property. While it was a new direction in physics, in other important scientific discourses of that time—specifically Huttonian geology and Darwinian evolution, similar approaches were taken in which a system's development followed principles, but did so in a way that both provided a direction of time and allowed for non-deterministic, though rule-based, time evolution. Boltzmann referred to both of these theories, especially the work of Darwin, frequently. The possibility that Darwin influenced Boltzmann's thought in physics can be seen as being supported by Boltzmann's later writings.

Keywords: Ludwig Boltzmann; entropy; Charles Darwin; evolution; model

1. Introduction

The things that you are liable
To read in the Bible,
It ain't necessarily so.

—Ira Gershwin

There is no 10,000 kg sphere of pure gold anywhere in the universe. But that fact's universality does not make it into a law of nature because while there does not happen to be such a sphere, there could be. Physical laws seem to require a special sort of necessity, what philosophers of science term "nomological necessity", that tells us what must or cannot happen [1]. Ludwig Boltzmann's formulation of the Second Law of Thermodynamics [2] was contentious in part because it challenged the traditional understanding of the nature of that necessity in physical law.

Contemporaries like James Clerk Maxwell, Josef Lochschmidt, and Ernst Zermelo objected to Boltzmann's approach in order to save the traditional account of natural law. If Boltzmann's proposal clashes with the commitment to traditional necessity, they held, then the proposal should be jettisoned. Boltzmann, on the other hand, opted to revise our understanding of what should be expected from physical laws.

What accounts for Boltzmann's willingness to be so philosophically radical? The reason may lay in a combination of two factors: (1) Boltzmann's brand of realism—he was an "entity realist", believing that atoms do exist, but not a "nomological realist", believing our best scientific theories merely provide a *Bild*, a useful picture that should not be seen as literally true; and, (2) Boltzmann may have employed Darwinian evolution to provide a *Bild* to understand the microscopic world in which random changes are responsible for the time evolution of a system that is time-irreversible in practice, but not in principle.

Thermodynamic entropy famously emerges from Sadi Carnot's work on steam engines, the machines that powered the industrial revolution. They ran on coal, which had to be mined. Those coal mines not only provided fuel, but also unearthed the geological strata that led to modern geology. Within the layers of rock were fossils, often of animals different from their modern counterparts and in locations where the animals did not seem to belong. This set the stage for new biological theories of speciation, and ultimately, Darwinian natural selection.

Huttonian geology and Darwinian evolution are both scientific theories that differentiate temporal directions without the necessity traditionally required in physics, and both were big news in the scientific community when Boltzmann was working on entropy. Geology and biology thus may have provided models for Boltzmann's thinking about his statistical picture of thermodynamics.

While Boltzmann's overture is no smoking gun, that is, there is no direct reference to Darwin in Boltzmann's works in which he develops his understanding of entropy, there are plenty of places both during and after his seminal pieces on statistical mechanics where Boltzmann does not only refer to, but actively employs, Darwin's thought for a range of purposes. Based upon them, a circumstantial case can be made that Boltzmann used biological evolution as a *Bild* through which to understand the time evolution of thermodynamic systems.

2. Maxwell's Models and Boltzmann's *Bilder*

Maxwell and Boltzmann held complicated relations to scientific realism, the view that our current best theories reflect reality. Maxwell was famous for his mechanical models, physical analogies that were useful heuristics, but never intended to describe an underlying reality [3]. Playing a similar role in Boltzmann's understanding of the scientific method was his notion of *Bild*, that is, pictures or models used to make sense of systems, but not necessarily to provide accurate accounts. Indeed, he held that scientists should pursue a range of these models, as each may be fruitful in a different way.

Yet, while both employed explicitly anti-realist approaches in doing science, both also harbored realist ambitions. If a notion employed in the models developed by scientists, say, of molecules, was sufficiently successful in accounting for a wide enough swath of phenomena without glaring anomalies, then we would have warrant for considering that notion to refer to an actual component of the universe. In both holding most scientific work to be of mere instrumental value, while still allowing that sufficient predictive and explanatory success provided metaphysical license, Henk de Regt [4] calls Maxwell and Boltzmann "flexible realists".

For Maxwell, absolute truth was restricted to the Divine [5]; only God could know the truths of the universe with certainty. Humans could, at best, develop a rough intuitive sense of the way the world worked, and this was aided by our ability to construct mental metaphors, models that frame the system under investigation in terms of an analogy to another system we understood better. "By a physical analogy I mean that partial similarity between the laws of one science and those of another which makes each of them illustrate the other (quoted in [6] (p. 208))". Maxwell (and many who followed him) dedicated great effort to constructing cognitively and actually building mechanical models of abstract entities like the magnetic field and the luminferous aether.

As those models became able to account for increasingly greater numbers of observable phenomena, and to predict new ones that had not been previously suspected, the question of the realistic interpretation of the models naturally emerged.

> "The question of the reality of analogies in nature derives most of its interest from its application to the opinion, that all phenomena of nature, being varieties of motion, can only differ in complexity, and therefore the only way of studying nature, is to master the fundamental laws of motion first, and then examine what kind of complication of these laws must be studied in order to obtain true views of the universe. If this theory be true, we must look for indications of these

> fundamental laws throughout the whole range of science, and not least among those remarkable products of organic life, the results of cerebration (commonly called 'thinking'). In this case, of course, the resemblances between the laws of different classes of phenomena should hardly be called analogies, as they are only transformed identities (quoted in [5] (p. 76))".

Maxwell thus argues that when a model is sufficiently successful, we cannot but begin to see it as being actually descriptive of the underlying reality. We should never forget that it is a model and not take the model to be a full and complete description, but we must make some limited inference in our grasping of things as they really are.

Boltzmann was deeply influenced by Maxwell's physics, but also his epistemology. Enticed by the success of Maxwell's method of theorizing, he, too, hews to a scientific methodology that employs a sort of model at its heart. But where those models were almost exclusively mechanical for Maxwell, Boltzmann moves to a notion that was reflective of what was happening in Austrian culture at the time.

The notion of "*Bild*" was very much in the air around Boltzmann [7]. On the one hand, it is the term that was used for a photograph, a new technology that led to a wave of philosophical conversation in the culture around reality and representation. Are the properties of the photograph, for example, absolute truths of the world? Is it perspectival or absolutely objective? Could it be used as evidence in a court? In science? What can be inferred about the subject of the photograph from the image itself and with what degree of certainty?

At the same time, it was also the root of the term "*bildung*", which referred to the process of self-creation through education and culture. In a class-conscious society, as the urban centers of the late Austro-Hungarian Dynasty were, there was a deep connection of the notion of "*Bild*" to what one should believe as a result of a process of discovery. It connoted a relation of the proper orientation of self to the social world.

These senses are embedded in Boltzmann's Austrian twist on Maxwell's notion of a mechanical model, turning it into more abstract notion of a scientific *Bild*. While Boltzmann uses the term in a multiplicity of ways [7], it is central to his approach to scientific methodology. A *Bild* is a model that includes elements beyond the observable, mentally developed cognitive constructs. These additional theoretical elements from the mind of the theorist create an explanatory construct which can be used not only to visualize what a system might look like that gives rise to the observed phenomena in the way a Maxwellian mechanical model does, but also to suggest future phenomena that might be accounted for as well.

In "On the Development of the Methods of Theoretical Physics in Recent Times", Boltzmann points to Wilhelm Weber's electro-magnetic theory which, although later discredited by Maxwell's theory, nonetheless suggested the hitherto undiscovered Hall effect [8]. Because even a *Bild* like Weber's that turned out not to work beyond the phenomenon it was designed to account for and yet could have such progressive elements, and because a *Bild* should be thought of as a heuristic tool and not an accurate description of the underlying reality, there is an advantage in having theoreticians create a multiplicity of different *Bilder*. In "On the Fundamental Principles and Equations of Mechanics" [9], Boltzmann writes, "If in this way we have grasped the task of thought in general and of science in particular, we obtain conclusions that are at first sight striking. We shall call an idea about nature false if it misrepresents certain facts or if there are obviously simpler ideas that represent these facts more clearly and especially if the idea contradicts generally confirmed laws of thought; however, it is still possible to have theories that correctly represent a large number of facts but are incorrect in other aspects, so that they have a certain relative truth. Indeed, it is even possible that we can construct a system of pictures of experience [*Bildern der Erscheinungen*] in several different ways (pp. 105–106)".

This multiplicity of distinct models from the same initial set of observations should be seen metaphorically along the lines of mutations in a Darwinian context. Different alterations will have potentially different advantages in terms of the "selection pressures" that are generated by additional experimental discoveries. Indeed, Rosa et al. [10] try to develop a Darwinian epistemology along this line. As such, we see in Boltzmann, as with Maxwell, that a sufficiently successful model should be taken to provide us with some sense that we are developing a picture that is in some limited way representative of the underlying reality.

Maxwell and Boltzmann both allow inferences from sufficiently successful models. The nature of that inference is what is called "entity realism". If there is an ineliminable element of a model that shows itself to be explanatory and predictive in a wide enough range of situations without significant failure, then there is reason to think that there is a correlate in the real system to that part of the model. In other words, models that are widely applicable without anomaly can give us reason to believe in something we cannot directly observe.

However, this realism does not extend to the model as a whole. While successful models can give us warrant for belief with respect to the furniture of the world, we must always remember that this is a model. Hence, the theory itself will always be at best an approximation, a mere analogy. The parts may point to real things, but the laws are not actual universal truths. While Maxwell and Boltzmann accepted a sort of entity-realism, they both rejected nomological realism, the view that our best current scientific theories tell us how the underlying reality actually operates.

3. The Reality of Molecules

This shared methodological approach led to a temporary disagreement between Maxwell and Boltzmann concerning an inference to the reality of molecules.

Michael Faraday was famously untrained in mathematics and thereby utilized mental pictures as analogies in working out his advances in electricity and magnetism. Maxwell, who was mathematically masterful, translated Faraday's insights into equations, much to Faraday's delight [5]. Maxwell continued to use Faraday's approach of mechanical models prolifically, developing a range of intricate physical analogies. He treated electricity as an incompressible fluid and the ether as a set of cogs.

But perhaps most importantly, in a series of papers beginning in 1860 with "Illustrations of the Dynamical Theory of Gases" [11], Maxwell constructed mechanical models of gases, treating them as collections of particles that began as impenetrable spheres that interacted only by contact but which became decreasingly idealized as he went on. With each new element added to the mechanical model, more empirical thermodynamic phenomena could be derived.

Maxwell was under no illusion that he was proving anything, but rather saw himself as providing evidence in favor of the kinetic theory of heat [6]. This evidence was the result of our ability to account for an increasing number of observable phenomena and regularities and suggested that this progress would likely continue with the further development of the molecular model. Maxwell famously wrote [11] "If the properties of such a system of bodies [as he assumed] are found to correspond to those of gases, an important physical analogy will be established, which may lead to more accurate knowledge of the properties of matter (p. 377)".

The success of Maxwell's research program impressed Boltzmann, who himself began to contribute to it. While both were thoroughly committed to the mechanical theory of heat, in 1871 Maxwell was more reticent to attribute reality to the molecules as a result of an anomaly neither could account for: specific heat.

The ratio of specific heat at constant volume to specific heat at constant pressure could be experimentally determined. Maxwell's initial attempt with his simplified picture of molecules failed. This was unsurprising, but when he made the model more realistic, accounting for rotational and translational energies, even while considering the molecules to be polyatomic, Maxwell still could not solve the problem. Boltzmann followed with his attempt, but to no avail [4].

Their inability to account for the correct measured values of specific heats led Maxwell to withhold a realist understanding of the work, referring to it as "the greatest difficulty the molecular theory has yet encountered (quoted in [4] (p. 212))". Boltzmann contended that the overall progress of the research program justified a realistic interpretation of atoms, where the problem made Maxwell pull back.

Ultimately, Boltzmann did solve the problem by creating the "dumbbell model" of diatomic molecules, which allowed for the rotational and translational motion, but because of the bond connecting the atoms, eliminated one degree of freedom. This changed the theoretical predictions in a way that allowed it to match the measured values. At that point, Maxwell, Boltzmann, and most others—with notable exceptions like Ernst Mach and Henri Poincaré—accepted the existence of atoms as having been demonstrated. His atomic *Bild* gave warrant for an assertion of the reality of its central entity. As Boltzmann would later write, "[C]ontemporary atomism provides a perfectly accurate picture [*vollkommen zutreffendes Bild*] of all mechanical phenomena and, because of the self-contained nature of this field, there is little expectation that we will discover phenomena that do not fit in the frame of this picture [*Rahmen des Bildes*] [12] (p. 150)".

4. How the H Did Boltzmann Decide to Move to a Statistical Law?

Boltzmann accepted the reality of atoms as he worked on the problem of specific heats, also focusing on making sense of other aspects of macroscopic thermodynamics in terms of the mechanical model, specifically, the concept of entropy. How could the macroscopic quantity governing the conversion of energy to work in engines be understood in the microscopic context? There was a macroscopic principle, the Second Law of Thermodynamics, that needed accounting for in terms of Boltzmann's *Bild*.

Starting in 1872 with his paper "Further Studies on the Thermal Equilibrium of Gas Molecules" [13], Boltzmann sought to model entropy on molecular systems governed by Newtonian mechanics and thereby develop a notion of microscopic entropy that would mirror the macroscopic notion by increasing in systems not in equilibrium and remain constant for those that were. The 1872 paper proposes a theorem involving the quantity that he would term H, which is the negation of entropy.

Boltzmann followed with a series of papers working on this problem, leading up to the 1877 paper "On the Relationship between the Second Fundamental Theorem of the Mechanical Theory of Heat and Probability Calculations Regarding the Conditions for Thermal Equilibrium" [2], in which he derives what is known as the Boltzmann distribution. In fact, he derives it twice in the paper: once using discrete mathematical means and again employing the continuous means of differential equations. The former, found in Section I of the paper, begins with the "chunking" that Boltzmann included, dividing the molecules into velocity classes and assuming every molecule in that class to have the same velocity, allowing him to deal with a finite number of velocity-classes. (This move, most famously, influenced Max Planck with its idea of quantification which led to his solution of the blackbody radiation problem). As Kim Sharp and Franz Matschinky, in their introduction to their translation of [2], note, "This assumption does not correspond to any realistic model, but it is easier to handle mathematically (p. 1976)". In Section II of the paper, Boltzmann then repeats the process, but with continuous energy distributions.

This redundancy might seem curious, but as Nadine de Courtenay [14] argues, "Boltzmann was one of the first physicists to recognize that mathematical language was not an inconsequential means of expressing physical processes (p. 50)".

Ernst Mach's insistence on differential equations in physics, for example, was not an innocent choice, Boltzmann argues, but begs the question in favor of his anti-atomism. Mach's positivism, his epistemological view that we should only believe that which is observable, led him to reject the existence of atoms as well as other in principle unobservable concepts like Isaac Newton's absolute space. Mach insisted that we replace the sorts of explanations in science that smuggle in unobservable entities and simply see differential equations as the last word, a method that is metaphysically empty, but scientifically full.

But this embrace of differential equations as the language of all physics was not philosophically neutral. Mach may have claimed to be trying to rid physics of metaphysical baggage, but he was actually stacking the philosophical deck by choosing to work only in smooth mathematical universes in making his models of the world [14]. Constraining physical laws to the form of differential equations invisibly imported an anti-atomistic metaphysic. Mathematical language is not mere metaphysically neutral formalism, according to Boltzmann, and Mach's use made him the metaphysician he railed against.

In choosing to work in both discrete and continuous languages, Boltzmann in [2] was dong two things. First, he was playing it metaphysically honestly, showing that his result was not dependent on a particular picture of the underlying world. This is not to say that he was not committed to atomism.

He was, but in showing that the result was not dependent upon discrete or continuous foundations, he had a second goal: to show that the physics was capable of being a bridge to connect the macroscopic world, in which the microscopic could be treated as if it were smooth, and the microscopic world, which could not. As such, the behavior of the resulting notion of entropy, and the statistical tools he would build around it, would provide a picture that holds true in both frames of reference. It worked for the macroscopic world that was seen as if it was continuous and for the microscopic world that had to be treated as discrete. By having a single treatment that is invariant under the change from discrete to smooth, the work had to be seen as bridging the intellectual chasm between the macroscopic thermodynamic and the microscopic atomistic.

But there was an aspect of his result that undermined this bridge. The resulting view of entropy was statistical. What began in the 1872 paper as an epistemic probability, a statistical generalization resulting from our inability to account for the multitude of atoms in a small amount of a gas, turned into a fully stochastic approach by 1877 in which the entropy was no longer tied to properties of the individual atoms in the gas, but rather now became a measure of the accessibility of abstract ensemble states. "Boltzmann's explanation was to consider the increase in entropy as a result of a statistical process, involving the calculation of probabilities, and not as a result of a dynamic process described only by mechanics [10]".

Boltzmann did not give up his commitment to the atomic hypothesis, but moved his thinking from the Maxwellian approach of deriving the macroscopic directly from mechanical aspects of the microscopic to a higher-level picture. In shifting the object of the law from standardly physical quantities like duration and velocity to properties of the constructed phase space, Boltzmann was radically revising how to think of the system and how to understand the macroscopic law governing entropy. The Second Law of Thermodynamics was no longer deterministic, but a statistical generalization that made what we see not necessary, but highly likely. That move, of course, generated serious objections.

5. Defending Necessity

So, according to Boltzmann's approach, entropy tends to increase, thereby distinguishing future from past, but does not always necessarily increase, and that probability is intrinsic to the system, not a mere result of our ignorance. Critics, notably Maxwell, Josef Lochschmidt, and Ernst Zermelo, objected to different parts of the position. Maxwell to the nature of the probability, Lochschmidt to the lack of time-reversibility, and Zermelo to the lack of necessity.

Maxwell did not object to the statistical move per se, but tried to show that the system is not stochastic. The uncertainty involved remains of the epistemological sort. His eponymous demon is a tiny intelligence in control of a valve. The demon is capable of determining the velocity of all molecules of a gas and capable of opening or shutting the vale at will. By choosing to only open the valve for high-velocity molecules, the demon thus becomes capable of sorting the molecules and thereby creating increased order, that is, of decreasing entropy.

Entropy thereby does not necessarily increase, but our statistical sense that it generally does is a result of our not being demonic, that is, of our having less cognitive capacity than the fanciful being. Because its ability to alter the amount of entropy in the system is accomplished through the intelligent processing of information, then, the appeal to probability is epistemic and not metaphysical. In other words, the statical nature of the Second Law would be a result of our ignorance, not a feature of the world itself [7] (p. 1278).

Josepf Lochschmidt, was a dear and close friend of Boltzmann. They both accepted the existence of atoms and the supposition that they were governed by Newtonian mechanics. But if this is true, Lochschmidt contended, then the properties of the laws governing the small would have to be the same as those governing the large since the ensemble is nothing more than a collection of necessarily determinable states. The rules governing the parts—which are deterministic and time-reversible—should not differ from those governing the whole.

Yet, Boltzmann's statistical approach creates an asymmetry between future and past. This, Lochschmidt argues, is problematic. "[I]n any system, the entire course of events becomes retrograde if at a certain moment the velocities of all its elements is reversed (quoted in [15] (p. 200))". Given that Newtonian mechanics is time-reversible, reflecting the velocities of all molecules would follow Newton's laws and decrease entropy, but this reversed system is as much a model of Newton law as the non-reversed one. As Flamm points out of the reflection, "This procedure is equivalent to time reversal", that is, the backward running film of the universe decreases entropy, but still fits the underlying mechanics Boltzmann uses.

It should be noted that this was also a line that was used against Darwin's theory of evolution. Critic William Hopkins [16] wrote, "a phenomenon is properly said to be explained, more or less perfectly, when it can be proved to be the necessary consequent of preceding phenomena, or more especially, when it can be clearly referred to some recognized cause; and any theory which enables us to be to do this may be said in a precise and logical sense, to explain the phenomenon in question. But Mr. Darwin's theory can explain nothing in this sense, because it cannot possibly assign any necessary relation between the phenomena and the causes to which it refers them (p. 267)".

Zermelo's objection took a different route. If the entropy of a system is based upon the distribution of positions and velocities, and these obey Newtonian principles, then given enough time, the system will eventually find itself back in the original orientation. But, since there was an increase in entropy moving away from the initial state, there would have to have been a decrease in order for the original state to recur.

The Second Law of Thermodynamics is a purported law of nature. Laws of nature are more than mere rules of thumb. They say what must happen. Making a law of nature into a statistical generalization is to undermine it as a law of nature. In physics, during that period, Hermann Bondi [17] contends, "it was widely thought that the perfect predictability of Newton's solar system 'clockwork' was what any physical science, nay what any human endeavor should aim to achieve (p. 159)". Boltzmann's understanding of entropy denied it and therefore, to many, it seemed not to be good science. As Bondi reported Ernest Rutherford as having said, "If you get a statistical answer, you asked the wrong question".

Boltzmann wrestled with all three of these concerns, acknowledging that they were, indeed, concerns that needed to be taken seriously. And yet, he did not waver from his approach. Formulating different versions of responses to them, he always stayed true to his approach. He was willing in the move from the micro- to the macroscopic to accept a deeply stochastic picture of nature which differentiated future from past, but which did not possess the sort of necessity that had been considered an essential aspect of physical law. What accounts for this digging in of his intellectual heels in the face of strong objections to the different aspects of his view from people he respected?

One possibility is another successful *Bild*, another cognitive framework that allowed him to make sense of this unusual world in a way that made sense to him. That heuristic could have been provided by Darwin's theory of natural selection.

6. Darwin in Germany and Austria around Boltzmann

Darwin's *Origin of Species* first appeared in German translation in April 1860, translated by Germany's most prominent paleontologist, Heinrich Georg Bronn, whose own research had been proceeding parallel to Darwin's [18]. However, Bronn passed away two years after the translation's publication. As a result, Darwinism in Germany required someone else to act as its spokesperson. Stepping into this role was the young, handsome, and charismatic Ernst Haeckel who famously spoke to the German Society of Naturalists and Physicians at Stettin in September of 1863, and with an address of historical significance, the twenty-nine year old zoologist launched the era of Darwinism in Germany. "With the translator Bronn", he told the assembled crowd, "I see in Darwin's direction the only possible way to come close to understanding the great law of development that controls the whole organic world [18] (p. 157)".

The approach that Haeckel develops, and which becomes greatly influential in the German-speaking world during Boltzmann's time, however, is less Darwinian and more in line with that of Jean Baptiste Lamarcke. We see this sort of Lamarckian understanding of Darwin in the writings of Hermann von Helmholtz [19]. Boltzmann studied with Helmholtz in Berlin in 1871, just before Boltzmann's initial work on the H theorem, and we find in Helmholtz's personal notebook from that period (notes which he used to give his lectures) passages like the following:

> "Recapitulation of Darwin's hypotheses: The law of heredity demonstrates a range of variations in all classes of organisms although the fewest differences arise among species in the wild. What remains to be determined is the limits of this change. This is where the following comes into consideration: that a much greater influence is to be expected from natural breeding than from artificial selection. The hypothesis of an independent origin of each species becomes incredibly unlikely as a result of (a) the homology among different species (b) Metamerism within the same species (c) Paleontological developments, and (d) the geographical affinity of like organisms to live together". [20]

The reference here to "metamorism" shows the ways in which Helmholtz saw Darwin as connected to his own research program on color perception and thereby would have likely been a topic Helmholtz would have been thinking about during the period when Boltzmann was in contact with him.

Boltzmann left Berlin for a position back in Vienna which he occupied until 1876, just before the publication of his statistical approach to entropy. At that time, his colleague was Ernst Mach, who had been installed in the newly created chair in "the history and philosophy of the inductive sciences". In this position, which Boltzmann himself would later occupy, Mach became a public spokesperson for his philosophical position of positivism. While the scientific and epistemological disagreements between Mach and Boltzmann are extremely well-trod ground, one point of agreement between them was their mutual enthusiasm for Darwinism (see [21]). Indeed, Mach asserts that it must have its imprint on all of human thought. For example, in his inaugural address "On Transformation and Adaptation in Scientific Thought" [22] for his position at Prague (the post he held before moving to Vienna), Mach writes, that "[K]nowledge, too, is a product of organic nature. And although ideas, as such, do not comport themselves in all respects like independent organic individuals, and although violent comparisons should be avoided, still, if Darwin reasoned rightly, the general imprint of evolution and transformation must be noticeable in ideas also (quoted in (pp. 217–218))"

While we can trace Boltzmann's public and published references to Darwin at least as far back as 1886, nine years after his transformative approach to entropy, it is clear that Boltzmann would have been surrounded by discussion of Darwin's theory from the wider culture and from important intellectual figures with whom he was interacting at the time he was working on this new understanding of the Second Law.

7. Darwin in Boltzmann

There is no smoking gun. We do not have Boltzmann in a correspondence or a footnote citing Darwin's influence on his thought in developing his statistical understanding of the Second Law of Thermodynamics. However, what we do have is a range of testimony from Boltzmann about a wide range of other ways that Darwin influenced his thinking at the time and afterward, some closer and others farther from this topic. We can also find analogues in the treatment and evidence of natural selection that mirror the arguments raised against Boltzmann's approach to the Second Law of Thermodynamics, and in the analogous situations, the concerns cease to be serious concerns. Putting this together, we may not be able to demonstrate that Boltzmann was thinking of the time evolution of ensembles of atoms along the lines of the adaptations of species in ecosystems, but it would fit comfortably within the larger confines of how we do know that Boltzmann dealt with other issues in epistemology and science.

Indeed, Darwin occupied a privileged place in the science of this time. In 1886, nine years after the statistical turn and while Boltzmann is working on the argument addressing Loschmidt's objection in print in his article "Neuer Beweis zweier Sätze über das Wärmegleichgewicht unter mehratomigen Gasmolekülen", he is called to give a talk to the Austro-Hungarian Imperial Academy on the Second Law of Thermodynamics [23]; he began by reflecting on the centrality of science in general on human progress.

> "If we regard the apparatus of experimental natural science as tools for obtaining practical gain, we can certainly not deny its success. Unimagined results have been achieved, things that the fancy of our forebears dreamt in their fairy tales, outdone by the marvels that science in concert with technology has realised before our astonished eyes. By facilitating the traffic of men, things and ideas, it helped to raise and spread civilization in a way that in earlier centuries is paralleled most nearly by the invention of the art of printing. And who is to set a term to the forward stride of the human spirit! The invention of a dirigible airship is hardly more than a question of time. Nevertheless I think that it is not these achievements that will put their stamp on our century: if you ask me for my innermost conviction whether it will one day be called the century of iron, or steam, or electricity, I answer without qualms that it will be named the century of the mechanical view of nature, of Darwin [23] (p. 15)".

Boltzmann thought Darwin more important than air travel.

It should not be lost that this is in a paper discussing the notion of entropy and that instead of calling Darwin's work "the theory of evolution" or "speciation by natural selection", instead Boltzmann chooses a name that explicitly parallels the phrase he used for the kinetic theory of gases—the mechanical view of heat. Indeed, Boltzmann's discussion seeks to erase the distinction between the explanatory sciences like physics and the merely descriptive historical sciences. "Since the mighty upswing of geology, physiology and so on, but above all since the general acceptance of the ideas of Darwin, these sciences boldly undertake to explain the forms of minerals and of organic life [23] (p. 16)". With the "mighty upswing" of geology and the great success of Darwin, there were examples of a mechanical approach to the world in which the development of a scientific system is the result of accidental, not deterministic factors, a fact that was not lost on Boltzmann.

Boltzmann not only admired the theory of evolution; he was clearly reading Darwin closely. Boltzmann reflects on the most profound questions of humanity "whence do we come, whither shall we go" and asserts that "essential and undeniable progress" has been made in "the present century, thanks to most careful studies and comparative experiments on the breeding of pigeons and other domestic animals, on the coloring of flying and swimming animals, by means of researches into the striking similarity of harmless to poisonous animals, through arduous comparisons of the shape of flowers with that of the insects that fertilize them (p. 14)". This is a list, almost chapter by chapter, of the evidence that Darwin sets out for his theory in *The Origin of Species*. Clearly, Boltzmann not only read the book, but knew its structure and argumentation intimately.

The influence of Darwin can be seen in multiple ways in the thought of Boltzmann: two epistemological, one metaphysical, and two scientific.

7.1. Darwin as an Answer to Kantianism

The structure of the human mind was a lively question among physicists of Boltzmann's time. The advancements of science required a foundation in epistemology. We had to know what knowledge was in order to continue to increase it. The prevailing trend in Continental philosophy in the 19th century moved away from the scientific and toward the romantic. The idealism of Friedrich Hegel ruled supreme in German-speaking philosophy departments. As such, philosophers seemed of little use when facing the challenge of non-Euclidean geometry or the rise of atomism, which posited the existence of the unobservable.

Some, like Boltzmann's contemporary at Vienna, Mach, took on a thoroughgoing empiricism, arguing that only what could be observed should play a part in our understanding of the world. Diametrically opposed to the *Naturphilosophie*, this meant the elimination of God and other purely speculative metaphysical entities like the spirit. Influenced by the work of Gustav Fechner, Mach [24] contended that we could do away not only with the idea of the soul, but with the notion of mind altogether, replacing it with a purely materialistic psychophysics. This monism stood as the key move in establishing a general stance: "all metaphysical elements are to be eliminated as superfluous and as destructive of the economy of science (xxxviii)". Having done away with the mind/soul, Mach also sought to cleanse science of Newton's absolute space and absolute time, atoms, and any other theoretical construct that was not reducible to sense perceptions.

Others did not take Mach's positivist route, but sought the last major philosophical figure who did take science seriously, which sparked a rebirth of Kantianism in Germany. Immanuel Kant argues in his *Critique of Pure Reason* [25] that certain mathematical and scientific notions, such as Euclidean geometry and Newtonian mechanics, are what he terms "synthetic a priori". The logical distinction between analytic and synthetic propositions distinguishes between those like "bachelors are unmarried" and "bachelors are slobs", wherein analytic sentences like the former have a predicate that is contained in the meaning of the subject, whereas synthetic sentences like the latter have a predicate that is not a part of the meaning of the subject. The a priori/a posteriori distinction is epistemological in separating those propositions that could be known without perception from those that require observation.

Before Kant, it was supposed that all analytic propositions are a priori and all synthetic propositions are a posteriori, but Kant argues that there is a subset of propositions that are synthetic a priori, i.e., that we know without experience, but which are not merely definitional. Consider the Euclidean axioms. We know that you can draw a circle of any size around any point, but this truth is not merely one that is arrived at through unpacking the notions of circle and point. It is a synthetic proposition in possessing content beyond the definition of the words making it up.

Kant's question, then, is "How are a priori synthetic judgements possible?" His answer is that they are an innate part of the structure of the human mind. They function as the categories by which the mind takes the raw manifold of perception, such as the blur of colors taken in by the eyes, and out of them constructs complex perceptions. The sensory organs provide the content of what we observe, but the necessary categories of the mind provide the form and when the activity of the mind acts upon the raw data fed in from the senses, the result is the observations we make of the world around us.

Because they are innate, they are universal. All humans begin with the same conceptual foundations. Observation can add to it, but all people start with the same set of categories to construct the world. And since these are the structural elements implicit within our observations, no possible experience could contradict them. They are apodictic, viz., necessarily true and in principle unfalsifiable. Because these are the ideas that build the world out of our perceptions, no perception could possibly falsify it, since it built those perceptions.

A number of major figures who championed a revived version of the Kantian synthetic a priori as the basis for their justification of scientific truths were scientists Boltzmann held in the greatest respect. Heinrich Hertz, for example, writes in the prefatory note to his book *The Principles of Mechanics: Presented in a New Form* [26]:

> "The subject-matter of the first book is completely independent of experience. All the assertions made are a priori judgments in Kant's sense. They are based upon the laws of the internal intuition of, and upon the logical forms followed by, the person who makes the assertions; with his external experience they have no other connection than these intuitions and forms may have (p. 45)".

Indeed, Hertz and Boltzmann had a lively correspondence with Hertz contending with Kant that the laws of thought must be synthetic a priori [27].

Boltzmann's response in [28] is to counter this brand of neo-Kantianism, which he thinks is so absurd as to be a joke, with Darwin.

> "What then will be the position of the so-called laws of thought in logic? Well, in light of Darwin's theory they will be nothing else but inherited habits of thought. Men have gradually become accustomed to fix and combine the words through which they communicate and which they rehearse in silence when they think, as well as the memory pictures of those words and everything in the way of internal ideas used for the denoting of things, in such a manner as to enable them always to intervene in the world of phenomena in the way intended, and in inducing others to do likewise, that is to communicate with them. These inventions are greatly promoted by storing and suitable ordering of memory pictures and by learning and practicing speech, and this promotion is the criterion of truth. This method for putting together and silently rehearsing mental images as well as spoken words became increasingly perfect and has passed into heredity in such a way that fixed laws of thought have developed...One can call these laws of thought a priori because through many thousands of years of our species' experience they have become innate to the individual, but it seems to be no more than a logical howler of Kant's to infer their infallibility in all cases...According to Darwin's theory this howler is perfectly explicable. Only what is certain has become hereditary; what was incorrect has been dropped. In this way these laws of thought acquired such a semblance of infallibility that even experience was believed to be answerable to their judgement. Since they were called a priori, it was concluded that everything a priori was infallible (pp. 194–195)".

Boltzmann is adopting an early version of the view that would later be famously held by Noam Chomsky, that a result of evolution is an inherent linguistic faculty that is innate and results in a particular intrinsic grammar beneath all human language-use [29]. This explains the correct elements of the Kantian synthetic a priori, Boltzmann contends, without committing us to the flawed apodictic aspect.

Boltzmann's epistemological approach was pragmatic, seeing the forming of beliefs based on observations as evidence as a successful adaptation of our ancestors. The human mind is an artifact of evolution and so, therefore, must the structures by which it determines what we ought to believe based upon what it is fed through the sense organs. Darwin, instead of Kant, forms the basis for his understanding of why the human mind is capable of scientific theorizing and testing.

What is most important in the current context, though, and should be emphasized, is that the move Boltzmann is making here epistemologically is to undermine the apodictic nature of Kantian synthetic a priori truths, that is, the necessity of these propositions, preferring instead a view in which they are the result of a Darwinian evolutionary process by which they arise accidentally through a historical process. This is very much homeomorphic to the move Boltzmann makes in 1877 with respect to the Second Law of Thermodynamics, taking it from a proposition whose necessity must be asserted to the result of probabilistic historical processes.

7.2. Darwin as a Metaphor for the Scientific Method

Darwin provides not merely the basis on which we should understand why humans are capable of science, but also provides a way of understanding the means by which scientific results become rationally accepted by the scientific community. Boltzmann writes in [30] "[N]o theory is absolutely true, and equally hardly any absolutely false either, but that each must gradually be perfected, as organisms must according to Darwin's theory. By being strongly attacked, a theory can gradually shed inappropriate elements while the appropriate residue remains (p. 153)".

Evolution is not only a theory, but a theory that provides an image for how theories emerge, survive, and grow. Rosa et al. in [10], expand upon this metaphorically Darwinian approach in their article "Constructivism and Realism in Boltzmann's Thermodynamic Atomism".

7.3. Darwin as Support for the Materialist Worldview

Boltzmann's atomistic worldview not only provided him with a picture of the microscopic workings of thermodynamic systems, but provided a thoroughgoing ontology (that is, a catalogue of everything that exists in reality). Heat had been considered a substance from Aristotle to the advocates of phlogiston theory. The mechanical theory of heat allowed for the simplification of our catalogue of things in the universe, and Boltzmann saw the mechanical theory of heredity as doing the same.

In the German-speaking world of the 19th century, one of the most important scientists connected with evolution was Haeckel, most remembered for his aphorism "Ontogeny recapitulates phylogeny". Coming out of the *Naturphilosophie* movement which leaned heavily on research in embryology and morphology paired with a robust religious metaphysic, Haeckel developed an explicitly evolutionary theory, albeit one that was more in line with that of Lamarck than Darwin's approach. But Haeckel [31] added a German philosophical twist to his view, placing Geist (spirit or soul) at the heart of the biological.

> "No reproach is more frequently made against the science of to-day, especially against its most hopeful branch, the study of development, than that it degrades living Nature to the level of a soulless mechanism, banishes from the world the ideal, and kills all the poetry of existence. We believe that our unprejudiced, comparative, genetic study of soul-life gives the lie to that unjust accusation. For if our uniform or monistic conception of Nature is rightly founded, all living matter has a soul, and that most wondrous of all natural phenomena that we usually designate by the word spirit or soul is a general property of living things. Far other than believing in a crude, soulless material, after the manner of our adversaries, we must rather suppose that the primal elements of soul-life, the simple forms of sensibility, pleasure, pain, the simple forms of motion, attraction, and repulsion, are in all living matter, in all protoplasm. But the grades of the up-building and composition of this soul vary in different living beings, and lead us gradually upwards from the quiescent cellsoul through a long series of ascending steps to the conscious and rational soul of man (p. 173)".

Contrary to the more mechanical picture of Darwin, Haeckel is explicit in his insertion of the metaphysical into the material. Life is a combination of matter and soul, body and spirit.

Boltzmann rejected the metaphysical dualism inherent in the position, asserting that all phenomena from astronomy to psychology could be accounted for in terms of the behavior of the material constituents of the world. The biggest challenge for this sort of metaphysical materialism, of course, is human consciousness. But Boltzmann [12] sees Darwin as having given us the tools for that.

> "The brain we view as the apparatus or organ for producing word pictures [Bilder], an organ which because of the pictures' great utility for the preservation of the species has, comfortably with Darwin's theory, developed in man to a degree of particular perfection, just as the neck of the giraffe and the bill of the stork have developed to an unusual length (p. 69)".

The human mind is no more mysterious than any other of a range of biological curiosities. Humans are just animals and our intelligence is just one more example of a notable adaptation. From [32], "From a Darwinian point of view we can grasp furthermore what is the relation of animal instinct to human intellect. The more perfect an animal, the more it shows incipient traces of intellect alongside instinct (p. 138)".

7.4. A Darwinian Explanation of the Development of Photosynthesis, Mind, and Life Itself

Life requires decreasing entropy, as such an essential element of life is the collection of free energy for the purpose of doing the work needed to maintain life. Using his understanding of radiative entropy, Boltzmann was able to make sense of the process according to Englebert Broda [21]. Boltzmann contends that plant photosynthesis arose and developed during a Darwinian struggle for improved supply of free energy:

> "The dependence of plant life on light had been discovered in London by Jan Ingen-Housz, a Dutchman mostly living in Vienna, in 1779. However, Ingen-Housz did not know why exactly light is needed. Indeed he could not know it as conservation of energy was unknown in his time, and so the need for a particular source of energy did not seem to exist. The second step had been taken by the discoverer of energy conservation (First Law of Thermodynamics), Julius Robert Mayer, in 1845. He wrote "The plants absorb a force, light, and produce a force: chemical difference". However, in his ignorance of the Second Law Mayer could not make a distinction between useful and useless forms of energy. This distinction, in respect to photosynthesis, was left to the physicist who understood the Second Law better than anybody else, and who explained it in atomistic terms, to Ludwig Boltzmann. In 1884 he had introduced the notion of the entropy of radiation (p. 62)".

This sort of combination of his approach to thermodynamics and Darwin's natural selection, Broda shows, is not limited to photosynthesis, but also gives an account of the arising of other bioenergetic processes such as fermentation and respiration. Indeed, Broda points to passages in Boltzmann's later lectures in which he makes this form of argument for the arising of consciousness and life itself.

7.5. Darwin as a Model for Thermodynamics

Again, while we do not have Boltzmann asserting the connection between Darwinian evolution and the time evolution of thermodynamic systems as a part of his thinking in the 1877 move to a statistical understanding of entropy, we do have instances of Boltzmann publicly connecting the two after the fact. Boltzmann employs Darwin's thought as a *Bild*, a heuristic model, for the development of statistical mechanics. There is significant textual evidence that the two are connected in Boltzmann's thought after the development of his statistical notion of entropy, which allows for an inference that Darwinian evolution may have served as a structural model for Boltzmann in developing his statistical understanding of thermodynamics.

In [12], Boltzmann connects mental phenomena, which he understands as material interactions that are the result of human evolutionary history, with the physical phenomena of electricity and, more importantly, for this point, heat. "Mental phenomena may well be much more remote from material ones than thermal or electric from purely mechanical ones, but to say that the two former are qualitatively while the latter three are only quantitatively different seems to me mere prejudice (72)". While Boltzmann used the term "mechanical" in different ways in various contexts, here, he is clearly meaning something along the lines of being governed by the laws of mechanics (as opposed to Maxwell's equations); as such, he is not creating a mechanical picture of mind in the sense that the mind is a result of Newton's laws of motion, but he is creating a mechanical theory of mind which does not require any sort of non-material soul.

It is mere prejudice, a fallacy to be eliminated, to distinguish between mental and thermal systems. They are not identical, but of the same sort, Boltzmann contends. As such, the type of arguments that support evolutionary outcomes would not be of a different epistemic species from the sort from those we give in making arguments about thermodynamics.

Indeed, we see in his "Reply to a Lecture on Happiness Given by Professor Ostwald" [33], an evolutionary discussion that maps very closely to a discussion of molecules in a gas. "As regards the concept of happiness, I derive it from Darwin's theory. Whether the in the course of aeons the first protoplasm developed 'by chance' in the damp mud of the vast waters on the Earth, whether egg cells, spores, or some other germs in the form of dust or embedded in meteorites once reached Earth from outer space, is here a matter of indifference. More highly developed organisms will hardly have fallen from the skies. To begin with there were thus only very simple individuals, simple cells or particles of protoplasm. Constant motion, so-called Brownian molecular motion, happens with all small particles as is well-known; growth by absorption of similar constituents and subsequent multiplication by division is likely explicable by purely mechanical means. It is equally understandable that these rapid motions were influenced and modified by their surroundings. Particles in which the change occurred in such a way that on average (by preference) they moved to regions where there were better materials to absorb (food), were better able to grow and propagate so as soon to overrun all the others (176)".

The approach to evolution reduces the generally macro-level interaction of members of species with their environment, giving rise to selection pressures which affect the relative success of mutations, and places it in a microscopic framework in which atomic-level interactions, like Brownian motion, are in play and driven "by chance". Framing the process stochastically while speaking of the average speed of small particles in rapid motion is directly parallel to the sort of calculations Boltzmann was undertaking in understanding the time evolution of gases with notions like the mean free path. The convergence of vocabularies and concerns between the two is not only striking, it is clearly intentional. Boltzmann is explicitly connecting evolution with thermodynamics.

It should be noted that the Professor Ostwald to whom Boltzmann is responding is Wilhelm Ostwald, a Nobel laureate who was one of the last major opponents of atomism, preferring a system based on energy as a foundational concept. From his [34], "What we hear originates in work done on the ear drum and the middle ear by vibrations of the air. What we see is only radiant energy which does chemical work on the retina that is perceived as light. When we touch a solid body we experience mechanical work performed during the compression of our fingertips...From this standpoint, the totality of nature appears as a series of spatially and temporally changing energies, of which we obtain knowledge in proportion as they impinge upon the body and especially upon the sense organs which fashioned for the reception of the appropriate energies (159)".

Ostwald was seeking to do away with particles, making energy the primary metaphysical constituent of the universe. In his evolutionary example, he is taking organisms—a clear instance of things—and intuitively elevating them so that the evolutionary example rhetorically supports his atomism.

It should be noted that Boltzmann uses in his example here, in the influence of Brownian motion. Ironically, it was Jean Perrin's experimental work on Brownian motion that led Ostwald to finally relent and reject his energetics theory for atomism. Perrin's experiments verified the theoretical work of Albert Einstein [35]. Boltzmann could not have known at the time that Brownian motion would have this effect on Ostwald or that Einstein would explain it in terms of atomic interaction several years later. Indeed, as John

Blackmore [36] persuasively argues, it is unlikely that Boltzmann was aware of Einstein's work on his matter.

We see a similar move on the part of Boltzmann in [32]: "We must mention also that most splendid mechanical theory in the field of biology, namely the doctrine of Darwin. This undertakes to explain the whole multiplicity of plants and the animal kingdom from the purely mechanical principle of heredity, which like all mechanical principles of genesis remains of course obscure (p. 132)".

Note the phrasing "that most splendid mechanical theory in the field of biology". Boltzmann is creating a class of mechanical theories, of which the mechanical theory of heat is the most obvious example, but then including with a celebratory nod a member of the group in biology.

But the most important element comes a couple of pages later when Boltzmann highlights an unexpected element of the evolutionary process: the expected unexpected, that is, the instances in which random mutations give rise to unfit organisms. "It is well-known that Darwin's theory explains by no means merely the appropriate character of human and animal bodily organs, but also gives an account of why often inappropriate and rudimentary organs or even errors of organization could and must occur (136)".

Darwinian evolution is driven by random mutations (although they did not understand the genetic force driving it at the time). That randomness will not always give rise to more fit organisms, but will be expected to create dead ends at an expected statistical rate. Yes, we generally focus on the developments that drive the process forward, but given that it is a random process there will also be expected cases of the undesirable and indeed the unexpected.

Evolution can run backward. The eye can develop, but then if a subpopulation takes to inhabiting caves, selection pressures could undo what eons of selection processes had constructed. It is the general case, but not absolutely necessary, that members of species become more complex to fit their environment.

This is an instance of Darwinian evolution enacting its own version of Zermelo's objection. It is perfectly possible that a strange set of circumstances could take a species and do and then undo all of the changes. Evolution is a time-oriented theory, yet there is a miniscule possibility that it will return that on which it works to its original stage. It is highly unlikely, but theoretically conceivable. That does not mean that evolution has not been working. It is just an unusual possibility given the statistical nature of the process.

But, if we accept this likelihood in the biological sense, Boltzmann seems to be saying, why would we have any problem with the analogue in the thermodynamic case? Clearly, we should not, and if Boltzmann was using Darwinian evolution, the mechanical theory of inheritance, as a *Bild* underlying his construction of statistical mechanics, then we have a perfectly reasonable explanation for why none of the concerns and objections raised would have seemed to be concerning to Boltzmann.

8. Conclusions

In 1877, Ludwig Boltzmann made a stunning shift. His new version of the Second Law of Thermodynamics was a statistical regularity, lacking the standard sort of nomological necessity that had been traditionally asserted as an essential property for a law of nature. It was a bold move that was novel in physics, but not in science writ large. Huttonian geology and especially Darwinian evolution, two theories that were tremendously prevalent in scientific discourse at the time, possessed structural similarities. In the decades after his shift in the understanding of entropy, Boltzmann refers to both of these theories.

This is especially true with respect to Darwin's work, of which Boltzmann was a vocal supporter. Indeed, we see Boltzmann model elements of his approach to the scientific method on Darwin. Evolution plays a central role in his understanding of the acquisition of human knowledge, of the emergence of life and consciousness, and Boltzmann considers ways in which his work on entropy must be a part of biological evolution itself.

All of this was after 1877, but given Boltzmann's time in Berlin with Hermann von Helmholtz, whose work bridged the physical and biological and who was thinking and speaking about Darwin, there is reason to believe that Darwin was on Boltzmann's mind just before his bold new understanding of entropy emerged.

After his view solidified, Boltzmann used Darwinian analogies to explain his work to the public. Boltzmann saw the connection between his approach and Darwin's to be similar enough to be able to use the latter as a *Bild*, a conceptual picture capable of making the latter more clear to the mind.

As a result of all of this circumstantial evidence, there seems to be warrant for considering the possibility that Darwin helped Boltzmann make the switch that appears in 1877. Again, there is no smoking gun here, but we do have a variety of different sorts of evidence garnered around a hypothesis. The inference based upon that is similar to Darwin's own approach to argumentation in *The Origin of Species*, an inductive method of inference he learned from his own teacher William Whewell [37], who called such an approach to reasoning "consilience".

What strikes the contemporary mind as odd about this claim is that a biological system could be used as a model for a physical system. Going back to August Comte and through the Logical Positivists (who develop in part from the ideas of Philipp Frank, Boltzmann's own student), the reductionist picture of science has psychology reducing to biology, which is just complicated chemistry, which in turn is nothing but physics. As such, when we look at the origin of ideas in the most basic of the sciences, physics, then surely the only influences are to be found in the discourse among physicists. But recall what Boltzmann himself thought his own era was the age of—the age of Darwin. While it may seem to those who have internalized the reductionist scheme to put the intellectual cart before the scientific horse, Boltzmann's own epistemology freed the scientist to find an appropriate *Bild* wherever one could.

If this inference is correct, then, Darwin and Boltzmann become the scientific Porgy and Bess with the reworked Gershwin's lyric: "The phase space sectors where you'd find a system's vectors. . .it ain't necessarily so".

Funding: This research received no external funding.

Institutional Review Board Statement: Not applicable.

Data Availability Statement: No new data were created or analyzed in this study. Data sharing is not applicable to this article.

Conflicts of Interest: The author declares no conflicts of interest.

References

1. Skyrms, B. Nomological Necessity and the Paradoxes of Confirmation. *Philos. Sci.* **1966**, *33*, 230–249. [CrossRef]
2. Boltzmann, L. On the Relationship between the Second Fundamental Theorem of the Mechanical Theory of Heat and Probability Calculations regarding the Conditions for Thermal Equilibrium. (trans.) Kim Sharp and Franz Matchinsky. *Entropy* **2015**, *17*, 1971–2009.
3. Hunt, B. *The Maxwellians*; Cornall University Press: Ithaca, NY, USA, 1991.
4. de Regt, H. Scientific Realism in Action: Molecular Models and Boltzmann's Bildtheorie. *Erkenntnis* **2005**, *63*, 205–230. [CrossRef]
5. Tolstoy, I. *James Clerk Maxwell: A Biography*; Cannongate: Edinburgh, Scotland, 1981.
6. Achinstein, P. *Particles and Waves: Historical Essays in the Philosophy of Science*; Oxford University Press: New York, NY, USA; Oxford, UK, 1991.
7. Lambert, R.; Steven, G. *Bild*-ing Science: The Multiplicity of *Bild*-Types in Boltzmann. *Found. Sci.* **2023**. [CrossRef]
8. Boltzmann, L. "On the Development of the Methods of Theoretical Physics in Recent Times," (1899). In *Theoretical Physics and Philosophical Problems*; McGuinness, B., Ed.; D. Reidel: Dordrecht, The Netherlands, 1974.
9. Boltzmann, L. "On the Fundamental Principles and Equations of Mechanics," (1899). In *Theoretical Physics and Philosophical Problems*; McGuinness, B., Ed.; D. Reidel: Dordrecht, The Netherlands, 1974.
10. Rosa, L.P.; Andrade, E.; Picciani, P.; Faber, J. Constructivism and Realism in Boltzmann's Thermodynamic Atomism. *Found. Phys.* **2020**, *50*, 1270–1293. [CrossRef]
11. Maxwell, J.C. *Illustration of the Dynamic Theory of Gases. The Scientific Papers of James Clerk Maxwell, Volume I*; Niven, W.D., Ed.; Dover: New York, NY, USA, 1965.

12. Boltzmann, L. On the Indispensability of Atomism in Natural Science. In *Theoretical Physics and Philosophical Problems*; McGuinness, B., Ed.; D. Reidel: Dordrecht, The Netherlands, 1974.
13. Boltzmann, L. "Further Studies on the Thermal Equilibrium of Gas Molecules" (1872). In *The Kinetic Theory of Gases: An Anthology of Classic Papers with Historical Commentary*; Hall, N., Ed.; World Scientific: London, UK, 2003.
14. de Cortenay, N. The Epistemological Virtues of Assumptions: Toward a Coming of Age of Boltzmann and Meinong's Objections to 'the Predjudice of the Actual'? *Stud. Hist. Philos. Sci. Part A* **2010**, *41*, 41–57. [CrossRef]
15. Flamm, D. *Four Papers by Loschmidt on the State of Thermal Equilibrium*; Springer: New York, NY, USA, 1997.
16. Hopkins, W. Physical Theories of the Phenomena of Life. In *Darwin and His Critics: The Reception of Darwin's Thery of Evolution by the Scientific Community*; Hull, D., Ed.; University of Chicago Press: Chicago, IL, USA, 1973.
17. Bondi, H. *The Arrow of Time: Statistics and the Universe. Pioneering Ideas for the Physical and Chemical Sciences: Josef Loschmidt's Contributions and Modern Developments in Structural Organic Chemistry, Atomistics, and Statistical Mechanics*; Fleischhacker, W., Schönfeld, T., Eds.; Plenum: New York, NY, USA, 1997.
18. Gliboff, S. *HG Bronn, Ernst Haeckel, and the Origins of German Darwinism: A Study in Translation and Transformation*; MIT Press: Cambridge, UK, 2008.
19. Fullinwider, S.P. Darwin Faces Kant: A Study in Nineteenth Century Physiology. *Br. J. Hist. Sci.* **1991**, *24*, 21–44. [CrossRef]
20. Helmholtz, H. *Notizbuch: Biologischer Theil*. Helmholtz Archive. Translated by Author. Available online: https://www.helmholtz.de/fileadmin/user_upload/Handschrift/Helmholtz_Handschrift_Hermann_web.pdf (accessed on 27 February 2024).
21. Geissler, E.; Scheler, W. *Darwin Today: The 8th Kühlungsborn Colloquium on Philosophical and Ethical Problems of Biosciences. Kühlungsborn 8–12 November 1981*; Antonov, A., Frolov, I.T., Eds.; DeGruyter: Berlin, Germany, 1982.
22. Mach, E. On Transformation and Adaptation in Scientific Thought. In *Popular Scientific Lectures*; 1885. Translated by Thomas McCormack; Open Court: Chicago, IL, USA, 1898.
23. Boltzmann, L. "The Second Law of Thermodynamics." (1886). In *Theoretical Physics and Philosophical Problems*; McGuinness, B., Ed.; D. Reidel: Dordrecht, The Netherlands, 1974.
24. Mach, E. *The Analysis of Sensations and the Relation of the Physical to the Psychical*; Open Court: La Salle, IL, USA, 1984.
25. Kant, I. *Critique of Pure Reason*; St. Martin's: New York, NY, USA, 1965.
26. Hertz, H. *The Principles of Mechanics: Presented in a New Form*; Dover: New York, NY, USA, 1956.
27. D'Agostino, S. Boltzmann and Hertz on the Bild-Conception of Physical Theory. *Hist. Sci.* **1990**, *28*, 380–398. [CrossRef]
28. Boltzmann, L. "On a Thesis of Schopenhauer's." (1905). In *Theoretical Physics and Philosophical Problems*; McGuinness, B., Ed.; D. Reidel: Dordrecht, The Netherlands, 1974.
29. Chomsky, N. *Aspects of the Theory of Syntax*; MIT Press: Cambridge, UK, 1965.
30. Boltzmann, L. "An Inaugural Lecture on Natural Philosophy." (1903). In *Theoretical Physics and Philosophical Problems*; McGuinness, B., Ed.; D. Reidel: Dordrecht, The Netherlands, 1974.
31. Haeckel, E. Cell Souls and Soul Cells. In *The Pedigree of Man and Other Essays*; Translated by Edward Aveling; A & H.B. Bonner: London, UK, 1903.
32. Boltzmann, L. "On the Principles of Mechanics." (1897). In *Theoretical Physics and Philosophical Problems*; McGuinness, B., Ed.; D. Reidel: Dordrecht, The Netherlands, 1974.
33. Boltzmann, L. "Reply to a Lecture on Happiness Given by Professor Ostwald." (1904). In *Theoretical Physics and Philosophical Problems*; McGuinness, B., Ed.; D. Reidel: Dordrecht, The Netherlands, 1974.
34. Ostwald, W. *Vorlessungen über Naturphilosophie*; Verlag Von Veit: Leipzig, Germany, 1902.
35. Einstein, A. "On the Motion of Small Particles Suspended in a Stationary Liquid, as Required by the Molecular Kinetic Theory of Heat," (1905). In *The Collected Papers of Albert Einstein, Volume 2*; Stachel, J., Ed.; Princeton University: Princeton, NJ, USA, 1987.
36. Blackmore, J. Boltzmann and Epistemology. *Synthese* **1999**, *119*, 157–189. [CrossRef]
37. Whewell, W. *Novum Oraganum Renovatum*; John W. Parker & Son: London, UK, 1858.

entropy

Article

A Numerical Study of Quantum Entropy and Information in the Wigner–Fokker–Planck Equation for Open Quantum Systems

Arash Edrisi [1], Hamza Patwa [1,2] and Jose A. Morales Escalante [1,2,*]

[1] Department of Physics & Astronomy, University of Texas at San Antonio, San Antonio, TX 78249, USA; arash.edrisi@my.utsa.edu (A.E.); hamza.patwa@my.utsa.edu (H.P.)
[2] Department of Mathematics, University of Texas at San Antonio, San Antonio, TX 78249, USA
* Correspondence: jose.morales4@utsa.edu

Abstract: Kinetic theory provides modeling of open quantum systems subject to Markovian noise via the Wigner–Fokker–Planck equation, which is an alternate of the Lindblad master equation setting, having the advantage of great physical intuition as it is the quantum equivalent of the classical phase space description. We perform a numerical inspection of the Wehrl entropy for the benchmark problem of a harmonic potential, since the existence of a steady state and its analytical formula have been proven theoretically in this case. When there is friction in the noise terms, no theoretical results on the monotonicity of absolute entropy are available. We provide numerical results of the time evolution of the entropy in the case with friction using a stochastic (Euler–Maruyama-based Monte Carlo) numerical solver. For all the chosen initial conditions studied (all of them Gaussian states), up to the inherent numerical error of the method, one cannot disregard the possibility of monotonic behavior even in the case under study, where the noise includes friction terms.

Keywords: kinetic theory; quantum information; Wigner–Fokker–Planck; Monte Carlo; Euler–Maruyama; open quantum systems; Wehrl entropy; quantum entropy; Husimi transform

1. Introduction

Open quantum systems is an area of great importance in both Computational and Applied Mathematics and Physics due to the relevance of its applications in topics such as quantum optics, semiconductors, and lately, quantum computing and information science. In particular, quantum computing and information sciences are experiencing a booming development given the recent advances (circa 2019 by Google and IBM) in the experimental implementation of quantum computers of the order of 100 qubits in the current NISQ (Noisy Intermediate Scale Quantum) era. This reflects the need to have a holistic understanding of quantum information and quantum entropy, particularly in the present NISQ era, since, to this day, quantum computing devices are error-prone due to the effects of environmental noise. Therefore, it is also of fundamental importance to conduct an interdisciplinary study of open quantum systems encompassing this physical phenomenon, their mathematical modeling, and a computational model of their mathematical abstraction.

There is a direct connection between quantum computing and information sciences and Wigner–Fokker–Planck models. The Wigner–Fokker–Planck equation is completely equivalent to the more common model for open quantum systems in quantum computing and information, namely the Lindblad master equation, in the case where the related variables of the system are continuous (specifically position and momentum, suitable then for a phase space formulation), as opposed to spin variables for example. Open quantum systems are fundamental to the study of quantum computing and information, because, in the NISQ era, environment noise is the fundamental challenge to the preservation of information in quantum computers. Most importantly, recent techniques [1] actually make use of environment noise to get the state indeed to the ground state, from which to then start the quantum computation. These techniques are called "ground-state preparation via Lindbladians" and have been very popular in the last year in the domain of current mathematical and computational challenges in quantum computing and information.

The purpose of this work is to perform a numerical study of the behavior of quantum entropy (as a measure of quantum information) for specific computational methods of open quantum systems models, namely, Monte Carlo stochastic solvers based on Euler–Maruyama techniques. Of the different models for open quantum systems, we particularly draw our attention to the Wigner–Fokker–Planck equation, which in the case of a harmonic potential $V(x) = x^2/2$ (taking units such that $\hbar = m = 1$) is as follows:

$$w_t + k \cdot \nabla_x w - x \cdot \nabla_k w = \nabla_{(x,k)} \cdot (D\nabla_{(x,k)} w) + \gamma \nabla_k \cdot (wk). \tag{1}$$

where $w(x,k,t)$ is the Wigner quasi-probability density function (quasi-PDF, since it might be negative in some phase space regions) defined over a position-momentum phase space $(x,k) \in \mathbb{R}^2$ at time t, where the momentum is formally given by $p = \hbar k$, D is the diffusion matrix, and γ is a friction coefficient. This model represents a quantum phase-space picture via the quasi-PDF w for a small subsystem in which the Hamiltonian transport is being perturbed by Markovian noise, introduced by a larger environment through energy exchanges with the subsystem. Noise is represented in this Equation then through the diffusion operator and the friction terms. For completeness, we mention that the Wigner–Fokker–Planck equation in the general case of an arbitrary potential $V(x)$ has the following form:

$$w_t + k \cdot \nabla_x w + \Theta[V]\{w\} = \Delta_k w + \nabla_k \cdot (kw) + \Delta_x w, \quad \text{where}$$
$$\Theta[\,V\,]\{w\} = -i/(2\pi)^d \int \delta V(x,\eta)\; w(x,k',t) \exp\{i\eta \cdot (k-k')\}\; dk' d\eta \;\text{ on } \mathbb{R}^{2d}, \tag{2}$$

where $\delta V(x,\eta) = V(x+\eta/2) - V(x-\eta/2)$. This more general case that includes non-harmonic potentials can be represented through the pseudo-differential integral operator $\Theta[\,V\,]\{w\}$ [2] above acting on the given potential $V(x)$. However, the harmonic case will be the focus of our numerical studies, since the Wigner–Fokker–Planck equation converges to an analytically known steady-state solution [3] for potentials of this form, which can be further interpreted as capturing the effects of decoherence due to the environment.

The current state of the research field, encompassing different kinds of numerical methods for open quantum systems, in general, can be stated in the following way. Several types of numerical methods have been used in the computational modeling of open quantum systems. Their mathematical modeling through Wigner–Fokker–Planck equations and their subsequent numerical solution via Monte Carlo stochastic simulations (in the context of Quantum Optics, for example) have been reported [4], as well as discrete velocity numerical methods for a stationary Wigner Equation [5]. It is known though that there is a natural stochastic error in the numerical solution by Monte Carlo methods, where this error will decrease as $N^{-1/2}$ by increasing the number of samples N. Therefore, the computational cost, inferred through the number of sample points, needs to grow quadratically to reduce the error linearly. There are previous works as well on the numerical simulation of Wigner models for quantum tunneling phenomena [6], as well as literature on operator splitting type methods for the Wigner–Poisson system [7], and also on the semi-discrete analysis of the Wigner Equation [8].

The phenomena of open quantum systems need a numerical description that reflects inherently the physics of quantum transport and noise (represented by diffusion for Markovian interactions). In Monte Carlo methods for open quantum systems models, such as the Wigner–Fokker–Planck equation [4], this is achieved with the combination of finite differences representing the transport plus the random sampling from normal distributions that model the diffusion processes (via the well-known connection between random walks and diffusion in the Feynman-Kac formulation). The combination of these two is called the Euler–Maruyama method (to first order), and it is a known numerical procedure for the solution of stochastic differential equations [9]. Extensions of the Monte Carlo method described are possible for Wigner functions that have a negative part, such as signed-particle methods [10,11]. This would be achieved by splitting the Wigner function into the sum of two parts, a positive part w_+ and a negative part w_-, both of which on their own can formally be considered probability densities. The distributions associated with w_+ and w_- would be related: when one loses a particle, the other would gain one, and vice versa. On the other hand, Discontinuous Galerkin (DG) methods can be designed to mimic numerically convection-diffusion problems, as is the case for Local DG or Nonsymmetric Interior Penalty Galerkin methods. For example, Ref. [2] focused on an adaptable DG scheme for Wigner–Fokker–Planck, where a Nonsymmetric Interior Penalty Galerkin method was used as a numerical method. There is literature on the use of Discontinuous Galerkin Methods in equations of the Quantum–Liouville type [12], as well as numerical modeling of a Quantum Liouville–Poisson system [13]. However, the aforementioned equations in the last sentence consider only quantum transport in the problem, because diffusion does not appear in Liouvillian transport. Previous works where the noise due to the environment in an open quantum system is modeled in a DG setting for master equations have been in reported in [14,15], for example.

In the particular context of mathematical studies of entropy in open quantum systems, Ref. [3] uses classical (mathematically speaking) entropy methods from kinetic theory, which consider relative entropies:

$$e_\varphi(f|g) = \int_{\mathbb{R}^n} \varphi(f/g)g(dx), \tag{3}$$

(where e_φ is the relative entropy function of f with respect to g, and φ the generating function) such as the logarithmic relative entropy, related to:

$$\varphi_1(\alpha) = \alpha \ln(\alpha) - \alpha + 1, \quad \alpha \in \mathbb{R}^+, \tag{4}$$

or the quadratic relative entropy, related to:

$$\varphi_2(\alpha) = K(\alpha - 1)^2, \quad \alpha \in \mathbb{R}, \quad K > 0, \tag{5}$$

to establish the conditions that imply the existence of a thermal equilibrium state, proving an exponential decay towards it. In [16], the free energy (or Kullback relative entropy of ρ with respect to $\exp(-V)$) is presented for the study of Fokker–Planck equations. In [17], entropy methods for diffusive PDEs are presented in general, and in particular as well for Fokker–Planck equations which might be possibly nonlinear.

Though the physical entropy in quantum Mechanics for a density matrix representation, the Von Neumann entropy:

$$S = -\mathrm{Tr}(\hat{\rho}\log(\hat{\rho})) \tag{6}$$

which is well known and constitutes the quantum extension of the classical Boltzmann/Shannon entropy:

$$H = -\int_{\Omega_x}\int_{\Omega_p} f\log(f)dxdp. \tag{7}$$

The definition of entropy in the Wigner formulation is not as direct as an extension since $\log(w)$ might not be able to be evaluated as a real-valued function for cases where $w < 0$ (which can happen; however, for Gaussian states, their Wigner function $w(x,p,t) > 0$ is nonnegative). However, one can use the Wehrl entropy [18], which is essentially the Boltzmann/Shannon H entropy abovementioned applied to the nonnegative Husimi transform of the Wigner function. Previous mathematical work [19] has also indicated that the only physically relevant quantum Fokker–Planck equation that makes quantum entropy grow monotonically (for all admissible initial conditions) is the frictionless one ($\gamma = 0$). However, mathematical studies of the quantum relative entropy in the Wigner–Fokker–Planck equation are indeed helpful and necessary to establish the convergence of its initial value problem to the steady-state solution in the harmonic problem precisely through entropy methods [3] as abovementioned. The result provided in [3] proves mathematically the convergence to the steady state.

Naturally, the abovementioned relative entropy results in [3] are crucial for showing convergence to a steady-state solution for the Wigner–Fokker–Planck initial value problem (IVP) under a harmonic potential, via entropy methods. There will be two distinctions between the nature of this result and our numerical studies though. The first one is that we will focus on the absolute entropy, not the relative entropy, where, as we have mentioned, it is known that monotonic behavior of the physical Von Neumann entropy is only guaranteed for the frictionless case $\gamma = 0$.

Second, our studies will not include the Von Neumann entropy but focus on the Wehrl entropy (absolute), where rather than the Wigner function it is the Husimi function the one in Wehrl's entropy argument. Therefore, though there are similarities in the functional form of the relative logarithmic entropy for the Wigner function w (or rather their positive and negative parts $w^{\pm}$) and the absolute Wehrl entropy for the Husimi function $h(w)$; in reality, they are analytically different. On the other hand, a generalized study of Fokker–Planck equations indicates that interpreting them as gradient flows in a Wasserstein metric (via a variational formulation of them) [20] is helpful to find exponential convergence of them by entropy production studies. Works related to this research direction have been performed in [16] and [21] for the Fokker–Planck equation, for the Lindblad equation in [22–24]. There is work as well on the gradient flows of the entropy for finite Markov Chains [25], as well as on the exponential decay of Rényi divergence under Fokker–Planck equations [26].

Regarding numerical studies of entropy in kinetic equations, there are previous numerical studies on kinetic equations for different applications, where the numerical entropy is monitored during the time evolution of the respective problem. Work on this regard has been done for the Vlasov–Boltzmann–Poisson system for electron transport in semiconductors [27], for the Vlasov–Poisson system [28], as well as for Fokker–Planck–Landau Type Equations for plasmas [29], and for the coupled Vlasov–Poisson Fokker–Planck–Landau equation [30], describing transport plasma models for Coulomb, Maxwell type, and hard-sphere particle interactions. In the particular application of open quantum systems, however, to the best of our knowledge, a numerical study of the absolute quantum entropy for the Wehrl entropy in the case of a Wigner–Fokker–Planck equation is lacking in the literature. Our contribution to the discipline is to fill this gap by studying precisely these numerical aspects in the particular case of a harmonic potential, which is the main aim of our work. We conclude from our numerical studies that, for most of the chosen initial conditions (all of them Gaussian states), we have observed monotonic behavior of the Wehrl entropy, whereas for one case, though the behavior might seem nonmonotonic without including the inherent error of the stochastic method used, one cannot disregard the possibility of monotonic behavior when considering the uncertainty error natural to our stochastic Monte Carlo method.

2. Materials and Methods

In this work, we utilize stochastic numerical methods of the Monte Carlo type (based on the Euler–Maruyama method) appropriate for the solution of the convective-diffusive Wigner–Fokker–Planck equation under a harmonic potential. Given the known initial and steady states, we compute the time evolution using the Monte-Carlo Euler–Maruyama Method, monitoring as well the L_2-type distance to the steady state and the Wehrl entropy applied to the Husimi transform as a sanity check. We solve analytically for the Husimi transform of a Gaussian state of the Wigner–Fokker–Planck, finding the expected initial and final values of the Wehrl entropy. We study two particular cases of Gaussian states: the harmonic groundstate (since we will choose it as the initial condition of our initial value problem) and the steady-state analytical solution to which we expect to converge numerically, as well as Gaussian states (Gaussian Wigner functions) whose covariance matrix is proportional to one of the two previously mentioned cases.

2.1. Monte-Carlo Solver: Euler–Maruyama Method

We will be using the ground state of the quantum harmonic oscillator in its Wigner representation as our initial condition in the numerical solution of the Wigner–Fokker–Planck Equation [31]:

$$w_0(x,p) = \frac{2}{h} e^{-\frac{a^2p^2}{\hbar^2} - \frac{x^2}{a^2}}. \tag{8}$$

To tackle the numerical solution of the Wigner–Fokker–Planck equation, we will employ the Euler–Maruyama method. This well-known stochastic numerical technique is particularly well-suited for simulating the dynamics of quantum systems in the presence of noise, namely open quantum systems. The Euler–Maruyama method is used to discretize the evolution of the Wigner function by following the trajectory of points in phase space, randomly sampling from the initial condition as a probability distribution (since it is a nonnegative function given that it is a Gaussian state, and therefore, its Wigner function is a Gaussian) to generate our sample of points constituting a point distribution. To choose successive states for our Wigner–Fokker–Planck kinetic equation in our simulations, we discretize both the transport (via a forward Euler time-step) and diffusion (via a random walk simulation of it by sampling from a Gaussian distribution with covariance related to the diffusion matrix representing the environment noise) processes over small time steps, incorporating in this way both deterministic transport and random diffusion components of the dynamics. Therefore, the phase-space point that represents the state of the system at the successive time $t + \Delta t$ for the i-th trajectory in particular is given by:

$$(x_i, p_i)(t + \Delta t) = (x_i, p_i)(t) + \Delta t(p_i, -x_i - \gamma p_i)(t) + \vec{E}_i, \tag{9}$$

where $\vec{E}_i \in \mathbb{R}^2$ is a random variable sampled from a Gaussian with covariance matrix $2D\Delta t$. More information about this is given in Section 2.2 and Equation (11). This approach allows us to mimic discretely the continuous evolution of the quantum state considering as well the random fluctuations induced by the environment. The Euler–Maruyama method is a numerical technique that falls under the category of Monte Carlo methods, specifically within the realm of stochastic differential Equations (SDEs). Monte Carlo methods involve the use of random sampling to obtain numerical results, and in the context of SDEs, these methods are employed to simulate the evolution of stochastic processes. The Euler–Maruyama method is particularly suited for solving stochastic differential equations of the form below:

$$dX_t = a(X_t)dt + b(X_t)dW_t, \tag{10}$$

where X_t is the state of the system at time t, $a(X_t)$ and $b(X_t)$ are deterministic functions, dt is the differential time step, and dW_t is the differential increment of a Wiener process/Brownian motion. The stochastic term $dW_t = W_{n+1} - W_n$ denotes one step in the random walk. However, we approximate this difference as $2D\Delta t\mathcal{N}(0,1)$, where $\mathcal{N}(0,1)$ represents the normal distribution with a mean of zero and a variance of one [9], and we have denoted by D the value of the diffusion matrix in the numerical simulations as well. The specifics are explained below.

2.2. Pseudo-Code and Methodology Description

The Wigner–Fokker–Planck equation can be formally considered to be related to an SDE of the form (10). As abovementioned, Monte Carlo simulation methods have been proposed for solving equations of this type, in the context of quantum optics [4]. Let the Wigner function at time $t = 0$ be $w(x,p,0)$. We then obtain an initial collection of N points $(q_i(0), p_i(0)) \equiv \vec{z}_i(0)$ in phase space by randomly sampling from $w(x,p,0)$. To each point $\vec{z}_i(t)$, a transport and diffusion part are applied to evolve the point to the next time step $t + \Delta t$, as per the Euler–Maruyama method:

$$\vec{z}_i(t + \Delta t) = \vec{z}_i(t) + \vec{\alpha}_i(t)\Delta t + \vec{E}_i, \tag{11}$$

where $\vec{\alpha}_i = (p_i, -q_i - \gamma p_i)$ is the transport vector for Equation (1), and $\vec{E}_i \in \mathbb{R}^2$ is a random variable with covariance matrix $2D\Delta t$. The term with $\vec{\alpha}_i$ represents the transport process (deterministic) and the term with $\vec{E}_i$ represents the diffusion (random) process. These processes are applied iteratively to all points at each time step. At the final time T, we obtain a final set of points $\vec{z}_i(T)$. This set of points is equivalent to a distribution obtained by sampling N points from the analytical solution of the WFP equation at time T.

It is important to note that this method applies in principle to nonnegative Wigner functions since Monte Carlo methods such as the Euler–Maruyama algorithm use probability densities, which are nonnegative by definition. Though one can use decompositions of the Wigner function into the difference $w^+ - w^-$ of two nonnegative Wigner functions w^+ and w^- as in the so-called signed-particle method [32–34], we only consider in this work specific initial conditions and potentials (namely, Gaussian states and harmonic potentials, respectively) for the method of choice that have the property $w(x,p,0) \geq 0 \implies w(x,p,t) \geq 0 \;\; \forall t > 0$. This is because it is known that for the case when $w(x,p,0)$ is a Gaussian state and the potential is harmonic, the time-evolution induced by the Wigner–Fokker–Planck dynamics (1) keeps the state Gaussian at all times, thus rendering a nonnegative Wigner function [35]. This justifies our use of the harmonic oscillator potential and its ground state (8) (which is a Gaussian state) as the initial condition.

The following pseudo-code (see Algorithm 1) depicts the implementation of our method (since in our units $\hbar = 1$, we have denoted the momentum as $p = \hbar k$).

Algorithm Specifics for the Monte Carlo Solver of Wigner–Fokker–Planck

1. Parameter Definitions
   - L: Length of the domain in position space.
   - σ_q and σ_p: Diagonal matrix elements of the covariance matrix of the Wigner function at the initial time.
   - δ_t: Time step for the simulation.
   - μ: Mean value vector for the initial distribution.
   - μ_1 and μ_2: Components of the mean value vector for the position and momentum, respectively.
   - D_{qq} and D_{pp}: Diffusion coefficients for position and momentum, respectively.
   - D: Diffusion matrix (assumed having zero off-diagonal terms).
   - γ: Friction coefficient.
2. Array Definitions
   - q: Array to store position values for each particle at each time step.
   - p: Array to store momentum values for each particle at each time step.
3. Initial Conditions
   - Gaussian Sampling: Initialize the position and momentum of each particle at the first time step using normal random number generation with mean components μ_1, μ_2 and standard deviations σ_q, σ_p.
4. Time Evolution Loop
   - Use nested loops to iterate over each time step i and each particle j.
   - Generate random noise ϵ using a multivariate normal distribution with mean μ and $2D\Delta t$ as covariance matrix for position and momentum.
   - Update the position $q[i+1,j]$ and momentum $p[i+1,j]$ of each particle using the Euler–Maruyama stochastic method.
5. Simulation Output
   - After the completion of the time evolution loop, the arrays q and p contain the simulated trajectories of position and momentum for each particle over time.

Algorithm 1 Euler–Maruyama for the Wigner–Fokker–Planck equation (harmonic potential)

Define:
$L \leftarrow 1.0$
$\sigma_q \leftarrow \frac{L}{\sqrt{2}}$
$\sigma_p \leftarrow \frac{1}{\sqrt{2}L}$
$\delta_t \leftarrow 0.01$
$Total\_Time \leftarrow 50.$
$NumOfTimeStep \leftarrow \text{round}\{\frac{Total\_Time}{\delta_t}\}$
$NumOfParticles \leftarrow 10^4$
$\mu_1 \leftarrow 0.$
$\mu_2 \leftarrow 0.$
$\mu \leftarrow \begin{bmatrix} \mu_1 & \mu_2 \end{bmatrix}$
$D_{qq} = 1., \quad D_{pp} = 1.$
$\gamma = 1.$
$D \leftarrow \begin{bmatrix} D_{qq} & 0 \\ 0 & D_{pp} \end{bmatrix}$
Arrays Initialization:
$q \leftarrow \text{zeros}[NumOfTimeStep,\ NumOfParticles]$
$p \leftarrow \text{zeros}[NumOfTimeStep,\ NumOfParticles]$
Initial Conditions:
$q[1,:] \leftarrow \text{normrnd}(\mu[1],\ \sigma_q,\ [1,\ NumOfParticles])$
$p[1,:] \leftarrow \text{normrnd}(\mu[2],\ \sigma_p,\ [1,\ NumOfParticles])$
Update:
for each $i \in NumOfTimeStep$ **do**
  for each $j \in NumOfParticles$ **do**
    $\epsilon \leftarrow \text{mvnrnd}(\mu,\ 2D\delta_t)$
    $q[i+1,\ j] \leftarrow q[i,j] + p[i,j]\delta_t + \epsilon[1]$
    $p[i+1,\ j] \leftarrow p[i,j] + (-q[i,j] - \gamma p[i,j])\delta_t + \epsilon[2]$
  end for
end for

The algorithm simulates the stochastic evolution of a system governed by a Langevin equation, incorporating random noise to account for the effects of an external environment. This type of simulation is commonly used in the study of open quantum systems, where the Euler–Maruyama method provides a computationally efficient approach to capture the stochastic dynamics of the system. The generated trajectories allow researchers to analyze the behavior of the system and study phenomena, such as decoherence and dissipation in quantum systems. Langevin dynamics is a mathematical and computational framework used to model the motion of particles in a physical system subject to random forces and friction. It is commonly applied in various scientific fields, including physics, chemistry, and biology, to describe the stochastic behavior of particles in a medium. The Langevin equation, named after the French physicist Paul Langevin, is a stochastic differential equation that incorporates both deterministic and random components to simulate the dynamics of a particle. The equation is often used in the context of Brownian motion, where particles experience random forces due to collisions with surrounding molecules or other particles.

We have written a code, in MATLAB and also its equivalent version in Python, both available in the GitHub repository https://github.com/phjame/StochasticWFP (accessed on 29 February 2024), for the computational implementation of our Euler–Maruyama-based Monte Carlo solver. For both languages, we have used their respective MATLAB and Python statistical toolboxes to fit a 2D Gaussian distribution onto each set of phase-space points for every time step. Subsequently, we utilize the resulting covariance matrix at each time step to compute the respective Wehrl quantum entropy, as well as to monitor the

L_2-norm between our numerical Wigner function at each time step and the steady-state solution μ:

$$\mu(x,k) = \frac{1}{2\pi\sqrt{5}}\, e^{-(\frac{1}{5}|x|^2+\frac{1}{5}x\cdot k+\frac{3}{10}|k|^2)} \tag{12}$$

The inequality condition for L_2-norm [2] is as follows:

$$\left\| \frac{w-\mu}{\sqrt{\mu}} \right\|_{L^2(\mathbb{R}^{2d})} \leq e^{-\sigma t} \left\| \frac{w_I-\mu}{\sqrt{\mu}} \right\|_{L^2(\mathbb{R}^{2d})} \tag{13}$$

with σ such that the Hessian of the quadratic form inside the argument of the Gaussian representing the steady-state solution satisfies:

$$Hess\left(\frac{1}{5}|x|^2 + \frac{1}{5}x\cdot k + \frac{3}{10}|k|^2\right) \geq \sigma\mathbb{I}. \tag{14}$$

More about these norm considerations can be found in Appendix A.

3. Results

3.1. Wigner–Fokker–Planck Model: Gaussian States under Harmonic Potential

3.1.1. Husimi Transform of a Gaussian State (Gaussian Wigner Function)

We want to find the Husimi transform of a Gaussian state, given by the Wigner function:

$$w(x,k) = \frac{1}{\sqrt{|\Sigma|(2\pi)^d}} \exp\left(-\frac{1}{2}(x,k)\Sigma^{-1}\begin{pmatrix} x \\ k \end{pmatrix}\right) = \frac{1}{2\pi\sqrt{|\Sigma|}} \exp\left(-\frac{1}{2}(x,k)\Sigma^{-1}\begin{pmatrix} x \\ k \end{pmatrix}\right)$$

or in a position-momentum representation, since $p = \hbar k$:

$$w(x,p) = \frac{1}{2\pi\sqrt{|\Sigma|}} \exp\left(-\frac{1}{2}(x,p/\hbar)\Sigma^{-1}\begin{pmatrix} x \\ p/\hbar \end{pmatrix}\right)$$

To calculate the Wehrl entropy:

$$H = -\int h \log(h) dx dp,$$

we need to calculate the Husimi function, obtained by applying the Husimi transform to our Wigner function above. Namely, we have:

$$h(x,p) = \int\int w(x',p')(\pi\hbar)^{-1} \exp\left(-(x'-x)^2/2s^2\right) \exp\left(-(p'-p)^2(2s^2/\hbar^2)\right) dx'dp'.$$

Thus, concretely, we have that the Husimi function for a Gaussian state is:

$$h = \frac{(\pi\hbar)^{-1}}{2\pi\sqrt{|\Sigma|}} \int\int \exp\left(-\frac{(x',\frac{p'}{\hbar})}{2}\Sigma^{-1}\begin{pmatrix} x' \\ \frac{p'}{\hbar} \end{pmatrix}\right) \exp\left(-\frac{(x'-x)^2}{2s^2}\right) \exp\left(-\frac{(p'-p)^2 2s^2}{\hbar^2}\right) dx'dp',$$

or if we stick with wave-numbers k instead of momentum, we have:

$$h(x,k) = \frac{(\pi\hbar)^{-1}\hbar}{2\pi\sqrt{|\Sigma|}} \int\int \exp\left(-\frac{(x',k')}{2}\Sigma^{-1}\begin{pmatrix} x' \\ k' \end{pmatrix}\right) \exp\left(-\frac{(x'-x)^2}{2s^2}\right) \exp\left(-2s^2(k'-k)^2\right) dx'dk',$$

equivalent to:

$$h(x,k) = \frac{\int\int \exp\left(-\frac{(x',k')}{2}\Sigma^{-1}\begin{pmatrix} x' \\ k' \end{pmatrix}\right)\exp\left(-\frac{(x'-x,k'-k)}{2}\begin{pmatrix} \frac{1}{s^2} & 0 \\ 0 & 4s^2 \end{pmatrix}\begin{pmatrix} x'-x \\ k'-k \end{pmatrix}\right)dx'dk'}{2\pi^2\sqrt{|\Sigma|}}.$$

If we use the notation:

$$\Sigma^{-1} = \begin{pmatrix} \Sigma^{-1}_{11} & \Sigma^{-1}_{12} \\ \Sigma^{-1}_{21} & \Sigma^{-1}_{22} \end{pmatrix} = \begin{pmatrix} a & b \\ c & d \end{pmatrix}$$

then one can prove that:

$$h = \frac{1}{\pi\sqrt{|\Sigma|}}\frac{\exp\left(\frac{\frac{c^2k^2}{2}s^4 - 2dk^2s^2(1+as^2) - 2cks^2x + \frac{c^2x^2}{8} - adx^2/2 - 2as^2x^2 + b^2(\frac{k^2s^4}{2} + \frac{x^2}{8}) + b(ck^2s^4 - 2ks^2x + \frac{cx^2}{4})}{d(1+as^2)+s^2(4-\frac{(b+c)^2}{4}+4as^2)}\right)}{\sqrt{4 - \frac{(b+c)^2}{4} + 4as^2 + ad + d/s^2}}.$$

Furthermore, considering that the covariance matrix is positive semi-definite and, therefore, symmetric, then we have $\Sigma^{-1}_{12} = b = c = \Sigma^{-1}_{21}$, so:

$$h(x,k) = \frac{1}{\pi}\frac{\sqrt{|\Sigma^{-1}|}\exp\left(\frac{-2s^2(1+as^2)dk^2 - 4bs^2kx - \frac{1}{2}a(d+4s^2)x^2 + \frac{1}{2}b^2(4k^2s^4+x^2)}{d(1+as^2)+s^2(4-b^2+4as^2)}\right)}{\sqrt{4-b^2+a(4s^2+d)+\frac{d}{s^2}}},$$

which is equivalent to:

$$h(x,k) = \frac{1}{\pi}\frac{\sqrt{|\Sigma^{-1}|}}{\sqrt{|\Sigma^{-1}|+4(1+as^2)+\frac{d}{s^2}}}\exp\left(-\frac{1}{2}\frac{(\frac{|\Sigma^{-1}|}{s^2}-4a)x^2 + 8bkx + 4(d+|\Sigma^{-1}|s^2)k^2}{[|\Sigma^{-1}|+4(1+as^2)+\frac{d}{s^2}]}\right)$$

or in a component-wise form:

$$h(x,k) = \frac{1}{\pi}\frac{\sqrt{|ad-b^2|}\exp\left(\frac{[b^2-a(d+4s^2)]x^2 - 8bs^2kx + 4s^2[b^2s^2-(1+as^2)d]k^2}{2[d(1+as^2)+s^2(4-b^2+4as^2)]}\right)}{\sqrt{4-b^2+a(4s^2+d)+\frac{d}{s^2}}}$$

which we can also write as:

$$h(x,k) = \frac{1}{\pi}\frac{\sqrt{|ad-b^2|}}{\sqrt{4-b^2+a(4s^2+d)+\frac{d}{s^2}}}\exp\left(-\frac{1}{2}(x,k)S^{-1}\begin{pmatrix} x \\ k \end{pmatrix}\right)$$

with the new inverse covariance matrix:

$$S^{-1} = \frac{1}{d(1+as^2)+s^2(4[1+as^2]-b^2)}\begin{pmatrix} a(d+4s^2)-b^2 & 4bs^2 \\ 4bs^2 & 4s^2[(1+as^2)d-b^2s^2] \end{pmatrix}.$$

The covariance matrix S is the one we will use in our calculations of the Wehrl entropy for the Husimi function of a Gaussian state.

3.1.2. Husimi Transform of Steady-State Solution for Harmonic Benchmark Problem

We only need to find specifically the covariance matrix for this case. Since we have in general that:

$$S^{-1} = \frac{1}{d(1+as^2)+s^2(4[1+as^2]-b^2)}\begin{pmatrix} a(d+4s^2)-b^2 & 4bs^2 \\ 4bs^2 & 4s^2[(1+as^2)d-b^2s^2] \end{pmatrix}$$

where:

$$\Sigma^{-1} = \begin{pmatrix} \Sigma_{11}^{-1} & \Sigma_{12}^{-1} \\ \Sigma_{21}^{-1} & \Sigma_{22}^{-1} \end{pmatrix} = \begin{pmatrix} a & b \\ c & d \end{pmatrix},$$

so for the inverse covariance matrix of the steady-state solution, we have:

$$\Sigma_\infty^{-1} = \begin{pmatrix} 2/5 & 1/5 \\ 1/5 & 3/5 \end{pmatrix} = \begin{pmatrix} 0.4 & 0.2 \\ 0.2 & 0.6 \end{pmatrix},$$

since:

$$\Sigma_\infty = \begin{pmatrix} 3 & -1 \\ -1 & 2 \end{pmatrix}.$$

Choosing the parameter value $s = 1/\sqrt{2} \iff 2s^2 = 1$, then we have:

$$S_\infty^{-1} = \frac{1}{d + a + ad/2 - b^2/2 + 2} \begin{pmatrix} a(d+2) - b^2 & 2b \\ 2b & [(2+a)d - b^2] \end{pmatrix}$$

$$S_\infty^{-1} = \frac{1}{\frac{3}{5} + \frac{2}{5} + \frac{2}{5}\frac{3}{5}/2 - \frac{1}{5}^2/2 + 2} \begin{pmatrix} \frac{2}{5}(\frac{3}{5}+2) - \frac{1}{5}^2 & 2\frac{1}{5} \\ 2\frac{1}{5} & [(2+\frac{2}{5})\frac{3}{5} - \frac{1}{5}^2] \end{pmatrix}$$

$$S_\infty^{-1} = \frac{1}{1 + \frac{1}{10} + 2} \begin{pmatrix} \frac{2}{5}(\frac{3+10}{5}) - \frac{1}{5}^2 & 2\frac{1}{5} \\ 2\frac{1}{5} & [(\frac{10+2}{5})\frac{3}{5} - \frac{1}{5}^2] \end{pmatrix}$$

$$S_\infty^{-1} = \frac{1}{3 + \frac{1}{10}} \begin{pmatrix} \frac{2}{5}(\frac{13}{5}) - \frac{1}{5}^2 & \frac{2}{5} \\ \frac{2}{5} & [(\frac{12}{5})\frac{3}{5} - \frac{1}{5}^2] \end{pmatrix}$$

$$S_\infty^{-1} = \frac{1}{15 + \frac{1}{2}} \begin{pmatrix} \frac{2}{5}(13) - \frac{1}{5} & 2 \\ 2 & [(12)\frac{3}{5} - \frac{1}{5}] \end{pmatrix}$$

$$S_\infty^{-1} = \frac{2}{31} \begin{pmatrix} \frac{25}{5} & 2 \\ 2 & \frac{35}{5} \end{pmatrix} = \frac{2}{31} \begin{pmatrix} 5 & 2 \\ 2 & 7 \end{pmatrix} = \frac{1}{31} \begin{pmatrix} 10 & 4 \\ 4 & 14 \end{pmatrix},$$

whose determinant is:

$$S_\infty^{-1} = \frac{140 - 16}{31^2} = \frac{124}{31^2} = \frac{4}{31}.$$

Therefore, we have that:

$$S_\infty = \frac{31}{4} \times \frac{1}{31} \begin{pmatrix} 14 & -4 \\ -4 & 10 \end{pmatrix} = \begin{pmatrix} 7/2 & -1 \\ -1 & 5/2 \end{pmatrix},$$

for which:

$$\frac{\ln(|S_\infty|)}{2} = \frac{\ln(31/4)}{2} \approx 1.0238.$$

3.1.3. Husimi Transform of Harmonic Groundstate

We will get the covariance matrix of the Gaussian associated with the Husimi transform of the harmonic ground state. Again, we have that:

$$S^{-1} = \frac{1}{d(1 + as^2) + s^2(4[1 + as^2] - b^2)} \begin{pmatrix} a(d + 4s^2) - b^2 & 4bs^2 \\ 4bs^2 & 4s^2[(1 + as^2)d - b^2s^2] \end{pmatrix}$$

where:

$$\Sigma^{-1} = \begin{pmatrix} \Sigma^{-1}_{11} & \Sigma^{-1}_{12} \\ \Sigma^{-1}_{21} & \Sigma^{-1}_{22} \end{pmatrix} = \begin{pmatrix} a & b \\ c & d \end{pmatrix}$$

so for the inverse covariance matrix of the harmonic ground state, we have:

$$\Sigma^{-1}_0 = \begin{pmatrix} 2 & 0 \\ 0 & 2 \end{pmatrix}$$

since choosing a position domain of length $L = 1$, we have that:

$$\Sigma_0 = \begin{pmatrix} 1/2 & 0 \\ 0 & 1/2 \end{pmatrix}$$

Choosing the parameter value $s = 1/\sqrt{2} \iff 2s^2 = 1$, then we have:

$$S^{-1}_0 = \frac{1}{d+a+ad/2-b^2/2+2} \begin{pmatrix} a(d+2)-b^2 & 2b \\ 2b & [(2+a)d-b^2] \end{pmatrix}$$

$$S^{-1}_0 = \frac{1}{2+2+4/2-0+2} \begin{pmatrix} 2(2+2)-0 & 0 \\ 0 & [(2+2)2-0] \end{pmatrix}$$

$$S^{-1}_0 = \frac{1}{8} \begin{pmatrix} 8 & 0 \\ 0 & 8 \end{pmatrix} = \begin{pmatrix} 1 & 0 \\ 0 & 1 \end{pmatrix} = I_d,$$

therefore, $S_0 = I_d$ and $\det(S_0) = 1$, for which:

$$\frac{\ln(|S_0|)}{2} = \frac{\ln(1)}{2} = 0.$$

3.1.4. Wehrl Entropy of a Gaussian State through Its Husimi Transform

Gaussian states in the Wigner formulation are represented simply by Gaussians. They are the only states for which their Wigner functions $w(x,k) \geq 0$ are nonnegative over all their domain. Moreover, the harmonic potential will transform Gaussian states into other Gaussian states until the steady-state solution is achieved [35].

We can evaluate their Wehrl entropy by the Husimi function $h(x,k)$. The Husimi function, obtained by applying a Husimi transform to our Wigner function, will be another Gaussian since w is Gaussian, though in general different from $w(x,k)$ in its analytical form. However, h will still be represented by a formula such as:

$$h(x,k,t) = \frac{1}{\sqrt{|\Sigma|(2\pi)^d}} \exp\left(-\frac{1}{2}(x,k)\Sigma^{-1}\begin{pmatrix} x \\ k \end{pmatrix}\right) = \frac{1}{2\pi\sqrt{|\Sigma|}} \exp\left(-\frac{1}{2}(x,k)\Sigma^{-1}\begin{pmatrix} x \\ k \end{pmatrix}\right),$$

where $d = 2$.

The Wehrl's entropy is obtained by calculating the following:

$$H = -\int h \log(h) dx dk,$$

$$H = -\int \frac{1}{2\pi\sqrt{|\Sigma|}} \exp\left(-\frac{1}{2}(x,k)\Sigma^{-1}\begin{pmatrix} x \\ k \end{pmatrix}\right) \log\left(\frac{1}{2\pi\sqrt{|\Sigma|}} \exp\left(-\frac{1}{2}(x,k)\Sigma^{-1}\begin{pmatrix} x \\ k \end{pmatrix}\right)\right) dx dk =$$

$$H = -\int \frac{1}{2\pi\sqrt{|\Sigma|}} \exp\left(-\frac{1}{2}(x,k)\Sigma^{-1}\begin{pmatrix} x \\ k \end{pmatrix}\right) [\log\left(\frac{1}{2\pi\sqrt{|\Sigma|}}\right) - \frac{1}{2}(x,k)\Sigma^{-1}\begin{pmatrix} x \\ k \end{pmatrix}] dx dk =$$

$$-\log\left(\frac{1}{2\pi\sqrt{|\Sigma|}}\right) [\frac{1}{2\pi\sqrt{|\Sigma|}} \int \exp\left(-\frac{1}{2}(x,k)\Sigma^{-1}\begin{pmatrix} x \\ k \end{pmatrix}\right) dx dk] +$$

$$\frac{1}{2\pi\sqrt{|\Sigma|}}\int \exp\left(-\frac{1}{2}(x,k)\Sigma^{-1}\left(\begin{array}{c} x \\ k \end{array}\right)\right)\frac{1}{2}(x,k)\Sigma^{-1}\left(\begin{array}{c} x \\ k \end{array}\right)dxdk =$$

$$-\log\left(\frac{1}{2\pi\sqrt{|\Sigma|}}\right) + \frac{1/2}{2\pi\sqrt{|\Sigma|}}\int \exp\left(-\frac{1}{2}(x,k)\Sigma^{-1}\left(\begin{array}{c} x \\ k \end{array}\right)\right)(x,k)\Sigma^{-1}\left(\begin{array}{c} x \\ k \end{array}\right)dxdk,$$

since our Gaussians are normalized to 1. Since Σ is a positive definite matrix, this means it can be diagonalized into a matrix with positive eigenvalues:

$$\Sigma = Q^{-1}DQ, \quad D = \mathrm{diag}(d_i), \quad d_i > 0, \quad i \in 1,2.$$

Since then, its inverse Σ^{-1} is also positive definite, one can perform the unique Cholesky factorization of Σ^{-1} via a nonsingular upper triangular matrix U:

$$\Sigma^{-1} = U^T U,$$

where it holds that:

$$|\Sigma^{-1}| = |U^T U| = |U|^2 \iff |\Sigma| = 1/|U|^2,$$

so:

$$H = \log\left(2\pi\sqrt{|\Sigma|}\right) + \frac{1/2}{2\pi\sqrt{|\Sigma|}}\int \exp\left(-\frac{1}{2}(x,k)U^T U\left(\begin{array}{c} x \\ k \end{array}\right)\right)(x,k)U^T U\left(\begin{array}{c} x \\ k \end{array}\right)dxdk =$$

$$H = \log\left(\frac{2\pi}{|U|}\right) + \frac{|U|/2}{2\pi}\int \exp\left(-\frac{1}{2}(x,k)U^T U\left(\begin{array}{c} x \\ k \end{array}\right)\right)(x,k)U^T U\left(\begin{array}{c} x \\ k \end{array}\right)dxdk,$$

and defining the coordinate transformation:

$$\left(\begin{array}{c} y \\ z \end{array}\right) = U\left(\begin{array}{c} x \\ k \end{array}\right) \iff (y,z) = (x,k)U^T,$$

which means that:

$$dydz = |U|dxdk \iff dxdk = dydz/|U|,$$

$$H = \log\left(\frac{2\pi}{|U|}\right) + \frac{|U|/2}{2\pi}\int \exp\left(-\frac{1}{2}(y,z)\left(\begin{array}{c} y \\ z \end{array}\right)\right)(y,z)\left(\begin{array}{c} y \\ z \end{array}\right)dydz/|U| =$$

$$H = \log\left(\frac{2\pi}{|U|}\right) + \frac{|U|/2}{2\pi}\int \exp\left(-\frac{y^2+z^2}{2}\right)(y^2+z^2)dydz/|U| =$$

$$H = \log\left(\frac{2\pi}{|U|}\right) + \frac{1/2}{2\pi}\int \exp\left(-\frac{y^2+z^2}{2}\right)(y^2+z^2)dydz =$$

$$H = \log\left(\frac{2\pi}{|U|}\right) + \frac{\langle y^2+z^2\rangle}{2},$$

one can notice that the second term is fixed, so it is a constant for all possible Gaussians, and therefore, the entropy is mostly defined by the first term. Since we have:

$$|\Sigma| = \frac{1}{|U|^2} \implies \sqrt{|\Sigma|} = \frac{1}{|U|},$$

then:

$$H = \log\left(2\pi\sqrt{|\Sigma|}\right) + \frac{\langle y^2+z^2\rangle}{2},$$

so, finally:

$$H = \frac{\log(|\Sigma|)}{2} + \log(2\pi) + \frac{\langle y^2 + z^2 \rangle}{2},$$

or equivalently:

$$H = \frac{\log(|\Sigma|)}{2} + C, \quad C = \log(2\pi) + \frac{\langle y^2 + z^2 \rangle}{2},$$

which, after diagonalization of the covariance matrix, can be equivalently expressed as:

$$H = \frac{\log(d_1) + \log(d_2)}{2} + C, \quad C = \log(2\pi) + \frac{\langle y^2 + z^2 \rangle}{2}.$$

Since solutions are known to converge to a steady state, this means that entropy will converge as well to its steady-state value. For the particular case of $D = I_d$, $\gamma = 1$, we have:

$$\Sigma_\infty = \begin{pmatrix} 3 & -1 \\ -1 & 2 \end{pmatrix}, \tag{15}$$

To complete the calculation, we simply recall that:

$$\frac{\langle y^2 + z^2 \rangle}{2} = \frac{1/2}{2\pi} \int \exp\left(-\frac{y^2 + z^2}{2}\right)(y^2 + z^2)dydz = \frac{\frac{2}{2}}{2\pi} \int dz \exp\left(-\frac{z^2}{2}\right) \int dy \exp\left(-\frac{y^2}{2}\right) y^2 =$$

$$\frac{1}{2\pi} \int dz \exp\left(-\frac{z^2}{2}\right)\left(-\int d[\exp\left(-\frac{y^2}{2}\right)]y\right) = [\frac{1}{\sqrt{2\pi}}]^2 \int dz \exp\left(-\frac{z^2}{2}\right) \int \exp\left(-\frac{y^2}{2}\right) dy = 1,$$

However, because there is a constant shift by C in the value of the Wehrl entropy H, in our numerical results section, we will limit ourselves to report the nonconstant part of this entropy, namely $\frac{\log(|\Sigma|)}{2}$ for Gaussian states.

Given our knowledge of the convergence in the Wigner–Fokker–Planck equation to a Gaussian steady-state in an exponential decay fashion [3], we notice that, indeed, the entropy might not strictly increase (one has freedom to choose, mathematically, the variances of a Gaussian representing the Gaussian state as higher or lower than the respective ones for the steady-state solution). The expected behavior when starting from a Gaussian state as the initial condition is that, under a harmonic potential, the entropy will converge to the steady-state entropy value.

The advantage of having a numerical solver of open quantum systems in the Wigner–Fokker–Planck formulation is that we can monitor closely (up to numerical error) the evolution of the entropy functional we chose to observe (the Wehrl entropy in our case), observing its behavior to determine numerically the type of behavior (either monotonic or not) for a given set of initial conditions, though one knows that the entropy will converge to a steady-state value in an attractor fashion. We monitor the Wehrl entropy then for Gaussian states under a harmonic potential in the Wigner–Fokker–Planck model for open quantum systems via Monte Carlo numerics, solving our convection-diffusion equation as a stochastic equation with an Euler–Maruyama methodology, as described in the previous subsections.

3.1.5. Numerical Results of Entropy Behavior

We present below the results of a numerical simulation of the Wigner–Fokker–Planck equation under a harmonic potential taking as initial condition a Gaussian state. In this case, the state is guaranteed to be Gaussian at all times, approaching in the limit as time goes to infinity for a known Gaussian steady-state analytical solution [3]. We specifically focus on the behavior of the Wehrl entropy (minus its constant shift) $\log(\det(\Sigma))/2$ applied to the Husimi transform of our Wigner function representing the Gaussian states taken as initial conditions. For reference, we include in our plots a straight line with the analytical value of the steady-state Wehrl entropy $\log(|\Sigma|)/2$ to which the entropies for the numerical solutions are expected to converge (which was indeed observed in our numerical simulations, as our subsequent figures will show).

Our first set of studies considers the evolution of an open quantum system whose initial condition is the harmonic ground state. After a long enough time, it converges numerically to the analytical steady-state solution. We study the numerical evolution of the Wehrl entropy at our different time iterations. In this case, the entropy increases from the initial value in the groundstate case towards the steady-state value, oscillating numerically around this limit value due to the inherent error in Monte Carlo methods (see Figure 1).

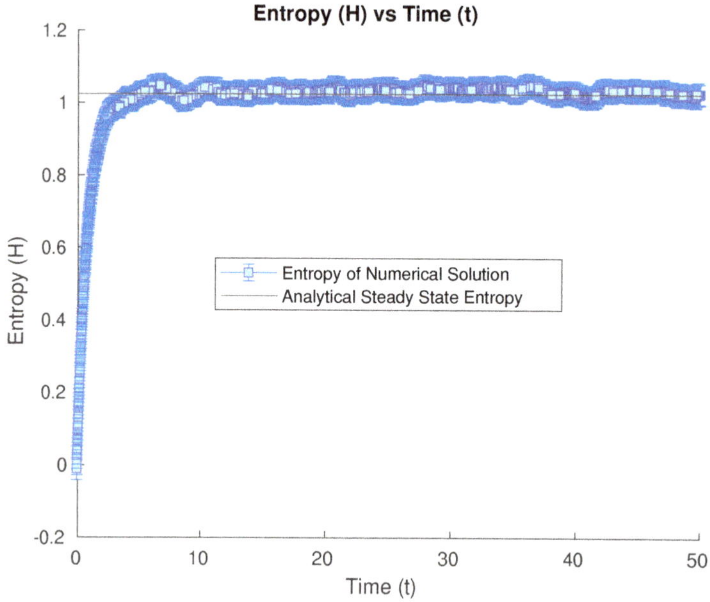

Figure 1. Numerical time evolution of the Wehrl entropy starting from a harmonic groundstate initial condition, until the steady state is achieved numerically.

Our second study considers the evolution of our open quantum system starting from a Gaussian state whose covariance matrix is $2.25^2 = 5.0625$ times the one for the harmonic oscillator. In this case, the entropy of the related Gaussian state is bigger than in the previous case, and it evolves by first sharply decreasing over a short time and then bouncing back up briefly (see Figure 2), until it decreases and oscillates numerically around the steady state. However, this brief nonmonotonic bounce is of the order of $\Delta H = 0.02$, which is below the inherent stochastic uncertainty of our Monte Carlo method (of the order of $\epsilon = 0.03$), obtained by observing the time evolution of our stochastic solver starting from the steady-state solution itself as initial condition (see Figure 3). Because the nonmonotonic bounce is below the uncertainty of the numerical method, up to the numerical error, we cannot deny the possibility of monotonic decay of the Wehrl's entropy being present in this plot within the uncertainty bounds.

For the third set of results, we will try different cases where the initial condition is related to a Gaussian state, whose covariance matrix is not diagonal. We will try first as a sanity check to start with the steady state as the initial condition. This will render the numerical simulation as simply noisy oscillations around the known equilibrium state with its given entropy value (see Figure 3), and in fact, this simulation will indicate the inherent uncertainty of our Euler–Maruyama method as it oscillates numerically around the steady state with the stochastic error of the method.

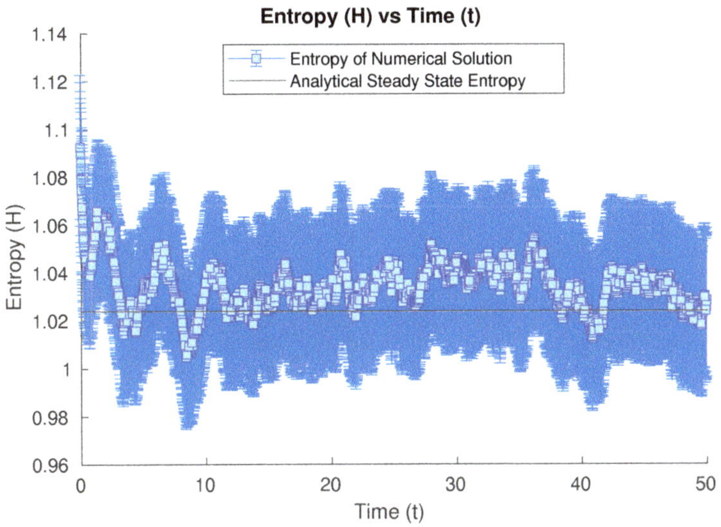

Figure 2. Numerical time evolution of the Wehrl entropy starting from a Gaussian state whose covariance is $2.25^2 = 5.0625$ times the harmonic groundstate one, until the steady state is achieved numerically. Error bars were included to also consider the inherent uncertainty of the Monte Carlo method in use.

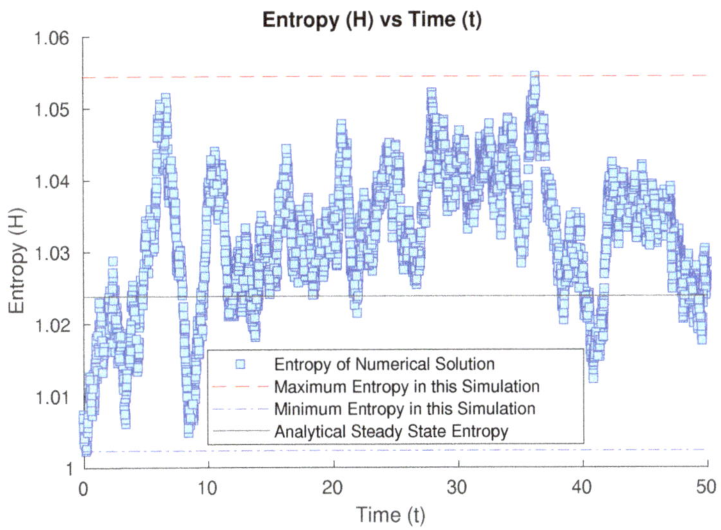

Figure 3. Numerical time evolution of the Wehrl entropy starting from the steady state as the initial condition, oscillating around it with the inherent numerical error of Monte Carlo methods.

We now will try as initial condition a Gaussian state, with the covariance matrix being a multiple of the steady state one, to study how the numerical solution will converge to the steady state (and particularly its entropy) from initial conditions with higher variances in the diagonalizing directions (more spread). We start then from a Gaussian state with a covariance matrix with values 1.5 times the ones of the steady state. Its entropy value is shown to converge to the steady-state entropy again (see Figure 4).

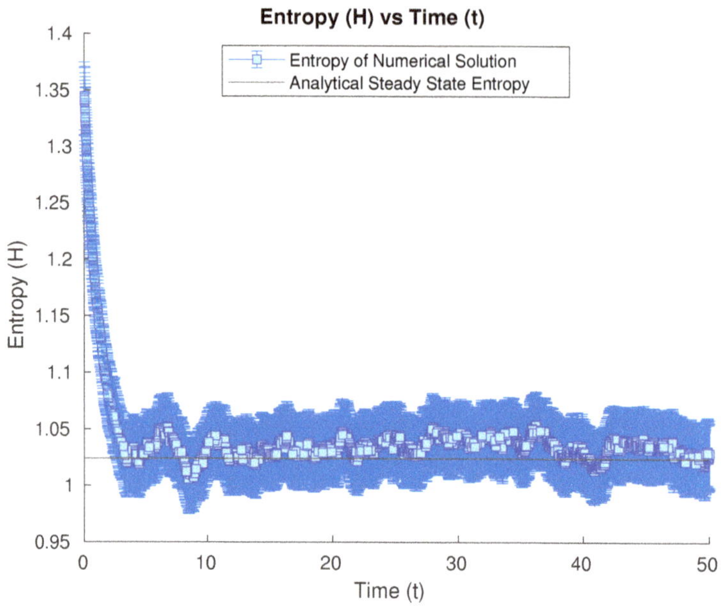

Figure 4. Numerical time evolution of the Wehrl entropy starting from a Gaussian state with a covariance matrix 1.5 times the steady-state one, having its entropy converge to the steady-state value.

Our last numerical examples start from initial conditions reflecting squeezed states centered at the origin with diagonal covariance matrices. One of them is such that (in units where $\hbar = 1 = m$) $\Delta x = 1 = \sigma_x$, $\Delta p = 1/2 = \sigma_p$ (Figure 5), and the other one is such that $\Delta x = 1/2 = \sigma_x$, $\Delta p = 1 = \sigma_p$ (Figure 6). Its entropy values converge to the steady state one during the numerical evolution time for both cases (see Figures 5 and 6).

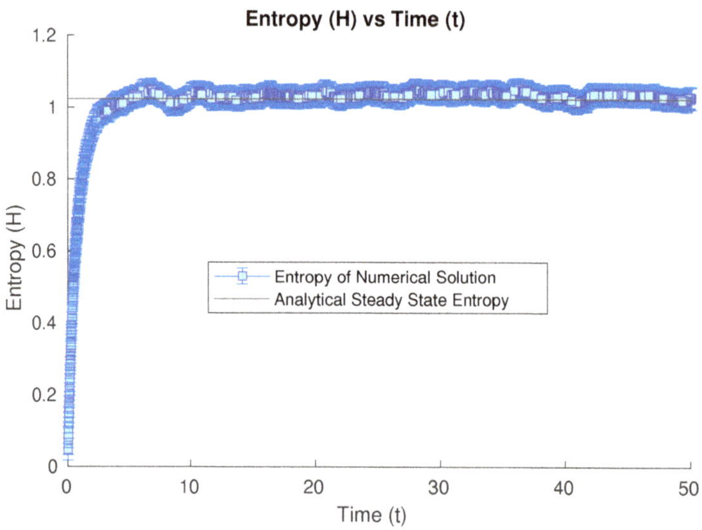

Figure 5. Numerical time evolution of the Wehrl entropy starting from a squeezed state such that $\Delta x = 1 = \sigma_x$, $\Delta p = 1/2 = \sigma_p$, having its entropy converge to the steady-state value.

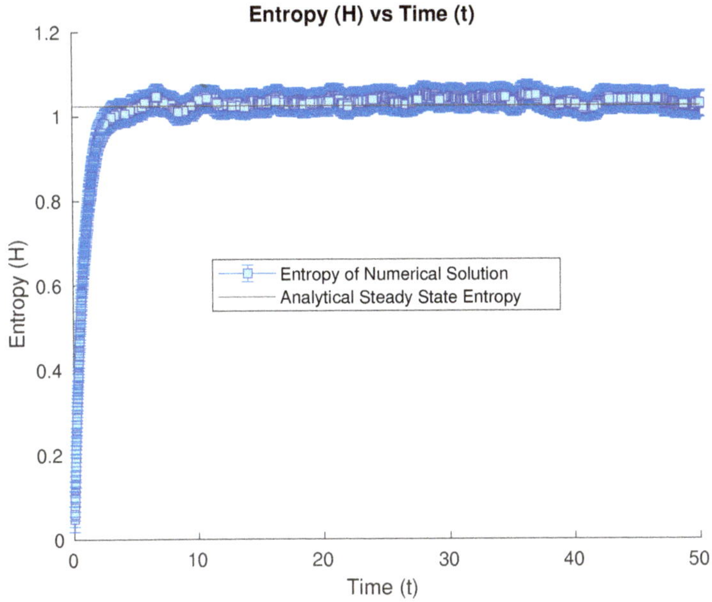

Figure 6. Numerical time evolution of the Wehrl entropy starting from a squeezed state such that $\Delta x = 1/2 = \sigma_x$, $\Delta p = 1 = \sigma_p$, having its entropy converge to the steady-state value as well.

4. Discussion

We find that the monotonic behavior of the entropy with time for the case with friction cannot be denied for any of the five initial conditions we picked. Without consideration of the inherent uncertainty in the numerical method of choice (Euler–Maruyama), one might feel tempted to interpret the inherent oscillations in the entropy evolution as nonmonotonic behavior. However, when including the natural stochastic error from the Euler–Maruyama-based method, one cannot disregard the possibility of monotonic behavior, since all oscillations are within the error bars of the numerical method. The neglect of these considerations would represent overfitting. Though we selected particular cases of Gaussian states as initial conditions, these were chosen based on their importance within the problem. Namely, the harmonic ground state is the paradigmatic case of a coherent state, studying as well a thermal state related to it, squeezed states are important as minimal uncertainty states with variances different from the coherent state ones, and the steady-state solution for the harmonic potential under noise is crucial to understand since all well-posed initial conditions converge to it. Though theoretical studies cannot guarantee monotonic behavior for a completely different entropy (the Von Neumann one), except for the frictionless case $\gamma = 0$, the numerical evidence seems to suggest that at least for Gaussian states one should explore the theoretical possibility of monotonic behavior of the Wehrl entropy in this case (or trying to find a counterexample perhaps related to the case where the average behavior without uncertainty considerations seems monotonic on a naive analysis).

5. Conclusions

We studied numerically the behavior of the Wehrl entropy related to the quantum information of a benchmark open quantum system problem, formulated in terms of the Wigner–Fokker–Planck equation. This benchmark problem is namely the case of a harmonic potential subject to Markovian noise producing diffusion and friction over the quantum transport of the problem. The main motivation to study the quantum entropy for an open quantum system via Monte Carlo methods based on Euler–Maruyama techniques is the absence of theoretical results that can guarantee either monotonicity or its opposite of absolute entropies of importance such as the Von Neumann or the Wehrl one (except for the Von Neumann entropy in the particular frictionless case $\gamma = 0$). Our numerical results for specific initial conditions, namely some particular Gaussian states related to the harmonic groundstate and the steady state (and some other Gaussian states with covariance matrices proportional to these two), seem to indicate that, within the error bounds of the numerical method, monotonicity could not be denied for the explored cases. These numerical results might motivate further theoretical work to study if one can guarantee for Gaussian states (or a subset of them) the monotonic behavior of Wehrl entropy in this benchmark case of a harmonic potential. On the other hand, one could also try to find a counterexample possibly related to the case that was closest yet unsuccessful (within the bounds of the numerical error of our method) to violating monotonicity from the particular cases of initial conditions selected. Further work will be considered in the extension of these numerical techniques for the case of non-Gaussian states or nonharmonic potentials, where the signed-particle method as in [11], for example, could be used.

Author Contributions: Conceptualization, J.A.M.E.; methodology, J.A.M.E.; software, J.A.M.E., A.E. and H.P.; validation, J.A.M.E., H.P. and A.E.; formal analysis, J.A.M.E., H.P. and A.E.; investigation, J.A.M.E., A.E. and H.P.; resources, J.A.M.E.; data curation, J.A.M.E.; writing—original draft preparation, J.A.M.E., H.P. and A.E.; writing—review and editing, J.A.M.E., H.P. and A.E.; visualization, J.A.M.E.; supervision, J.A.M.E.; project administration, J.A.M.E.; funding acquisition, J.A.M.E. All authors have read and agreed to the published version of the manuscript.

Funding: This research received no external funding. However, start-up funding support by the University of Texas at San Antonio is gratefully acknowledged by the last author. The APC was funded by the UTSA start-up funding support aforementioned.

Institutional Review Board Statement: Not applicable.

Informed Consent Statement: Not applicable.

Data Availability Statement: Code that produces the data supporting reported results can be found in the following GitHub repository: https://github.com/phjame/StochasticWFP (accessed on 29 February 2024).

Acknowledgments: We would like to acknowledge Maria Gabriela Boada Gutierrez, who participated in writing assistance, technical editing, language editing, and proofreading of the manuscript.

Conflicts of Interest: The authors declare no conflict of interest.

Appendix A. Weighted L_2-Distance between Wigner Function and Steady-State Solution for Harmonic Potential

We consider the following weighted L_2-distance between the Wigner function $w(x,p,t)$ at time t and the steady-state solution for the harmonic potential $\mu(x,p) = \lim_{t\to\infty} w(x,p,t)$ as in [2]:

$$\left|\left|\frac{w-\mu}{\sqrt{\mu}}\right|\right|_{L^2}. \tag{A1}$$

If we proceed to calculate this weighted distance, we have that:

$$\begin{aligned}\left|\left|\frac{w-\mu}{\sqrt{\mu}}\right|\right|_{L^2} &= \int dxdk(w-\mu)^2/\mu \\ &= \int dxdk(w^2/\mu - 2w + \mu) \\ &= \int dxdk(w^2/\mu) - 1,\end{aligned} \tag{A2}$$

the last line standing true because Wigner functions are normalized to 1 in phase space. Now, the integral in Equation (A2) will only converge if the resulting function w^2/μ is indeed a Gaussian distribution. If the subtraction of the inverse covariance matrices of w^2 and μ is not positive definite, then the related integral will diverge and the norm will blow up to infinity. This imposes constraints as to which Wigner functions will result in this L_2-distance being well defined. Specifically, let us take μ to be the normalized Gaussian centered at the origin with covariance matrix:

$$\Sigma_\mu = \begin{bmatrix} 3 & -1 \\ -1 & 2 \end{bmatrix}, \tag{A3}$$

and w to have the covariance matrix:

$$\Sigma_w(t) = \begin{bmatrix} A(t) & B(t) \\ B(t) & D(t) \end{bmatrix}, \tag{A4}$$

where $A(t)$, $B(t)$, and $D(t)$ are time-dependent parameters such that not only Σ_w is positive definite at all times but also that w^2/μ has a positive definite covariance matrix at all times. This is satisfied for w^2/μ if the condition $\det\left(2\Sigma_w^{-1} - \Sigma_\mu^{-1}\right)(t) > 0$ holds. One can show that this condition is equivalent to satisfying:

$$Q \equiv \frac{20 - 4B - B^2 + A(D-4) - 6D}{-6B^2 + A(6D-20)} < 0. \tag{A5}$$

In the case where the initial condition $w|_{t=0}$ is the harmonic oscillator ground state, the norm of $w|_{t=0} = w_0$ is well defined. The covariance matrix of w_0 is, after taking $\hbar = m = \omega = 1, \Sigma_0 = I/2$, where I is the identity matrix. Thus, $Q = -61/34$ for w_0, which is less than 0.

Now, let us consider the Wigner functions with covariance matrices $C_0\Sigma_0$ and $C_\mu\Sigma_\mu$, where C_0 and C_μ are positive real numbers. There exist valid values of C_0 and C_μ such that the norm (A1) is defined. Applying the condition (A5) to these covariance matrices, we obtain the following results:

$$C_0 \in (0, 10 - 2\sqrt{5}) \cup (\frac{20}{3}, 10 + 2\sqrt{5})), \tag{A6}$$

$$C_\mu < 2. \tag{A7}$$

These conditions are part of the motivations in our choice of coefficients in the analysis of the Wehrl entropy for our Monte Carlo numerical solutions presented in this work.

References

1. Ding, Z.; Chen, C.F.; Lin, L. Single-ancilla ground state preparation via Lindbladians. *arXiv* **2023**, arXiv:2308.15676.
2. Gamba, I.; Gualdani, M.P.; Sharp, R.W. An adaptable discontinuous Galerkin scheme for the Wigner-Fokker–Planck equation. *Commun. Math. Sci.* **2009**, *7*, 635–664. [CrossRef]
3. Sparber, C.; Carrillo, J.; Dolbeault, J.; Markowich, P. On the Long Time Behavior of the Quantum Fokker–Planck Equation. *Monatshefte Math.* **2004**, *141*, 237–257. [CrossRef]
4. D'ariano, G.; Macchiavello, C.; Moroni, S. On the Monte Carlo simulation approach to Fokker–Planck equations in quantum optics. *Mod. Phys. Lett. B* **1994**, *08*, 239–246. [CrossRef]
5. Frensley, W.R. Boundary conditions for open quantum systems driven far from equilibrium. *Rev. Mod. Phys.* **1990**, *62*, 745–791. [CrossRef]
6. Ringhofer, C. A Spectral Method for the Numerical Simulation of Quantum Tunneling Phenomena. *SIAM J. Numer. Anal.* **1990**, *27*, 32–50. [CrossRef]
7. Arnold, A.; Ringhofer, C. An Operator Splitting Method for the Wigner–Poisson Problem. *SIAM J. Numer. Anal.* **1996**, *33*, 1622–1643. [CrossRef]
8. Goudon, T. Analysis of a Semidiscrete Version of the Wigner Equation. *SIAM J. Numer. Anal.* **2002**, *40*, 2007–2025. [CrossRef]
9. Keener, J.; *Biology in Time and Space: A Partial Differential Equation Modeling Approach;* Pure and Applied Undergraduate Texts; American Mathematical Society: Providence, RI, USA, 2021.
10. Nedjalkov, M.; Kosina, H.; Selberherr, S.; Ringhofer, C.; Ferry, D.K. Unified particle approach to Wigner-Boltzmann transport in small semiconductor devices. *Phys. Rev. B* **2004**, *70*, 115319. [CrossRef]

11. Muscato, O. A benchmark study of the Signed-particle Monte Carlo algorithm for the Wigner equation. *Commun. Appl. Ind. Math.* **2017**, *8*, 237–250. [CrossRef]
12. Ganiu, V.; Schulz, D. Application of Discontinuous Galerkin Methods onto Quantum–Liouville type Equations. In Proceedings of the International Workshop on Computational Nanotechnology, Barcelona, Spain, 12–16 June 2023.
13. Suh, N.D.; Feix, M.R.; Bertrand, P. Numerical simulation of the quantum Liouville–Poisson system. *J. Comput. Phys.* **1991**, *94*, 403–418. [CrossRef]
14. Morales, J.; Boada-Gutierrez, M.G. *Discontinuous Galerkin schemes for Master Equations Modeling Open Quantum Systems*; NUMTA conference paper, Springer Lecture Notes in Computer Science; UTSA: San Antonio, TX, USA, 2024.
15. Morales Escalante, J.A. NIPG-DG schemes for transformed master equations modeling open quantum systems. *arXiv* **2023**, arXiv:2308.11580.
16. Markowich, P.A.; Villani, C. On the trend to equilibrium for the Fokker–Planck equation: An interplay between physics and functional analysis. *Mat. Contemp. (SBM)* **2000**, *19*, 1–2.
17. Jüngel, A. *Entropy Methods for Diffusive Partial Differential Equations*; Springer: Berlin/Heidelberg, Germany, 2016.
18. Wehrl, A. On the relation between classical and quantum-mechanical entropy. *Rep. Math. Phys.* **1979**, *16*, 353–358. [CrossRef]
19. Arnold, A.; López, J.L.; Markowich, P.A.; Soler, J. An analysis of quantum Fokker-Planck models: A Wigner function approach. *Rev. Matemática Iberoam.* **2004**, *20*, 771–814. [CrossRef]
20. Jordan, R.; Kinderlehrer, D.; Otto, F. The Variational Formulation of the Fokker–Planck Equation. *SIAM J. Math. Anal.* **1998**, *29*, 1–17. [CrossRef]
21. Arnold, A.; Markowich, P.; Toscani, G.; Unterreiter, A. On convex sobolev inequalities and the rate of convergence to equilibrium for Fokker–Planck type equations. *Commun. Partial. Differ. Equations* **2001**, *26*, 43–100. [CrossRef]
22. Temme, K.; Pastawski, F.; Kastoryano, M.J. Hypercontractivity of quasi-free quantum semigroups. *J. Phys. A Math. Theor.* **2014**, *47*, 405303. [CrossRef]
23. Müller-Hermes, A.; Franca, D.S. Sandwiched Rényi Convergence for Quantum Evolutions. *Quantum* **2018**, *2*, 55. [CrossRef]
24. Carlen, E.A.; Maas, J. Gradient flow and entropy inequalities for quantum Markov semigroups with detailed balance. *J. Funct. Anal.* **2017**, *273*, 1810–1869. [CrossRef]
25. Maas, J. Gradient flows of the entropy for finite Markov chains. *arXiv* **2011**, arXiv:1102.5238.
26. Cao, Y.; Lu, J.; Lu, Y. Exponential Decay of Rényi Divergence Under Fokker–Planck Equations. *J. Stat. Phys.* **2019**, *176*, 1172–1184. [CrossRef]
27. Cheng, Y.; Gamba, I.; Proft, J. Positivity-preserving discontinuous Galerkin schemes for linear Vlasov-Boltzmann transport equations. *Math. Comput.* **2012**, *81*, 153–190. [CrossRef]
28. Heath, R.; Gamba, I.; Morrison, P.; Michler, C. A discontinuous Galerkin method for the Vlasov–Poisson system. *J. Comput. Phys.* **2012**, *231*, 1140–1174. [CrossRef]
29. Pennie, C.A.; Gamba, I.M. Entropy Decay Rates for Conservative Spectral Schemes Modeling Fokker–Planck–Landau Type Flows in the Mean Field Limit. *arXiv* **2020**, arXiv:1910.03110.
30. Pennie, C.A.; Gamba, I.M. Decay of entropy from a conservative spectral method for Fokker–Planck–Landau type equations. *AIP Conf. Proc.* **2019**, *2132*, 020001. [CrossRef]
31. Case, W.B. Wigner functions and Weyl transforms for pedestrians. *Am. J. Phys.* **2008**, *76*, 937–946. [CrossRef]
32. Sellier, J.M.; Dimov, I. Wigner Functions, Signed Particles, and the Harmonic Oscillator. *J. Comput. Electron.* **2015**, *14*, 907–915. [CrossRef]
33. Ferry, D.K.; Nedjalkov, M. *The Wigner Function in Science and Technology*; IoP Publishing: Bristol, UK, 2018.
34. Benam, M.; Nedjalkov, M.; Selberherr, S. A Wigner Potential Decomposition in the Signed-Particle Monte Carlo Approach. In *Numerical Methods and Applications*; Nikolov, G., Kolkovska, N., Georgiev, K., Eds.; Springer: Cham, Switzerland, 2019; pp. 263–272.
35. Serafini, A. *Quantum Continuous Variables: A Primer of Theoretical Methods*; CRC Press, Taylor & Francis Group: Boca Raton, FL, USA, 2017.

entropy

Article

Exact Results for Non-Newtonian Transport Properties in Sheared Granular Suspensions: Inelastic Maxwell Models and BGK-Type Kinetic Model

Rubén Gómez González [1,†] and Vicente Garzó [2,*,†,‡]

[1] Departamento de Física, Universidad de Extremadura, Avda. de Elvas s/n, E-06006 Badajoz, Spain; ruben@unex.es
[2] Departamento de Física and Instituto de Computación Científica Avanzada (ICCAEx), Universidad de Extremadura, Avda. de Elvas s/n, E-06006 Badajoz, Spain
* Correspondence: vicenteg@unex.es
† These authors contributed equally to this work.
‡ https://fisteor.cms.unex.es/investigadores/vicente-garzo-puertos/.

Abstract: The Boltzmann kinetic equation for dilute granular suspensions under simple (or uniform) shear flow (USF) is considered to determine the non-Newtonian transport properties of the system. In contrast to previous attempts based on a coarse-grained description, our suspension model accounts for the real collisions between grains and particles of the surrounding molecular gas. The latter is modeled as a bath (or thermostat) of elastic hard spheres at a given temperature. Two independent but complementary approaches are followed to reach exact expressions for the rheological properties. First, the Boltzmann equation for the so-called inelastic Maxwell models (IMM) is considered. The fact that the collision rate of IMM is independent of the relative velocity of the colliding spheres allows us to exactly compute the collisional moments of the Boltzmann operator without the knowledge of the distribution function. Thanks to this property, the transport properties of the sheared granular suspension can be *exactly* determined. As a second approach, a Bhatnagar–Gross–Krook (BGK)-type kinetic model adapted to granular suspensions is solved to compute the velocity moments and the velocity distribution function of the system. The theoretical results (which are given in terms of the coefficient of restitution, the reduced shear rate, the reduced background temperature, and the diameter and mass ratios) show, in general, a good agreement with the approximate analytical results derived for inelastic hard spheres (IHS) by means of Grad's moment method and with computer simulations performed in the Brownian limiting case ($m/m_g \to \infty$, where m_g and m are the masses of the particles of the molecular and granular gases, respectively). In addition, as expected, the IMM and BGK results show that the temperature and non-Newtonian viscosity exhibit an S shape in a plane of stress–strain rate (discontinuous shear thickening, DST). The DST effect becomes more pronounced as the mass ratio m/m_g increases.

Keywords: granular suspensions; Boltzmann kinetic equation; inelastic Maxwell models; BGK-type kinetic model; non-Newtonian transport properties

1. Introduction

A very usual way of assessing the effect of the surrounding fluid on the dynamics properties of solid particles is through an effective fluid–solid force [1–4]. In some models, this force is simply proportional to the velocity particle (Stokes linear drag law) [5–12]. This type of force attempts to mimic the energy dissipated by grains due to their friction on the interstitial viscous gas. A more sophisticated model [13] also incorporates a Langevin-like stochastic term that accounts for the energy transferred to grains due to their "collisions" with particles of the background gas. However, although this coarse-grained approach has provided reliable results in the past, it would be desirable to consider a suspension model that takes into account the real collisions between grains and particles of the surrounding (molecular) gas. This sort of suspension model (which was inspired by a previous work of Biben et al. [14]) has been recently proposed [15]. In this model, granular particles are assumed to be sufficiently rarefied so that they do not disturb the state of the molecular (background) gas. As a consequence, the interstitial gas may be treated as a *thermostat* at the temperature T_g. Moreover, although the concentration (mole fraction) of grains is quite small, apart from the elastic collisions between solid and molecular gas particles, one has to consider the *inelastic* collisions between grains themselves. This model can be useful to analyze transport properties in particle-laden suspensions [16] where very dilute particles (like aerosols) are immersed in a fluid (like air).

The rheological properties of a granular suspension under simple (or uniform) shear flow (USF) have been recently determined [17]. In contrast to previous attempts [5–8,10–12,18], the results obtained in Ref. [17] were derived from the collisional model proposed in Ref. [15]. On the other hand, a limitation of these results is that they were *approximately* obtained by employing Grad's moment method [19], namely, a method based on the truncation of a series expansion of the velocity distribution function in (orthogonal) Sonine polynomials. The use of this approximate method is essentially motivated by the form of the collision rate for inelastic hard spheres (IHS) appearing inside the Boltzmann collision operator. The collision rate for IHS is proportional to the magnitude of the normal component of the relative velocity of the two spheres that are about to collide. This velocity dependence of the collision rate for IHS prevents the possibility of deriving exact expressions for the transport properties in the USF problem, even in the case of elastic collisions.

A possible way to overcome the technical difficulty of the hard-sphere kernel is to consider the so-called inelastic Maxwell models (IMM). As for the conventional Maxwell molecules [20–22], the collision rate of IMM is independent of the relative velocity of the two colliding spheres [23]. The use of IMM instead of IHS opens up the possibility of obtaining exact analytical results of the Boltzmann equation in some specific nonequilibrium situations, like the USF. In particular, the knowledge of the collisional moments of the Boltzmann equation for IMM enables a clear exploration of the impact of inelasticity on the non-Newtonian transport properties of the granular suspension without introducing uncontrolled approximations.

Another possible alternative for obtaining exact results is to consider a kinetic model that retains the relevant physical properties of the Boltzmann collision operator but turns out to be more tractable than the true kinetic equation. This route has been widely employed in the past in the case of molecular dilute gases where it has been shown that several exact solutions in far from equilibrium states agree very well with Monte Carlo simulations of the Boltzmann equation. Here, we will consider a kinetic model for granular suspensions [24] to complement the theoretical expressions obtained from the Boltzmann equation for IMM. Since this kinetic model is based on the well known Bhatnagar–Gross–Krook (BGK) model [22] for molecular gases, we will refer to it as a BGK-type kinetic model.

The objective of this paper is to determine the rheological properties of granular particles immersed in a bath of elastic hard spheres under USF. At a macroscopic level, the USF is characterized by constant number densities for solid and gas particles, a uniform temperature, and a (common) linear velocity profile $U_{g,x} = U_x = ay$, where a is the constant shear rate. Here, $\mathbf{U}_g$ and $\mathbf{U}$ denote the mean flow velocities of the molecular and granular gases, respectively. Since we are interested here in the steady state where the system admits a non-Newtonian hydrodynamic description, an external thermostat force (proportional to the peculiar velocity) must be introduced to keep the temperature T_g of the molecular gas constant.

The use of an IMM as well as a BGK-type kinetic model allows us to exactly compute the rheological properties of the granular suspension. These properties are expressed as nonlinear functions of the (reduced) shear rate $a^* = a/\gamma$ (where γ is a drift coefficient characterizing the friction of solid particles on the viscous gas), the coefficient of restitution, the (reduced) background temperature T_g^*, and the diameter σ/σ_g and mass m/m_g ratios.

Here, σ_g and m_g are the diameter and mass of the particles of the molecular gas, respectively, while σ and m are the diameter and mass of the solid particles, respectively.

A very interesting phenomenon appearing in gas–solid suspensions is the so-called discontinuous shear thickening (DST), where the non-Newtonian shear viscosity of the granular suspension drastically increases with the shear rate. In most of the cases, the DST phenomena occurs in highly concentrated suspensions of particles such as mixtures of cornstarch in water [25]. Although several mechanisms have been proposed in the granular literature [25] to explain this nonequilibrium discontinuous transition, the problem is still not understood.

As occurs for IHS [17], our results show that the kinetic granular temperature and the non-Newtonian viscosity exhibit a DST effect for sufficiently large values of the mass ratio m/m_g. In fact, in the Brownian limiting case ($m/m_g \to \infty$), the expressions of the rheological properties derived here reduce to those previously obtained [26] from a coarse-grained description based on the Fokker–Planck operator. This agreement justifies the use of this latter approach to analyze the DST effect in dilute granular suspensions [27]. It is important to remark that the DST effect found here occurs in a system with a simple structure (low-density granular gas immersed in a dilute molecular gas). This means that the origin of this phenomenon is essentially associated here with the *nonlinear* response of the granular suspension to the presence of large shear rates rather than to the complex structure of the system.

Apart from the transport properties (which are related to the second-degree velocity moments), the explicit forms of the higher-degree velocity moments as well as the velocity distribution function of the granular gas were also obtained from the BGK model. This is one of the main advantages of using a kinetic model instead of the true Boltzmann equation. Our results show, in particular, that the fourth-degree moments of the distribution function also exhibit a DST effect. With respect to the velocity distribution function, as expected, we find that its distortion from equilibrium is more significant as both the mass m/m_g and diameter σ/σ_g ratios depart from 1. In addition, a comparison between the BGK results and numerical solutions of the Boltzmann equation from the direct simulation Monte Carlo (DSMC) method [28] for IHS shows a generally good qualitative agreement between both approaches.

The plan of the paper is as follows. The Boltzmann kinetic equation for a granular gas immersed in a bath of elastic hard spheres under USF is presented in Section 2. The balance equations for the temperatures of the molecular and granular gases are also displayed. Section 3 deals with the calculations carried out for IMM of the rheological properties of the granular suspension. While a shear thinning effect is always found for the nonlinear shear viscosity of the molecular gas, the corresponding shear viscosity of the granular gas exhibits a DST effect for sufficiently large values of the mass ratio m/m_g. The results derived from the BGK-type kinetic model are provided in Section 4, while a comparison between the theoretical results obtained for IHS, IMM, and BGK model is displayed in Section 5 for several systems. Our results highlight a good agreement for the rheology between the three different approaches. Moreover, theoretical results obtained from the IMM and BGK model are also compared against computer simulations in the Brownian limit ($m/m_g \to \infty$), showing a good agreement. The paper is closed in Section 6 with some concluding remarks.

2. Boltzmann Kinetic Equation for Sheared Granular Suspensions

We consider a set of solid particles (granular gas) of mass m and diameter σ which are immersed in a solvent (molecular gas) constituted by particles of mass m_g and diameter σ_g. As usual, the granular gas is modeled as a gas of hard disks ($d = 2$) or spheres ($d = 3$) with inelastic collisions. In the simplest model, the inelasticity of collisions is characterized by a constant (positive) coefficient of normal restitution $\alpha \leq 1$, where $\alpha = 1$ refers to elastic collisions. On the other hand, collisions between solid particles and particles of the molecular gas are elastic. We also assume that the number density of grains is much smaller than that of solvent so that the state of the latter is not perturbed by the presence of the former. In these conditions, we can treat the molecular gas as a bath or *thermostat* at the temperature T_g (once the parameters of the system, specifically the shear rate, have been set). Moreover, although the granular gas is sufficiently rarefied, we take into account the collisions among grains in its corresponding Boltzmann kinetic equation.

We assume that the system (granular particles plus solvent) is subjected to USF. As said in the Introduction section, this state is characterized by constant densities n_g and n, uniform temperatures T_g and T, and by a (common) linear profile of the x component of the flow velocities along the y axis:

$$n_g \equiv \text{const}, \quad n \equiv \text{const}, \tag{1}$$

$$\nabla T_g = \nabla T = 0, \tag{2}$$

$$U_{g,i} = U_i = a_{ij} r_j, \quad a_{ij} = a\delta_{ix}\delta_{jy}, \tag{3}$$

with a being the *constant* shear rate. Here, n_g, $\mathbf{U}_g$, and T_g are the number density, the mean flow velocity, and the temperature, respectively, of the molecular gas. In terms of its one-particle velocity distribution function $f_g(\mathbf{r}, \mathbf{v}; t)$, these hydrodynamic fields are defined as

$$\left\{n_g, n_g\mathbf{U}_g, dn_gT_g\right\} = \int d\mathbf{v}\left\{1, \mathbf{v}, m_gV^2\right\}f_g(\mathbf{v}), \tag{4}$$

where $\mathbf{V} = \mathbf{v} - \mathbf{U}$ is the peculiar velocity. Note that in Equation (4), the Boltzmann constant $k_B = 1$. We will take this value throughout the paper for the sake of simplicity. In addition, in Equations (1)–(3), n, $\mathbf{U}$, and T denote the number density, the mean flow velocity, and the (granular) temperature, respectively, of the granular gas. They are defined as

$$\{n, n\mathbf{U}, dnT\} = \int d\mathbf{v}\left\{1, \mathbf{v}, mV^2\right\}f(\mathbf{v}). \tag{5}$$

Since the only spatial gradient present in the USF problem is the shear rate, the pressure tensor

$$\mathsf{P}_g = m_g \int d\mathbf{v}\; \mathbf{V}\mathbf{V}f_g(\mathbf{v}) \tag{6}$$

of the molecular gas, and the pressure tensor

$$\mathsf{P} = m \int d\mathbf{v}\; \mathbf{V}\mathbf{V}f(\mathbf{v}) \tag{7}$$

of the granular gas are the relevant fluxes in the problem. They provide information on the transport of momentum across the system. Our main target is to determine P_g and P for arbitrary shear rates.

One of the main advantages of the USF at a microscopic level is that it becomes a spatially homogeneous state when the velocities of the particles are referred to a Lagrangian frame moving with the linear velocity $U_i = a_{ij}r_j$. In this new frame and in the steady state, the distribution functions of the molecular and granular gases adopt the form

$$f_g(\mathbf{r}, \mathbf{v}) = f_g(\mathbf{V}), \quad f(\mathbf{r}, \mathbf{v}) = f(\mathbf{V}). \tag{8}$$

In addition, as the state of the solvent is not perturbed by the solid particles, the temperature T_g in the USF state increases in time due to the viscous heating term $-aP_{xy} > 0$. Thus, as usual in nonequilibrium molecular dynamics simulations [29], an external nonconservative force (thermostat) is introduced in the molecular gas to achieve a stationary state. Among the different possibilities, for simplicity, a force proportional to the particle velocity (Gaussian thermostat) of the form $\mathbf{F}_g = -m_g\xi\mathbf{V}$ is considered in this paper. The parameter ξ is chosen to be a function of the shear rate by the condition that T_g reaches a constant value in the long time limit. Analogously, the granular gas is also subjected to this kind of Gaussian thermostat (i.e., $\mathbf{F} = -m\xi\mathbf{V}$), where ξ is the same quantity for the solvent and the solid particles.

Under the above conditions, in the low-density regime, the distribution function $f_g(\mathbf{V})$ of the molecular gas obeys the nonlinear (closed) Boltzmann equation

$$-aV_y\frac{\partial f_g}{\partial V_x} - \xi\frac{\partial}{\partial \mathbf{V}}\cdot \mathbf{V}f_g = J_g[\mathbf{V}|f_g, f_g], \tag{9}$$

while the distribution function $f(\mathbf{V})$ of the granular gas obeys the kinetic equation

$$-aV_y\frac{\partial f}{\partial V_x} - \xi\frac{\partial}{\partial \mathbf{V}}\cdot \mathbf{V}f = J[\mathbf{V}|f, f] + J_{\text{BL}}[\mathbf{V}|f, f_g]. \tag{10}$$

Here, $J_g[f_g, f_g]$ and $J[f, f]$ are the nonlinear Boltzmann collision operators for the molecular and granular gases, respectively, and $J_{\text{BL}}[f, f_g]$ is the linear Boltzmann–Lorentz collision operator [30,31]. The balance equations for T_g and T can easily obtained by multiplying both sides of Equations (9) and (10) by $m_g V^2$ and mV^2, respectively, and integrating over velocity. The results are

$$-aP_{g,xy} = d\xi p_g, \tag{11}$$

$$-aP_{xy} - \frac{d}{2}p\zeta_g = d\xi p + \frac{d}{2}p\zeta, \tag{12}$$

where $p_g = n_g T_g$ and $p = nT$ are the hydrostatic pressures of the molecular and granular gases, respectively, and the partial production rates ζ and ζ_g are defined, respectively, as

$$\zeta = -\frac{m}{dnT}\int d\mathbf{v}\; V^2\; J[\mathbf{v}|f, f], \quad \zeta_g = -\frac{m}{dnT}\int d\mathbf{v}\; V^2\; J_{\text{BL}}[\mathbf{v}|f, f_g]. \tag{13}$$

The cooling rate ζ gives the rate of kinetic energy loss due to inelastic collisions between particles of the granular gas. It vanishes for elastic collisions ($\alpha = 1$). The term ζ_g gives the transfer of kinetic energy between the particles of the granular gas and the solvent. This term vanishes when the granular and molecular gases are at the same temperature ($T_g = T$). Equation (11) implies that, in the steady state, the viscous heating term ($-aP_{g,xy} > 0$) is exactly balanced by the heat extracted in the gas by the external thermostat. On the other hand, since ζ_g can be positive or negative, Equation (12) implies that, in the steady state, the term $-aP_{xy} - (d/2)p\zeta_g$ is exactly compensated for the cooling terms arising from collisional dissipation (ζp) and the thermostat term (ξp).

The USF state is, in general, a non-Newtonian state characterized by shear-rate-dependent transport coefficients. In particular, one can define the non-Newtonian shear viscosity of the molecular gas as

$$\eta_g = -\frac{P_{g,xy}}{a}. \tag{14}$$

Analogously, the non-Newtonian shear viscosity of the granular gas is given by

$$\eta = -\frac{P_{xy}}{a}. \tag{15}$$

In addition, beyond the Navier–Stokes domain, normal stress differences are expected in the USF. This means that $P_{g,xx} \neq P_{g,yy} \neq P_{g,zz}$ and $P_{xx} \neq P_{yy} \neq P_{zz}$.

It is quite evident that the evaluation of the rheological properties of the molecular and granular gases requires the knowledge of the pressure tensors P_g and P. The nonzero elements of these tensors can be obtained by multiplying by $m_g\mathbf{VV}$ and $m\mathbf{VV}$ both sides of Equations (9) and (10), respectively, and integrating over $\mathbf{V}$. However, to achieve explicit forms for P_g and P, one has to compute the collisional moments

$$\mathsf{A}_g = m_g \int d\mathbf{V}\mathbf{V}\mathbf{V} J_g[f_g, f_g], \tag{16}$$

$$\mathsf{B} = m \int d\mathbf{V}\mathbf{V}\mathbf{V} J[f, f], \quad \mathsf{C} = m \int d\mathbf{V}\mathbf{V}\mathbf{V} J_{\text{BL}}[f, f_g]. \tag{17}$$

In the case of IHS, the collisional moments A_g, B, and C cannot be exactly computed. As said in the Introduction section, a good estimate of them for IHS was made in Ref. [17] by means of Grad's moment method [19]. This method is based on the expansion of the distributions $f_g(\mathbf{V})$ and $f(\mathbf{V})$ in a complete set of orthogonal polynomials, with the coefficients being the corresponding velocity moments of those distributions. The above expansion generates an infinite hierarchy of moment equations that must be truncated at a given order. This truncation allows one to arrive at a closed set of coupled equations for the velocity moments that can be recursively solved. Thus, since the results derived in Ref. [17] for the rheological properties of molecular and granular gases are approximated, it is convenient to revisit the problem and determine the *exact* expressions for the non-Newtonian transport properties of the granular suspension. To achieve such exact forms, two independent approaches will be considered in this paper: (i) the Boltzmann kinetic equation for IMM; and (ii) a BGK-type kinetic model for IHS. This task will be carried out in the next two sections.

3. Rheology from Inelastic Maxwell Models

In this section, we will consider an IMM, namely, a collisional model where the collision rate of the two colliding spheres are independent of their relative velocity. In this case, the Boltzmann collision operator $J_g[f_g, f_g]$ (this is a simple version of the Boltzmann collision operator for Maxwell molecules) of the molecular gas can be written as [22]

$$J_g[\mathbf{v}_1|f_g,f_g] = \frac{\nu_g^{\mathrm{M}}}{n_g S_d}\int d\mathbf{v}_2 \int d\widehat{\sigma}\,\left[f_g(\mathbf{V}_1'')f_g(\mathbf{V}_2'') - f_g(\mathbf{V}_1)f_g(\mathbf{V}_2)\right], \tag{18}$$

where $S_d = 2\pi^{d/2}/\Gamma(\frac{d}{2})$ is the total solid angle in d dimensions and ν_g^{M} is an independent-velocity collision frequency. In Equation (18), the primes on the velocities denote the initial values $\{\mathbf{V}_1'', \mathbf{V}_2''\}$ that lead to $\{\mathbf{V}_1, \mathbf{V}_2\}$ following a binary collision:

$$\mathbf{V}_1'' = \mathbf{V}_1 - (\widehat{\sigma}\cdot\mathbf{g}_{12})\widehat{\sigma}, \quad \mathbf{V}_2'' = \mathbf{V}_2 + (\widehat{\sigma}\cdot\mathbf{g}_{12})\widehat{\sigma}. \tag{19}$$

The effective collision frequency ν_g^{M} can be seen as a free parameter of the model to be chosen to attain agreement with the properties of interest of the original Boltzmann equation for IHS. For instance, to correctly capture the velocity dependence of the original IHS collision rate, we can assume that the IMM collision rate is proportional to $\sqrt{T_g}$.

In the context of IMM, the inelastic Boltzmann collision operator $J[f,f]$ is [23,31]

$$J[\mathbf{V}|f,f] = \frac{\nu^{\mathrm{M}}}{nS_d}\int d\mathbf{V}_2 \int d\widehat{\sigma}\,\left[\alpha^{-1}f(\mathbf{V}_1'')f(\mathbf{V}_2'') - f(\mathbf{V}_1)f(\mathbf{V}_2)\right], \tag{20}$$

while the Boltzmann–Lorentz collision operator $J_{\mathrm{BL}}[f,f_g]$ is defined as [22]

$$J_{\mathrm{BL}}[\mathbf{V}|f,f_g] = \frac{\nu_0^{\mathrm{M}}}{nS_d}\int d\mathbf{V}_2 \int d\widehat{\sigma}\,\left[f(\mathbf{V}_1'')f_g(\mathbf{V}_2'') - f(\mathbf{V}_1)f_g(\mathbf{V}_2)\right]. \tag{21}$$

The relationship between $(\mathbf{V}_1'', \mathbf{V}_2'')$ and $(\mathbf{V}_1, \mathbf{V}_2)$ in Equation (20) is

$$\mathbf{V}_1'' = \mathbf{V}_1 - \frac{1+\alpha}{2\alpha}(\widehat{\sigma}\cdot\mathbf{g}_{12})\widehat{\sigma}, \quad \mathbf{V}_2'' = \mathbf{V}_2 + \frac{1+\alpha}{2\alpha}(\widehat{\sigma}\cdot\mathbf{g}_{12})\widehat{\sigma}, \tag{22}$$

while in Equation (21), it is

$$\mathbf{V}_1'' = \mathbf{V}_1 - 2\mu_g(\widehat{\sigma}\cdot\mathbf{g}_{12})\widehat{\sigma}, \quad \mathbf{V}_2'' = \mathbf{V}_2 + 2\mu(\widehat{\sigma}\cdot\mathbf{g}_{12})\widehat{\sigma} \tag{23}$$

where

$$\mu_g = \frac{m_g}{m+m_g}, \quad \mu = \frac{m}{m+m_g}. \tag{24}$$

In addition, as in the case of the collision frequency ν_g^{M}, the collision frequencies ν^{M} and ν_0^{M} for granular–granular and granular–molecular collisions, respectively, can be chosen to optimize the agreement with the results derived from IHS. We will choose them later.

As mentioned in previous works on IMM [26], the main advantage of computing the collisional moments of the Boltzmann operator for Maxwell models (both elastic and inelastic models) is that they can be exactly provided in terms of the velocity moments of the distribution functions without the explicit knowledge of the latter. This property has been exploited to compute the second-, third-, and fourth-degree collisional moments of IMM for monocomponent and multicomponent granular gases [31]. The exact knowledge of the second-degree collisional moments allow us to obtain exact expressions for the rheological properties of the molecular and granular gases. Let us separately evaluate the rheology of both gases.

3.1. Rheological Properties of the Molecular Gas

The pressure tensor of the molecular gas is defined by Equation (6). To obtain the nonzero elements of this tensor, one multiplies both sides of the Boltzmann equation (9) by $m_g V_i V_j$ and integrates over velocity. The result is

$$a_{ik}P_{g,kj} + a_{jk}P_{g,ki} + 2\xi P_{g,ij} = -\frac{2\nu_g^{\mathrm{M}}}{d+2}\left(P_{g,ij} - p_g\delta_{ij}\right), \tag{25}$$

where $p_g = n_g T_g$ is the hydrostatic pressure of the molecular gas. Upon obtaining Equation (25), use is made of the result [31]:

$$A_{g,ij} = m_g \int d\mathbf{V} V_i V_j J_g[f_g, f_g] = -\frac{2\nu_g^{\mathrm{M}}}{d+2}\left(P_{g,ij} - p_g\delta_{ij}\right). \tag{26}$$

The (reduced) elements of the pressure tensor $\mathsf{P}_g^* = \mathsf{P}_g/(n_g T_g)$ can be easily obtained from Equation (25). They are given by

$$P_{g,xx}^* = \frac{1}{1+2\widetilde{\xi}}\left[1 + \frac{2\widetilde{a}^2}{(1+2\widetilde{\xi})^2}\right], \quad P_{g,yy}^* = P_{g,zz}^* = \frac{1}{1+2\widetilde{\xi}}, \quad P_{g,xy}^* = -\frac{\widetilde{a}}{(1+2\widetilde{\xi})^2}. \tag{27}$$

Here, we introduce the quantities

$$\widetilde{a} = \frac{a}{\widetilde{\nu}_g^{\mathrm{M}}}, \quad \widetilde{\xi} = \frac{\xi}{\widetilde{\nu}_g^{\mathrm{M}}}, \quad \widetilde{\nu}_g^{\mathrm{M}} = \frac{2}{d+2}\nu_g^{\mathrm{M}}. \tag{28}$$

The constraint $P_{g,xx}^* + (d-1)P_{g,yy}^* = d$ leads to a cubic equation relating $\widetilde{\xi}$ and $\widetilde{a}$:

$$\widetilde{a}^2 = d\widetilde{\xi}(1+2\widetilde{\xi})^2. \tag{29}$$

The real root of Equation (29) gives the shear-rate dependence of $\widetilde{\xi}(\widetilde{a})$. It is given by

$$\widetilde{\xi}(\widetilde{a}) = \frac{2}{3}\sinh^2\left[\frac{1}{6}\cosh^{-1}\left(1 + \frac{27}{d}\widetilde{a}^2\right)\right]. \tag{30}$$

Comparison between the results derived here for $P_{g,ij}^*$ with those recently [17] obtained for IHS by means of Grad's moment method [19] shows that both results are identical if the effective collision frequency ν_g^{M} is given by

$$\nu_g^{\mathrm{M}} = \frac{4\pi^{(d-1)/2}}{\Gamma\left(\frac{d}{2}\right)} n_g \sigma_g^{d-1}\sqrt{\frac{T_g}{m_g}}. \tag{31}$$

Henceforth, we will take the choice (31) for ν_g^{M}.

From Equations (27), one can identify the (dimensionless) non-Newtonian shear viscosity $\eta_g^* = \nu_g^{\rm M}\eta_g/p_g = -P_{g,xy}^*/\widetilde{a}$ and the (dimensionless) normal stress difference $\Psi_g^* = P_{g,xx}^* - P_{g,yy}^*$ as

$$\eta_g^* = \frac{1}{(1+2\widetilde{\xi})^2}, \quad \Psi_g^* = \frac{2\widetilde{a}^2}{(1+2\widetilde{\xi})^3}. \tag{32}$$

Note that the results derived here for Maxwell molecules yield $P_{g,yy} = P_{g,zz}$. This result contrasts with the one obtained for hard spheres by numerically solving the Boltzmann equation by means of the DSMC method [28], where it has been shown that $P_{g,yy} \neq P_{g,zz}$. However, the difference $(P_{g,yy} - P_{g,zz})$ found in the simulations is, in general, quite small [9].

It is also important to note that in the case of Maxwell molecules there is an exact equivalence between the description with and without the drag force $-m\xi\mathbf{V}$. Nevertheless, for non-Maxwell molecules, this type of force does not play a neutral role in the transport properties of the system [32].

The shear-rate dependence of η_g^* and Ψ_g^* is plotted in Figure 1 for a three-dimensional system ($d=3$). As expected, the nonlinear viscosity η_g^* decreases with increasing (reduced) shear rate $\widetilde{a}$ (shear thinning effect). The opposite effect is observed for the normal stress difference function Ψ_g^* since it increases with the shear rate. Figure 1 also highlights the excellent agreement found between the theoretical results for the Maxwell molecules with those obtained by numerically solving the Boltzmann equation for hard spheres from the DSMC method [28].

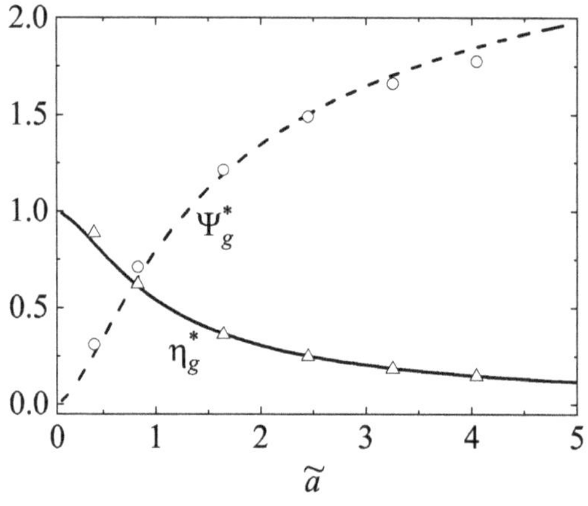

Figure 1. Plot of the nonlinear shear viscosity η_g^* and the normal stress difference Ψ_g^* for hard spheres ($d=3$) as functions of the (reduced) shear rate $\widetilde{a}$. Symbols refer to the DSMC results for hard spheres.

3.2. Rheological Properties of the Granular Gas

As in the case of the molecular gas, the rheology of the granular gas can be also determined by multiplying both sides of Equation (10) by mV_iV_j and integrating over $\mathbf{V}$. After some algebra, one achieves the result

$$a_{ik}P_{kj} + a_{jk}P_{ki} + 2\xi P_{ij} = -\nu^{\rm M}\Big[\nu_\eta^* P_{ij} + p\Big(\zeta^* - \nu_\eta^*\Big)\delta_{ij}\Big] - \frac{4}{d}\nu_0^{\rm M}\mu_g \\ \times\Big[\Big(1 - \frac{2}{d+2}\mu_g\Big)P_{ij} - \frac{2}{d+2}\mu\frac{n}{n_g}P_{g,ij} - \frac{d}{d+2}\mu_g n T_g\Big(\chi + \frac{m}{m_g}\Big)\delta_{ij}\Big], \tag{33}$$

where use is made of the results [31]:

$$B_{ij} = m\int d\mathbf{V} V_i V_j J[f,f] = -\nu^{\rm M}\Big[\nu_\eta^* P_{ij} + p\Big(\zeta^* - \nu_\eta^*\Big)\delta_{ij}\Big], \tag{34}$$

$$C_{ij} = m \int d\mathbf{V} V_i V_j J_{\rm BL}[f, f_g] = -\frac{4}{d}\nu_0^{\rm M}\mu_g\Big[\Big(1-\frac{2}{d+2}\mu_g\Big)P_{ij} - \frac{2}{d+2}\mu\frac{n}{n_g}P_{g,ij} - \frac{d}{d+2}\mu_g n T_g\Big(\chi+\frac{m}{m_g}\Big)\delta_{ij}\Big]. \tag{35}$$

In Equation (33),

$$\zeta^* = \frac{\zeta}{\nu^{\rm M}} = \frac{1-\alpha^2}{2d} \tag{36}$$

is the (reduced) cooling rate for the granular gas, $\chi = T/T_g$ is the temperature ratio, and

$$\nu_\eta^* = \frac{(d+1-\alpha)(1+\alpha)}{d(d+2)}. \tag{37}$$

The partial cooling rate ζ_g can be exactly obtained from Equation (34) as

$$\zeta_g = \frac{4\nu_0^{\rm M}}{d}\mu_g[1-\mu_g(1+\theta)], \tag{38}$$

where

$$\theta = \frac{mT_g}{m_g T} \tag{39}$$

is the ratio of the mean square velocities of granular and molecular gas particles. The forms (36) and (38) can be employed to fix the values of the free parameters $\nu^{\rm M}$ and $\nu_0^{\rm M}$. They are chosen under the criterion that ζ and ζ_g of IMM are the same as that of IHS of diameters σ and σ_0. In this latter case, the above cooling rates are estimated by using Grad's approximation [17]. In this approximation,

$$\zeta^{\rm IHS} = \frac{2\pi^{(d-1)/2}}{d\Gamma\left(\frac{d}{2}\right)} n\sigma^{d-1}\sqrt{\frac{T}{m}}(1-\alpha^2), \tag{40}$$

$$\zeta_g^{\rm IHS} = \frac{8\pi^{(d-1)/2}}{d\Gamma\left(\frac{d}{2}\right)} n_g\overline{\sigma}^{d-1}\mu_g\left(\frac{1+\theta}{\theta}\right)^{1/2}\sqrt{\frac{2T_g}{m_g}}[1-\mu_g(1+\theta)], \tag{41}$$

where $\overline{\sigma} = (\sigma+\sigma_g)/2$. Equations (38), (40), and (41) yield the identities

$$\nu^{\rm M} = \frac{4\pi^{(d-1)/2}}{\Gamma\left(\frac{d}{2}\right)} n\sigma^{d-1}\sqrt{\frac{T}{m}}, \quad \nu_0^{\rm M} = \frac{2\pi^{(d-1)/2}}{\Gamma\left(\frac{d}{2}\right)} n_g\overline{\sigma}^{d-1}\left(\frac{1+\theta}{\theta}\right)^{1/2}\sqrt{\frac{2T_g}{m_g}}. \tag{42}$$

To compare with the rheological properties of IHS [17], it is convenient at this level of the description to identify the friction (or drift) coefficient γ appearing in the Brownian limiting case ($m/m_g \to \infty$) when the molecular gas is at equilibrium. In fact, this limiting case is the situation considered when one employs a coarse-grained approach [1–4] to assess the impact of the interstitial gas on the dynamics properties of grains. In this limiting case, the expression (35) of the collisional moment C_{ij} reduces to

$$C_{ij}^{\rm Br} = -\frac{8\pi^{(d-1)/2}}{d\Gamma\left(\frac{d}{2}\right)} nT_g n_g\overline{\sigma}^{d-1}\left(\frac{m_g}{m}\right)^{1/2}\sqrt{\frac{2T_g}{m}}(P_{k\ell}^*-\delta_{k\ell}), \tag{43}$$

where $P_{ij}^* = P_{ij}/(nT_g)$, and we take into account that in the Brownian limit $\mu_g \to m_g/m$ and $(1+\theta)/\theta \to 1$ in the expression (42) of $\nu_0^{\rm M}$. The form of C_{ij} derived in Ref. [26] by replacing the Boltzmann–Lorentz collisional operator (21) by the Fokker–Planck operator

$$\gamma\frac{\partial}{\partial \mathbf{V}}\cdot\mathbf{V}f + \gamma\frac{T_g}{m}\frac{\partial^2 f}{\partial V^2} \tag{44}$$

is

$$C_{ij} = -2\gamma n T_g (P^*_{k\ell} - \delta_{k\ell}). \tag{45}$$

Comparison between Equations (43) and (45) allows us to identify γ for IMM as

$$\gamma = \frac{4\pi^{(d-1)/2}}{d\Gamma\left(\frac{d}{2}\right)} n_g \overline{\sigma}^{d-1} \left(\frac{m_g}{m}\right)^{1/2} \left(\frac{2T_g}{m}\right)^{1/2}. \tag{46}$$

The expression (46) for the friction coefficient γ for IMM is the same as the one obtained for IHS [15,17].

We are now in a position to determine the nonzero elements of the (reduced) pressure tensor P^*_{ij}. From Equations (33) and (46), one obtains the equation

$$\begin{aligned} a^*_{ik} P^*_{kj} + a^*_{jk} P^*_{ki} + 2\xi^* P^*_{ij} &= -\nu^{*\mathrm{M}} \left[\nu^*_\eta P^*_{ij} + \chi(\zeta^* - \nu^*_\eta)\delta_{ij}\right] - 2\mu\left(\frac{1+\theta}{\theta}\right)^{1/2} \\ &\quad \times \left(X\delta_{ij} + Y P^*_{ij} + Z P^*_{g,ij}\right), \end{aligned} \tag{47}$$

where $a^*_{ij} = a_{ij}/\gamma$, $\xi^* = \xi/\gamma$,

$$\nu^{*\mathrm{M}} = \frac{\nu^{\mathrm{M}}}{\gamma} = \frac{2^{d+1} d}{\sqrt{\pi}} \phi \sqrt{\chi T^*_g}, \tag{48}$$

and

$$X = -\frac{d}{d+2}\mu\left(\frac{1+\theta}{\theta}\right), \quad Y = 1 - \frac{2\mu_g}{d+2}, \quad Z = -\frac{2\mu}{d+2}. \tag{49}$$

Here,

$$\phi = \frac{\pi^{d/2}}{2^{d-1} d\Gamma\left(\frac{d}{2}\right)} n\sigma^d \tag{50}$$

is the solid volume fraction of the granular gas, $T^*_g = T_g/m\sigma^2\gamma^2$, and upon deriving Equation (48), use is made of the identity

$$\frac{n\sigma^{d-1}}{n_g \overline{\sigma}^{d-1}} = \frac{2^{d+\frac{3}{2}}}{\sqrt{\pi}} \left(\frac{m_g}{m}\right)^{1/2} \phi \sqrt{T^*_g}. \tag{51}$$

As occurs for the rheology of the molecular gas, Equation (47) shows that the diagonal elements of the pressure tensor P^*_{ij} orthogonal to the shear plane xy are equal to P^*_{yy} (i.e., $P^*_{yy} = P^*_{zz} = \cdots = P^*_{dd}$). This implies that the xx element is given by $P^*_{xx} = d\chi - (d-1)P^*_{yy}$. The yy and xy elements of the (reduced) pressure tensor can be written as

$$P^*_{yy} = \frac{\Omega_{yy}}{\nu_{yy}}, \quad P^*_{xy} = \frac{\Omega_{xy} - a^* P^*_{yy}}{\nu_{yy}}, \tag{52}$$

where

$$\nu_{yy} = 2\xi^* + \nu^{*\mathrm{M}} \nu^*_\eta + 2\mu\left(\frac{1+\theta}{\theta}\right)^{1/2} Y, \tag{53}$$

$$\Omega_{yy} = -\nu^{*\mathrm{M}} \chi\left(\zeta^* - \nu^*_\eta\right) - 2\mu\left(\frac{1+\theta}{\theta}\right)^{1/2} \left(X + Z P^*_{g,yy}\right), \tag{54}$$

$$\Omega_{xy} = -2\mu\left(\frac{1+\theta}{\theta}\right)^{1/2} Z P^*_{g,xy}. \tag{55}$$

Note that the elements of the pressure tensor $P^*_{g,yy}$ and $P^*_{g,xy}$ of the molecular gas must be expressed in terms of the (reduced) shear rate a^* and the (reduced) thermostat parameter

ξ^*. For this, one has to take into account the relationships between $\widetilde{a}$ and $\widetilde{\xi}$ with a^* and ξ^*, respectively. They are given by $\widetilde{a} = (\gamma/\widetilde{\nu}_g^{\rm M})a^*$ and $\widetilde{\xi} = (\gamma/\widetilde{\nu}_g^{\rm M})\xi^*$, where

$$\frac{\gamma}{\widetilde{\nu}_g^{\rm M}} = \frac{d+2}{\sqrt{2}d}\left(\frac{\overline{\sigma}}{\sigma_g}\right)^{d-1}\frac{m_g}{m}. \tag{56}$$

The equation defining the temperature ratio χ can be easily derived from Equation (47) as

$$\frac{2}{d}a^* P_{xy}^* + 2\xi^*\chi = -\nu^{*\rm M}\chi\zeta^* + 2\mu^2\left(\frac{1+\theta}{\theta}\right)^{1/2}(1-\chi). \tag{57}$$

From Equations (52) and (57), one finally obtains a^* in terms of the parameter space of the system:

$$a^* = \sqrt{\frac{d}{2}\frac{2\mu^2(1+\theta^{-1})^{1/2}(1-\chi) - (\nu^{*\rm M}\zeta^* + 2\xi^*)\chi}{\frac{\Omega_{xy}/a^*}{\nu_{yy}} - \frac{\Omega_{yy}}{\nu_{yy}^2}}}. \tag{58}$$

As happens in the case of IHS [17], the temperature ratio χ cannot be expressed in Equation (58) as an explicit function of the (reduced) shear rate and the remaining parameters of the system. On the other hand, for given values of the parameter space $\Xi \equiv \left(\chi, \alpha, \sigma/\sigma_g, m/m_g, \phi, T_g^*\right)$, the temperature ratio can be implicitly determined from the physical solution to Equation (58).

3.3. Brownian Limit

Before illustrating the shear-rate dependence of the rheological properties of the molecular gas for arbitrary values of the mass ratio m/m_g, it is convenient to check the consistency of the present results with those derived in Ref. [26] for IMM by using the Fokker–Planck operator (44). This consistency is expected to apply in the Brownian limit $m/m_g \to \infty$. In this limiting case, at a given value of the (reduced) shear rate a^*, $\theta \to \infty$, $\gamma/\nu_g^{\rm M} \propto m_g/m \to 0$, $\widetilde{a} \propto m_g/m \to 0$, $\widetilde{\xi} \propto \widetilde{a}^2 \propto (m_g/m)^2 \to 0$, and $\xi^* = \widetilde{\xi}(\widetilde{\nu}_g^{\rm M}/\gamma) \propto m_g/m \to 0$. Consequently, $P_{g,ij}^* = \delta_{ij}$ and the expressions of P_{yy}^*, P_{xy}^*, and a^* are

$$P_{yy}^* = \frac{2-\nu^{*\rm M}\chi\left(\zeta^* - \nu_\eta^*\right)}{2+\nu^{*\rm M}\nu_\eta^*}, \quad P_{xy}^* = -\frac{2-\nu^{*\rm M}\chi\left(\zeta^* - \nu_\eta^*\right)}{\left(2+\nu^{*\rm M}\nu_\eta^*\right)^2}a^*, \tag{59}$$

$$a^* = \sqrt{\frac{d}{2}\frac{\nu^{*\rm M}\zeta^* + 2(1-\chi^{-1})}{\nu^{*\rm M}\left(\nu_\eta^* - \zeta^*\right) + 2\chi^{-1}}}\left(2+\nu^{*\rm M}\nu_\eta^*\right). \tag{60}$$

Equations (59) and (60) are consistent with Equations (32), (33), and (35) of Ref. [26]. It is important to note that, to assess consistency with the Fokker–Planck results, the size ratio has been kept constant or proportional to the mass ratio so that $\xi^* \to 0$.

4. Rheology from a BGK-Type Kinetic Model of the Boltzmann Equation

We consider in this section the results derived for the USF from a BGK-type kinetic model of the Boltzmann equation [24,33]. In the problem for the granular suspension considered here, one has to replace the true Boltzmann operators $J_g[f_g, f_g]$, $J[f, f]$, and $J_{\rm BL}[f, f_g]$ by simpler relaxation terms that retain the relevant physical properties of those operators but are more tractable than the true kinetic equations. As in the case of IMM, let us separately determine the rheological properties of the molecular and granular gases by starting from these kinetic models.

4.1. Rheological Properties of the Molecular Gas

In the case of the molecular gas, the Boltzmann collision operator $J_g[f_g, f_g]$ is replaced by the conventional BGK kinetic model [22,34]:

$$J_g[f_g, f_g] \to -\nu_g\left(f_g - f_g^{\rm M}\right), \tag{61}$$

where ν_g is an effective velocity-independent collision frequency and $f_g^{\rm M}$ is the Maxwellian distribution

$$f_g^{\rm M}(\mathbf{V}) = n_g\left(\frac{m_g}{2\pi T_g}\right)^{d/2} \exp\left(-\frac{m_g V^2}{2T_g}\right). \tag{62}$$

Thus, according to Equation (9), the velocity distribution function $f_g(\mathbf{V})$ obeys the BGK kinetic equation

$$-aV_y\frac{\partial f_g}{\partial V_x} - \xi\frac{\partial}{\partial \mathbf{V}}\cdot \mathbf{V} f_g = -\nu_g\left(f_g - f_g^{\rm M}\right). \tag{63}$$

The nonzero elements of the pressure tensor $P_{g,ij}$ can be easily obtained from Equation (63) by multiplying both sides of this equation by $m_g V_i V_j$ and integrating over $\mathbf{V}$. The BGK expressions of the (reduced) elements of the pressure tensor $P_{g,ij}^*$ are given by Equations (27) with the replacement $\widetilde{\nu}_g^{\rm M} \to \nu_g$. As a consequence, the results derived from the BGK equation for the rheological properties agree with those obtained from the Boltzmann equation for hard spheres when

$$\nu_g = \widetilde{\nu}_g^{\rm M} = \frac{8\pi^{(d-1)/2}}{(d+2)\Gamma\left(\frac{d}{2}\right)} n_g \sigma_g^{d-1}\sqrt{\frac{T_g}{m_g}}. \tag{64}$$

4.2. Rheological Properties of the Granular Gas

In the case of the granular gas, we consider the kinetic model proposed by Vega Reyes et al. [24] for granular mixtures. This kinetic model is based on the equivalence between a system of elastic spheres subject to a drag force proportional to the peculiar velocity $\mathbf{V}$ with a gas of IHS [35]. Thus, the true inelastic Boltzmann collision operator is replaced by a relaxation term plus a drag force term. This (approximate) mapping between elastic hard spheres plus drag force with IHS allows us to extend known kinetic models of molecular gases to inelastic gases. Here, we consider the well-known Gross and Krook (GK) model for molecular (elastic) mixtures [36] for the corresponding relaxation terms. With this approach, the collision operators $J[f, f]$ and $J_{\rm BL}[f, f_g]$ are defined, respectively, as

$$J[f, f] \to -\nu'(\alpha)\left(f - f^{\rm M}\right) + \frac{\epsilon}{2}\frac{\partial}{\partial \mathbf{V}}\cdot \mathbf{V} f, \tag{65}$$

$$J_{\rm BL}[f, f_g] \to -\nu\left(f - \widetilde{f}_g\right). \tag{66}$$

In Equations (65) and (66),

$$f^{\rm M}(\mathbf{V}) = n\left(\frac{m}{2\pi T}\right)^{d/2} \exp\left(-\frac{mV^2}{2T}\right) \tag{67}$$

is the Maxwellian distribution of the granular gas,

$$\epsilon = \frac{2\pi^{(d-1)/2}}{d\Gamma\left(\frac{d}{2}\right)} n\sigma^{d-1}\sqrt{\frac{T}{m}}(1-\alpha^2), \tag{68}$$

and the reference distribution function $\widetilde{f}_g(\mathbf{V})$ is [36]

$$\widetilde{f}_g(\mathbf{V}) = n\left(\frac{m}{2\pi\widetilde{T}}\right)^{d/2} \exp\left(-\frac{mV^2}{2\widetilde{T}}\right). \tag{69}$$

Note that in the definition (68) of ϵ, the cooling rate ζ was evaluated by using the Maxwellian approximation for the distribution f. In this case, ζ is given by Equation (40).

In Equations (65) and (66), the quantities ν', ν, and $\widetilde{T}$ are chosen to optimize the agreement with some properties of interest of the Boltzmann equation for IHS. The usual method of obtaining the above parameters is to ensure that the kinetic model reproduces the collisional transfer equations of momentum and energy for elastic collisions ($\alpha = 1$). However, since $\mathbf{U} = \mathbf{U}_g$ in the USF, we only have one constraint (the one associated with the transfer of energy) instead of two, so that $\widetilde{T}$ and ν admit more than one form. Here, we propose a choice (see Appendix A for more technical details) that leads to an excellent agreement with the results obtained for IHS from Grad's moment method [17]. More specifically, we take the following values of $\widetilde{T}$ and ν:

$$\widetilde{T} = T_g, \quad \nu = \frac{8\pi^{(d-1)/2}}{d\Gamma\left(\frac{d}{2}\right)} n_g \overline{\sigma}^{d-1} \frac{m m_g}{(m+m_g)^2} \left(\frac{2T_g}{m_g} + \frac{2T}{m}\right)^{1/2}. \tag{70}$$

It still remains to completely define the model to chose the effective collision frequency ν'. It is defined here to reproduce the collisional moment

$$\int d\mathbf{v}\, mV_iV_jJ[f,f] \tag{71}$$

of the Boltzmann equation for IHS when one takes Grad's trial distribution for f [17]. This leads to the expression

$$\nu' = \frac{2\pi^{(d-1)/2}}{d(d+2)\Gamma\left(\frac{d}{2}\right)} n\sigma^{d-1} \sqrt{\frac{T}{m}} (1+\alpha)[d+1+(d-1)\alpha]. \tag{72}$$

Therefore, the BGK kinetic equation for the sheared granular gas is given by

$$-aV_y\frac{\partial f}{\partial V_x} - \xi\frac{\partial}{\partial \mathbf{V}}\cdot \mathbf{V}f = -\nu'\left(f - f^{\mathrm{M}}\right) + \frac{\epsilon}{2}\frac{\partial}{\partial \mathbf{V}}\cdot \mathbf{V}f - \nu\left(f - \widetilde{f}_g\right), \tag{73}$$

where ν and ν' are defined by Equations (70) and (72), respectively, and the Maxwellian distribution $\widetilde{f}_g$ is given by Equation (69) with $\widetilde{T} = T_g$.

The possibility of determining all the velocity moments of the distribution function is likely one of the main advantages of employing a kinetic model instead of the true Boltzmann equation. In the USF problem, it is convenient to define the general velocity moments

$$M_{k_1,k_2,k_3} = \int d\mathbf{V} V_x^{k_1} V_y^{k_2} V_z^{k_3} f(\mathbf{V}). \tag{74}$$

Although, here, we are essentially interested in the three-dimensional case, we will compute the velocity moments for $d = 3$ and $d = 2$. Note that for hard disks ($d = 2$), $k_3 = 0$ in Equation (74), since the z-axis is meaningless. The hierarchy of moment equations can be obtained by multiplying Equation (73) by $V_x^{k_1} V_y^{k_2} V_z^{k_3}$ and integrating over $\mathbf{V}$. The result is

$$ak_1 M_{k_1-1,k_2+1,k_3} + (\nu' + \nu + k\lambda) M_{k_1,k_2,k_3} = N_{k_1,k_2,k_3}, \tag{75}$$

where $\lambda = \xi + \epsilon/2$, $k = k_1 + k_2 + k_3$, and

$$N_{k_1,k_2,k_3} = n\left(\frac{2T_g}{m}\right)^{k/2} \left(\nu + \chi^{k/2}\nu'\right) M^L_{k_1,k_2,k_3}. \tag{76}$$

In Equation (75), for hard spheres ($d=3$)

$$M^L_{k_1,k_2,k_3}=\int d\mathbf{c}\, c_x^{k_1}c_y^{k_2}c_z^{k_3}e^{-c^2}=\pi^{-3/2}\Gamma\left(\frac{k_1+1}{2}\right)\Gamma\left(\frac{k_2+1}{2}\right)\Gamma\left(\frac{k_3+1}{2}\right), \tag{77}$$

if k_1, k_2, and k_3 are even, being zero otherwise. For hard disks ($d=2$),

$$M^L_{k_1,k_2,0}=\pi^{-1}\Gamma\left(\frac{k_1+1}{2}\right)\Gamma\left(\frac{k_2+1}{2}\right) \tag{78}$$

if k_1 and k_2 are even, being zero otherwise.

The solution to Equation (75) can be written as (see Appendix B of Ref. [26] for some details)

$$M_{k_1,k_2,k_3}=\sum_{q=0}^{k_1}\frac{k_1!}{(k_1-q)!}\frac{(-a)^q}{(\nu'+\nu+k\lambda)^{1+q}}N_{k_1-q,k_2+q,k_3}. \tag{79}$$

The nonzero elements of the pressure tensor P_{ij} can be easily obtained from Equation (79). In dimensionless form, the BGK expressions of the elements of $P^*_{ij}=P_{ij}/nT_g$ are

$$P^*_{yy}=P^*_{zz}=\frac{\nu+\nu'\chi}{\nu'+\nu+2\lambda},\quad P^*_{xy}=-\frac{\nu+\nu'\chi}{(\nu'+\nu+2\lambda)^2}a, \tag{80}$$

$$P^*_{xx}=\frac{\nu+\nu'\chi}{\nu'+\nu+2\lambda}\left[1+\frac{2a^2}{(\nu'+\nu+2\lambda)^2}\right]. \tag{81}$$

The (steady) temperature ratio $\chi=T/T_g$ can be obtained from the constraint $P^*_{xx}+(d-1)P^*_{yy}=d\chi$. This yields the implicit equation

$$a^*=\left(\nu'^*+\nu^*+2\lambda^*\right)\sqrt{\frac{d}{2}\chi\frac{\nu^*(1-\chi^{-1})+2\lambda^*}{\nu^*+\nu'^*\chi}}, \tag{82}$$

where $a^*=a/\gamma$,

$$\nu'^*=\frac{\nu'}{\gamma}=\frac{2^d}{d+2}\phi\sqrt{\frac{\chi T^*_g}{\pi}}(1+\alpha)[d+1+(d-1)\alpha], \tag{83}$$

$$\nu^*=\frac{\nu}{\gamma}=2\mu^2\left(\frac{1+\theta}{\theta}\right)^{1/2},\quad \lambda^*=\frac{\lambda}{\gamma}=\xi^*+\frac{\epsilon^*}{2}, \tag{84}$$

$$\xi^*=\frac{\sqrt{2}d}{d+2}\left(\frac{\sigma_g}{\overline{\sigma}}\right)^{d-1}\frac{m}{m_g}\widetilde{\xi},\quad \epsilon^*=2^d\phi\sqrt{\frac{\chi T^*_g}{\pi}}(1-\alpha^2). \tag{85}$$

Here, $\widetilde{\xi}$ is given by Equation (30) and we recall that $T^*_g=T_g/m\sigma^2\gamma^2$.

4.3. Brownian Limit

As in the case of IMM, it is quite interesting to consider the Brownian limiting case $m/m_g\to\infty$. In this case, $P^*_{g,ij}=\delta_{ij}$, $\mu\to 1$, $\theta\to\infty$, and $\xi^*\to 0$. Thus, following similar steps to those made for IMM, one obtains the expressions

$$P^*_{yy}=\frac{2+\nu'^*\chi}{2+\nu'^*+\epsilon^*},\quad P^*_{xy}=-\frac{2+\nu'^*\chi}{(2+\nu'^*+\epsilon^*)^2}a^*, \tag{86}$$

$$a^*=\left(2+\nu'^*+\epsilon^*\right)\sqrt{\frac{d}{2}\chi\frac{2(1-\chi^{-1})+\epsilon^*}{2+\nu'^*\chi}}. \tag{87}$$

4.4. Velocity Distribution of the Granular Gas

Apart from obtaining all the velocity moments, the use of kinetic models allows us, in some cases, to explicitly determine the velocity distribution function $f(\mathbf{V})$. The BGK-type Equation (73) reads

$$\left(1-d\widehat{\lambda}-\widehat{a}V_y\frac{\partial}{\partial V_x}-\widehat{\lambda}\mathbf{V}\cdot\frac{\partial}{\partial\mathbf{V}}\right)f(\mathbf{V})=f_{\rm R}(\mathbf{V}), \tag{88}$$

where $\widehat{a}=a/(\nu'+\nu)$, $\widehat{\lambda}=\lambda/(\nu'+\nu)$, and

$$f_{\rm R}(\mathbf{V})=\frac{\nu'}{\nu'+\nu}f^{\rm M}(\mathbf{V})+\frac{\nu}{\nu'+\nu}\widetilde{f}_g(\mathbf{V}). \tag{89}$$

The hydrodynamic solution to Equation (88) can be formally written as

$$\begin{aligned} f(\mathbf{V}) &= \left(1-d\widehat{\lambda}-\widehat{a}V_y\frac{\partial}{\partial V_x}-\widehat{\lambda}\mathbf{V}\cdot\frac{\partial}{\partial\mathbf{V}}\right)^{-1}f_{\rm R}(\mathbf{V}) \\ &= \int_0^\infty ds\, e^{-(1-d\widehat{\lambda})s}e^{\widehat{a}sV_y\frac{\partial}{\partial V_x}}e^{\widehat{\lambda}s\mathbf{V}\cdot\frac{\partial}{\partial\mathbf{V}}}f_{\rm R}(\mathbf{V}). \end{aligned} \tag{90}$$

The action of the velocity operators in Equation (90) on an arbitrary function $g(\mathbf{V})$ is [26]

$$e^{\widehat{a}sV_y\frac{\partial}{\partial V_x}}g(V_x,V_y,V_z)=g(V_x+\widehat{a}sV_y,V_y,V_z), \tag{91}$$

$$e^{\widehat{\lambda}s\mathbf{V}\cdot\frac{\partial}{\partial\mathbf{V}}}g(V_x,V_y,V_z)=g\left(e^{\widehat{\lambda}s}V_x,e^{\widehat{\lambda}s}V_y,e^{\widehat{\lambda}s}V_z\right). \tag{92}$$

The explicit form of the one-particle velocity distribution function can be explicitly obtained when one takes into account in Equation (90) the action of the velocity operators given by Equations (91) and (92). The result can be written as

$$f(\mathbf{V})=n\left(\frac{m}{2T}\right)^{d/2}\varphi(\mathbf{c}),\quad \mathbf{c}=\left(\frac{m}{2T}\right)^{1/2}\mathbf{V}, \tag{93}$$

where

$$\begin{aligned} \varphi(\mathbf{c}) &= \pi^{-d/2}\int_0^\infty ds\, e^{-(1-d\widehat{\lambda})s}\left\{\frac{\nu'}{\nu'+\nu}\exp\left[-e^{2\widehat{\lambda}s}\left(\mathbf{c}+s\widehat{\mathsf{a}}\cdot\mathbf{c}\right)^2\right]\right. \\ &\left.+\frac{\nu}{\nu'+\nu}\chi^{d/2}\exp\left[-\chi e^{2\widehat{\lambda}s}\left(\mathbf{c}+s\widehat{\mathsf{a}}\cdot\mathbf{c}\right)^2\right]\right\}. \end{aligned} \tag{94}$$

Here, we introduce the tensor $\widehat{a}_{ij}=\widehat{a}\delta_{ix}\delta_{jy}$.

In order to illustrate the shear-rate dependence of the distribution function, it is convenient to consider the marginal distribution for $d=3$:

$$\begin{aligned} \varphi_x(c_x) &= \int_{-\infty}^\infty dc_y\int_{-\infty}^\infty dc_z\,\varphi(\mathbf{c}) \\ &= \frac{1}{\sqrt{\pi}}\int_0^\infty ds\,\frac{e^{-(1-\widehat{\lambda})s}}{\sqrt{1+\widehat{a}^2s^2}}\left\{\frac{\nu'}{\nu'+\nu}\exp\left(-e^{2\widehat{\lambda}s}\frac{c_x^2}{1+\widehat{a}^2s^2}\right)\right. \\ &\left.+\frac{\nu}{\nu'+\nu}\chi^{1/2}\exp\left(-\chi e^{2\widehat{\lambda}s}\frac{c_x^2}{1+\widehat{a}^2s^2}\right)\right\}. \end{aligned} \tag{95}$$

In the Brownian limit, $\xi^*\to 0$, $\mu\to 1$, $\theta\to\infty$, and so $\nu^*\to 2$, $\lambda^*\to\epsilon^*/2$, and

$$\widehat{\lambda}\to\frac{\epsilon^*/2}{2+\nu'^*},\quad \widehat{a}\to\frac{a^*}{2+\nu'^*}. \tag{96}$$

Thus, when $m/m_g \to \infty$, Equation (95) becomes

$$\varphi_x(c_x) = \frac{1}{\sqrt{\pi}} \int_0^\infty ds \, \frac{e^{-(1-\widehat{\lambda})s}}{\sqrt{1+\widehat{a}^2 s^2}} \left\{ \frac{\nu'^*}{2+\nu'^*} \exp\left(-e^{2\widehat{\lambda}s} \frac{c_x^2}{1+\widehat{a}^2 s^2} \right) + \frac{2}{2+\nu'^*} \chi^{1/2} \exp\left(-\chi e^{2\widehat{\lambda}s} \frac{c_x^2}{1+\widehat{a}^2 s^2} \right) \right\}, \tag{97}$$

where $\widehat{\lambda}$ and $\widehat{a}$ are given by Equations (96).

5. Comparison between IMM and BGK Results

In Sections 3 and 4, we made use of the Boltzmann equation for IMM and the BGK-type kinetic model to investigate the shear-rate dependence of rheological properties in a sheared granular suspension. These properties are expressed in terms of the coefficient of restitution α, the reduced background temperature T_g^*, and the diameter σ/σ_g and mass m/m_g ratios. Additionally, there exists a residual dependence on density through the volume fraction ϕ. To avoid that, one could, for instance, reduce the shear rate using the effective collision frequencies $\nu^{\rm M}(T)$ or $\nu(T)$. However, for consistency with simulations and considering the background temperature T_g as a known quantity, we opted to employ $\gamma(T_g)$ as the reference frequency.

Given that in this section the second-degree moments of the distribution function are compared with molecular dynamics (MD) simulations for IHS in the Brownian limiting case [18], we set fixed values of $T_g^* = 1$ and $\phi = 0.0052$ for subsequent analysis. The selection of T_g^* as a free parameter imposes a constraint between the diameter σ/σ_g and mass m/m_g ratios [17]:

$$\frac{\sigma}{\sigma_g} = \left[\left(\frac{\sqrt{\pi}}{4\sqrt{2}} \frac{n}{n_g} \sqrt{\frac{m}{m_g}} \frac{1}{\phi \sqrt{T_g^*}} \right)^{1/(d-1)} - 1 \right]^{-1}. \tag{98}$$

This relation ensures convergence of results to those obtained via the Fokker–Planck equation as $m/m_g \to \infty$, since $\xi^* \to 0$. Furthermore, since we want to recover the results obtained in Ref. [17] derived from Grad's method, we take $n/n_g = 10^{-3}$ (rarefied granular gas).

The second-degree moments expressed through the reduced temperature χ, non-Newtonian shear viscosity η^*, and the normal stress difference Ψ^* are plotted in Figure 2 for $\alpha = 0.9$ and 1. Here, $\Psi^* = P_{xx}^* - P_{yy}^* = d\chi - dP_{yy}^*$. Equations (52) and (80) provide analytical expressions for IMM and BGK-type kinetic model, respectively, from which rheological properties are illustrated. Notably, there is nearly perfect agreement between Grad's solution for IHS, as obtained in Ref. [17], and both IMM and BGK-type results for any mass ratio, highlighting the ability of relatively simple models to capture essential properties of granular suspensions.

In particular, a DST transition characterized by S-shaped curves becomes more pronounced as the mass ratio m/m_g increases. Specifically, the non-Newtonian shear viscosity η^* exhibits a discontinuous transition (at a certain value of a^*) which intensifies as the particles of the granular gas become heavier. The theoretical results are validated with MD simulations [18] in the Brownian limiting case ($m/m_g \to 0$), showing generally good agreement despite slight discrepancies in the transition zone. Simulations suggest a sharper transition, likely due to the absence of molecular chaos in highly nonequilibrium situations. To address this, DSMC simulations for IHS are performed in the same limit, showing good agreement with theoretical results and further emphasizing a more pronounced transition. This phenomenon is likely attributable to a sudden growth of higher-order moments, resulting in a proportional increase in the deviation from theoretical predictions. As a result, the lower moments are also affected. Some technical details of the application of the DSMC method are available in the supplementary material of Ref. [15].

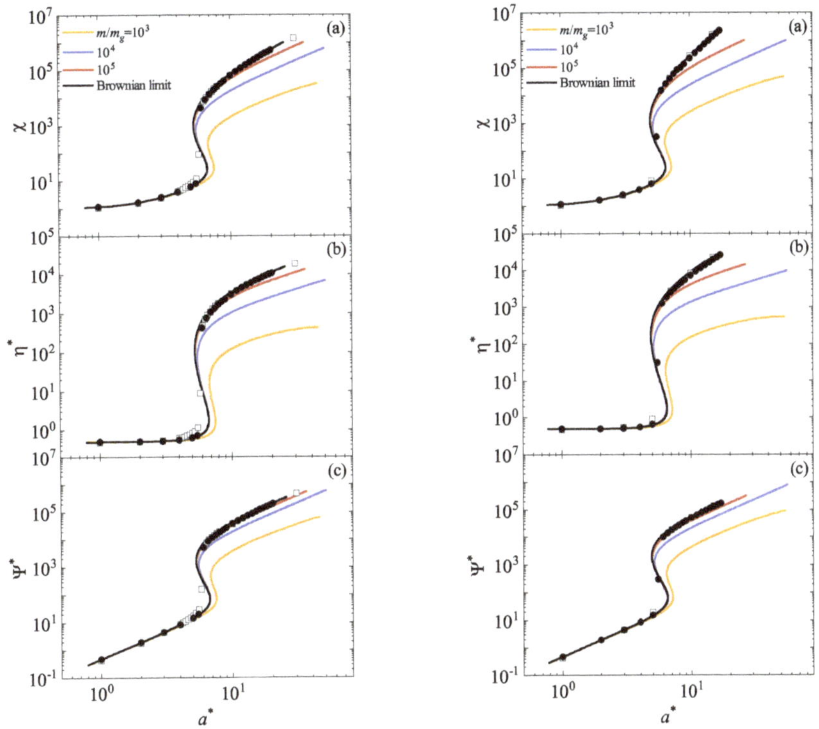

Figure 2. Plots of the (steady) granular temperature χ (**a**), the non-Newtonian shear viscosity η^* (**b**), and the normal stress difference Ψ^* (**c**) as functions of the (reduced) shear rate a^* for two different values of the coefficient of restitution α: 0.9 (**left panel**) and 1 (**right panel**). The graphs represent four distinct mass ratio values m/m_g: 10^3 (yellow lines), 10^4 (blue lines), 10^6 (red lines), and the Brownian limit (black lines). Here, $T_g^* = 1$, $d = 3$, and $\phi = 0.0052$. The solid lines correspond to the IMM results, the dashed lines are the BGK-like results, and the dotted lines refer to Grad's solution for IHS. Symbols denote computer simulation results performed in the Brownian limit: circles refer to the DSMC data obtained in this paper for IHS, while squares are MD results obtained in Ref. [18] for IHS.

The simplicity of the BGK and IMM models enables exploration beyond second-degree moments. Accordingly, we utilize the BGK-type kinetic equation to compute the fourth-degree moments. Although similar calculations could be performed in the case of IMM, we opt to omit them due to their extensive analytical effort. Additionally, drawing insights from the Fokker–Planck model [26] and dry granular gases, we anticipate potential divergences of the moments derived from IMM under certain shear rate conditions. We focus our efforts on calculating the following fourth-degree moments:

$$M_{4|0}^* = \frac{m^2}{nT_g^2} \int d\mathbf{V} V^4 f(\mathbf{V}), \tag{99}$$

$$M_{2|xy}^* = \frac{m^2}{nT_g^2} \int d\mathbf{V} V^2 V_x V_y f(\mathbf{V}). \tag{100}$$

Thus, in terms of the generic moments M_{k_1,k_2,k_3} and according to the expression (79), the moments $M^*_{4|0}$ and $M^*_{2|xy}$ are given by

$$M^*_{4|0} = \frac{m^2}{nT_g^2}(M_{4,0,0} + M_{0,4,0} + M_{0,0,4} + 2M_{2,2,0} + 2M_{2,0,2} + 2M_{0,2,2}), \tag{101}$$

$$M^*_{2|xy} = \frac{m^2}{nT_g^2}(M_{3,1,0} + M_{1,3,0} + M_{1,1,2}). \tag{102}$$

Figure 3a illustrates the ratio $M^*_{4|0}(a^*)/M^*_{4|0}(0)$ as a function of a^* for $\alpha = 0.9$ and 1 and three values of the mass ratio. We observe that variations in the mass ratio do not significantly alter the trends observed in the Brownian limiting case [26]. An abrupt transition in the higher-order moments is evident within a small region of a^*. Specifically, the kurtosis $M^*_{4|0}$ increases with the mass ratio m/m_g until it converges to the value obtained in the Brownian limit. Consistent with the conclusions drawn in Ref. [17], an increase in the mass of the particles of the granular gas results in an elevation of the granular temperature. Consequently, energy nonequipartition accentuates and moves the suspension away from equilibrium, leading to an increase in kurtosis as the distribution function deviates from its Maxwellian form. Regarding the influence of collisional dissipation, we observe that the effect of α on $M^*_{4|0}$ remains relatively discrete. Figure 3b illustrates the shear-rate dependence of the (reduced) moment $M^*_{2|xy}$. This moment vanishes in the absence of shear rate ($a^* = 0$). Similar conclusions to those made for the moment $M^*_{4|0}$ can be drawn.

Theoretical predictions for the fourth-degree moments are compared against DSMC simulations for IHS conducted in this paper in the Brownian limiting case. A qualitative agreement is observed, although simulations suggest a sharper transition. Some quantitative discrepancies are noticeable, which are mainly disguised by the scale. To assess the reliability of the BGK-type results, we focus on the region $0 < a^* < 1$, where all the fourth-degree velocity moments of IMM are well-defined functions of the shear rate. In addition, non-Newtonian effects are still significant in the range of values of the (reduced) shear rate $a^* \leq 1$. To this purpose, Figure 4 shows the (reduced) fourth-degree moments $M^*_{4|0}(a^*)/M^*_{4|0}(0)$ and $M^*_{2|xy}(a^*)$ for $\alpha = 0.7$ and 1. These moments are also illustrated as obtained for IMM in the Brownian limit [26]. It is worth noting that the results derived in Ref. [26] stem from considering an effective force modeling the interstitial gas, diverging from the limit of a Boltzmann–Lorentz operator modeled by a BGK-type equation as $m/m_g \to \infty$. Consequently, since DSMC simulations employ the exact Fokker–Planck operator, they perfectly align with the IMM results, while discrepancies emerge when compared with the BGK-type results. It is noteworthy that the BGK-type model slightly overestimates the deviation from the Newtonian situation ($a^* = 0$), a phenomenon also observed for molecular gases [37]. Moreover, non-Newtonian effects become apparent even at low values of a^*.

Finally, in Figure 5, the ratio $R_x(c_x) = \varphi_x(c_x)/(\pi^{-1/2}e^{-c_x^2})$ is plotted for $a^* = 1$ and four different values of the mass ratio. Here, the marginal distribution $\varphi_x(c_x)$ is given by Equation (95). It is evident that the deviation from equilibrium ($R_x \neq 1$) becomes more significant as the mass ratio m/m_g increases. Moreover, a comparison between theory and DSMC simulations reveals some disagreement in the BGK-type solution. Although the relative difference of these discrepancies is relatively small (it is about 8%), this contradicts what was observed in Ref. [9], where good agreement between the BGK solution and DSMC data was shown in the region of thermal velocities.

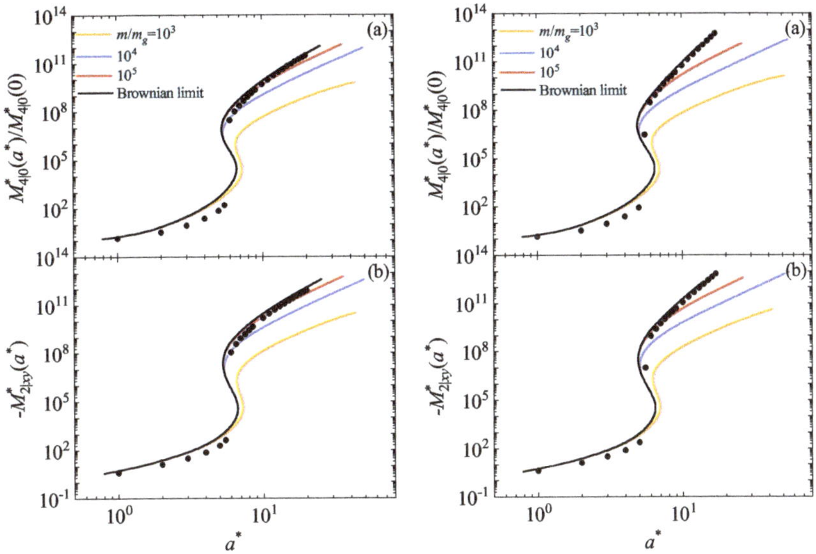

Figure 3. Plots of the (reduced) fourth-degree moments $M^*_{4|0}(a^*)/M^*_{4|0}(0)$ (**a**) and $-M^*_{2|xy}(a^*)$ (**b**) as functions of the (reduced) shear rate a^* obtained from the BGK-type equation for two different values of the coefficient of restitution α: 0.9 (**left panel**) and 1 (**right panel**). The graphs represent four distinct mass ratio values m/m_g: 10^3 (yellow lines), 10^4 (blue lines), 10^6 (red lines), and the Brownian limit (black lines). Here, $T^*_g = 1$, $d = 3$, and $\phi = 0.0052$. Symbols refer to the DSMC results obtained for IHS in this paper in the Brownian limit.

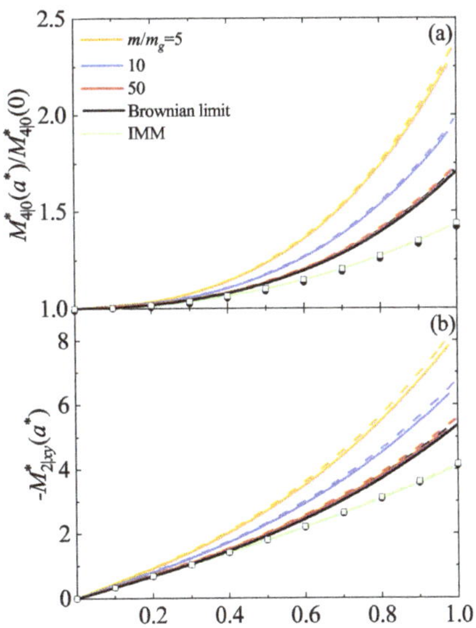

Figure 4. Plots of the (reduced) fourth-degree moments $M^*_{4|0}(a^*)/M^*_{4|0}(0)$ (**a**) and $-M^*_{2|xy}(a^*)$ (**b**) as functions of the (reduced) shear rate a^* obtained from the BGK-type equation for two different values of the coefficient of restitution α: 0.7 (solid lines) and 1 (dashed lines). The graphs represent four distinct mass ratio values m/m_g: 5 (yellow lines), 10 (blue lines), 50 (red lines), and the Brownian limit (black lines). Here, $T^*_g = 1$, $d = 3$, and $\phi = 0.0052$. Symbols refer to the DSMC results for IHS in the Brownian limit (squares for $\alpha = 1$ and circles for $\alpha = 0.7$). The green lines are the IMM results as obtained in Ref. [26] in the Brownian limit.

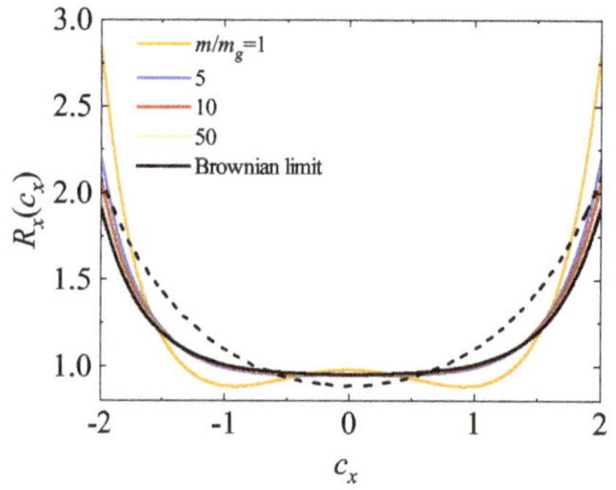

Figure 5. Plot of the ratio $R_x(c_x) = \varphi_x(c_x)/(\pi^{-1/2}e^{-c_x^2})$ for $a^* = 1$ as a function of the (reduced) velocity c_x for $\alpha = 0.9$ and five different values of the mass ratio m/m_g: 1 (yellow line), 5 (blue line), 10 (red line), 50 (green line), and the Brownian limit (black lines). Here, $T_g^* = 1$, $d = 3$, and $\phi = 0.0052$. The dashed line refers to the DSMC results for IHS in the Brownian limit.

6. Concluding Remarks

In our study, we explored the non-Newtonian transport properties of a dilute granular suspension subjected to USF using the Boltzmann kinetic equation. The particles are represented as d-dimensional hard spheres with mass m and diameter σ, immersed in an interstitial gas acting as a thermostat at temperature T_g. Various models for granular suspensions incorporate a gas–solid force to represent the influence of the external fluid. While some models consider only isolated body resistance via a linear drag law [5,6,8,10–12,38,39], others [13,40] include an additional Langevin-type stochastic term. In this paper, we consider a suspension model where the collisions between grains and particles of the interstitial (molecular) gas are taken into account. Thus, based on previous studies [14,41], we discretize the surrounding molecular gas, assigning individual particles with mass m_g and diameter σ_g, thereby accounting for *elastic* collisions between grains and background gas particles in the starting kinetic equation.

Under USF conditions, the system is characterized by constant density profiles n and n_g, uniform temperatures T and T_g, and a (common) flow velocity $U_x = U_{g,x} = ay$, where a denotes the shear rate. In agreement with previous investigations on uniform sheared suspensions, the mean flow velocity $\mathbf{U}$ is coupled to that of the gas phase $\mathbf{U}_g$. Consequently, the viscous heating term due to shear and the energy transferred by the grains from collisions with the molecular gas is compensated by the cooling terms derived from collisional dissipation, allowing the achievement of a steady state. A distinctive feature of the USF is that the one-particle velocity distribution function $f(\mathbf{r}, \mathbf{v})$ depends on space only through its dependence on the peculiar velocity $\mathbf{V} = \mathbf{v} - \mathbf{U}$. Consequently, the velocity distribution function becomes uniform in the Lagrangian reference frame, moving with $\mathbf{V}$. This means that $f(\mathbf{r}, \mathbf{v}) \equiv f(\mathbf{V})$. Based on symmetry considerations, the heat flux $\mathbf{q}$ vanishes, making the pressure tensor P the relevant flux. Therefore, to understand the intricate dynamics of granular suspensions under shear flow, it is imperative to focus on their non-Newtonian properties—derived from the pressure tensor. These include the (reduced) temperature $\chi = T/T_g$, the (reduced) nonlinear shear viscosity η^*, and the (reduced) normal stress difference Ψ^*.

Given that the most challenging aspect of dealing with the Boltzmann equation lies in the collision operator, it is reasonable to explore alternatives that render this operator more analytically tractable than in the case of IHS. Among the most sophisticated techniques in this regard is to consider the Boltzmann equation for IMM. As in the case of elastic collisions [20,22], the collision rate for IMM is independent of the relative velocity of the colliding particles. As a consequence, the collisional moments of degree k of the Boltzmann collisional operator can be expressed as a linear combination of velocity moments of degree less than or equal to k. To complement the results derived for IMM, we also considered in this paper the use of a BGK-type kinetic model where the true Boltzmann operator is replaced by a simple relaxation term. Here, we employed both approaches to compute the rheological properties of the sheared granular suspension. Thus, our objective was twofold. Firstly, we aimed to assess the reliability and compatibility of the proposed models with previous results [17] obtained for IHS using Grad's moment method. Additionally, DSMC simulations for IHS were performed as an alternative method to validate any potential discrepancies identified. Secondly, taking advantage of the capabilities provided by the BGK model, we endeavored to calculate the velocity distribution function and the higher-order moments that offer insights into its characteristics.

Before proceeding with the computation of rheological properties, it is necessary to understand the response of the molecular fluid to shear stress. This assessment was also conducted using both the Maxwell molecules and BGK-type kinetic model that were later used to model the granular gas. A novelty here is the application of a force (Gaussian thermostat) of the form $\mathbf{F} = -m\xi\mathbf{V}$ to compensate for the energy gained through viscous shear stresses. This force, by consistency, also applies to the granular gas, maintaining convergence to a steady state. As anticipated, the results agree well with those obtained through Grad's moment method [17] (see Equations (32) and (64)). Consequently, once the problem conditions (including the shear rate a) are defined, the molecular temperature T_g is determined, effectively serving as a thermostat for the granular gas.

After accurately describing the rheology of the molecular gas, we focused on modeling the granular gas. Using both the IMM and BGK-type model separately, we calculated the nonzero elements of the the pressure tensor. The knowledge of these elements allowed us to identify the relevant rheological properties of the granular suspension. As shown in Figure 2, these quantities are represented as functions of the coefficient of restitution α and the mass ratio m/m_g. In particular, we find that the theoretical results obtained from the Grad's method for IHS, IMM, and BGK-type model show remarkable agreement, with almost indistinguishable curves. This underlines the effectiveness of *structurally simple* models in capturing the complexities of sheared granular suspensions. We observe a DST-type transition starting at a certain value of a^*, which increases with the mass ratio m/m_g. Interestingly, similar to the MD simulations performed for IHS [18], the DSMC data suggest a more abrupt transition than predicted by theory. Given that the main divergences between Grad's (for IHS) and DSMC results arise from the form of the distribution function, the significance of investigating higher-order moments to assess the deviation of the distribution function from its Maxwellian reference is then justified.

Based on the previous literature where discrepancies in fourth-degree moments have been observed [26], and acknowledging the potential lengthiness of calculations, for the sake of simplicity, we decided to employ only the BGK-type model to compute higher-order moments. Specifically, we concentrated on the (symmetric) fourth-degree moments $M_{4|0}$ and $M_{2|xy}$. The shear-rate dependence of these moments is illustrated in Figure 3 for the same parameter values of α and m/m_g as those employed for the rheological quantities. Initially, we note that the fourth-degree moments also exhibit an abrupt transition at a value of a^* that increases with m/m_g until reaching the Brownian limit. We think that the DST behavior will also appear in all higher moments.

Furthermore, DSMC simulations in the Brownian limit qualitatively capture the profile of these fourth-degree moments, although some quantitative disparities are apparent. To ascertain the extent of these discrepancies, we narrowed our focus to the interval $0 < a^* < 1$, where non-Newtonian effects are apparent. Additionally, we included IMM results directly as obtained from an effective Fokker–Planck-type model [26]. Figure 4 illustrates that BGK results overestimate the deviation from the moments computed when no shear stress is applied compared to DSMC simulations and the results obtained using an effective force to model the interstitial gas. These disparities are also observed in the marginal distribution function φ_x.

The theoretical findings presented here motivate the comparison with computer simulations. Although the observed agreement in the Brownian limit is encouraging, there is scope to extend this agreement to scenarios with finite mass ratios. Our plan is to carry out simulations of this type in the near future, which we expect will further validate and improve our theoretical framework. In addition, we plan to extend our current findings to finite densities by exploring the Enskog kinetic equation, which will allow us to evaluate the involvement of density in the occurrence of these phenomena. Recent results [42] derived in the context of the Enskog equation by using the Fokker–Planck operator have shown that there is a transition from the discontinuous shear thickening (observed in dilute gases) to the continuous shear thickening for denser systems. We want to see if this behavior is also present for large but finite mass ratios. The above lines of research will be some of the main objectives of our upcoming projects.

Author Contributions: Conceptualization, R.G.G. and V.G.; software, validation, R.G.G. and V.G.; formal analysis, R.G.G. and V.G.; writing—original draft preparation, V.G.; writing—review and editing, R.G.G. and V.G. All authors have read and agreed to the published version of the manuscript.

Funding: The authors acknowledge financial support from Grant PID2020-112936GB-I00 funded by MCIN/AEI/10.13039/501100011033, and from Grant IB20079 funded by Junta de Extremadura (Spain) and by ERDF "A way of making Europe".

Institutional Review Board Statement: Not applicable.

Informed Consent Statement: Not applicable.

Data Availability Statement: The data that support the findings of this study are available from the corresponding author upon reasonable request.

Conflicts of Interest: The authors report no conflicts of interest.

Abbreviations

The following abbreviations are used in this manuscript:

IMM	Inelastic Maxwell model
IHS	Inelastic hard spheres
USF	Uniform shear flow
DST	Discontinuous shear thickening
DSMC	Direct simulation Monte Carlo
BGK	Bhatnagar–Gross–Krook kinetic model
GK	Gross–Krook kinetic model

Appendix A. Some Technical Details in the BGK-Type Kinetic Model

In this Appendix, we give some details on the determination of the parameters $\widetilde{T}$ and ν appearing in the relaxation term (66). To obtain them, we require the collisional transfer of energy of grains due to their elastic collisions with particles of the molecular gas to be the same as that obtained from the true Boltzmann–Lorentz collision operator. This implies that

$$\int d\mathbf{v}V^2 J_{\rm BL}[\mathbf{V}|f,f_g] = -\nu \int d\mathbf{v}V^2\Big[f(\mathbf{V}) - \widetilde{f}_g(\mathbf{V})\Big]. \tag{A1}$$

Given that the collisional moment involving the operator $J_{\rm BL}[f,f_g]$ cannot be exactly computed for IHS, one estimates this moment by replacing f and f_g with their Grad's solutions [19]. In this approximation, one achieves the results [17]

$$\int d\mathbf{v}V^2 J_{\rm BL}[\mathbf{V}|f,f_g] = -\frac{8\pi^{(d-1)/2}}{\Gamma\left(\frac{d}{2}\right)} n n_g \overline{\sigma}^{d-1}\frac{m_g}{(m+m_g)^2}\left(\frac{2T_g}{m_g}+\frac{2T}{m}\right)^{1/2}(T-T_g). \tag{A2}$$

Moreover,

$$\int d\mathbf{v}V^2\Big[f(\mathbf{V}) - \widetilde{f}_g(\mathbf{V})\Big] = -d\frac{n\nu}{m}\Big(T-\widetilde{T}\Big). \tag{A3}$$

From Equations (A2) and (A3), one obtains the identity

$$\frac{8\pi^{(d-1)/2}}{d\Gamma\left(\frac{d}{2}\right)} n_g \overline{\sigma}^{d-1} \frac{m m_g}{(m+m_g)^2} \left(\frac{2T_g}{m_g} + \frac{2T}{m}\right)^{1/2} (T - T_g) = \nu\left(T - \widetilde{T}\right). \tag{A4}$$

Equation (A4) allows us to make the identifications (70).

Finally, from Grad's moment method [19], the collisional moment (71) can be written as [17]

$$\int d\mathbf{v}\, m V_i V_j J[f,f] = -\nu^{\mathrm{M}} n T_g \left[\nu_\eta^{*\mathrm{IHS}} P_{ij}^* - \chi\left(\nu_\eta^{*\mathrm{IHS}} - \zeta^{*\mathrm{IHS}}\right)\delta_{ij}\right], \tag{A5}$$

where ν^{M} is defined in Equation (42) and

$$\nu_\eta^{*\mathrm{IHS}} = \frac{(1+\alpha)[3(1-\alpha)+2d]}{2d(d+2)}, \quad \zeta^{*\mathrm{IHS}} = \frac{1-\alpha^2}{2d}. \tag{A6}$$

The BGK-type kinetic model (65) yields the result

$$\int d\mathbf{v}\, m V_i V_j \left[-\nu'\left(f - f^{\mathrm{M}}\right) + \frac{\epsilon}{2}\frac{\partial}{\partial \mathbf{V}} \cdot \mathbf{V} f\right] = -n T_g \left[(\nu' + \epsilon) P_{ij}^* - \nu' \chi \delta_{ij}\right], \tag{A7}$$

where we recall that $P_{ij}^* = P_{ij}/nT_g$. Comparison between Equations (A5) and (A7) yields Equation (72) for $\nu'(\alpha)$.

References

1. Koch, D.L. Kinetic theory for a monodisperse gas-solid suspension. *Phys. Fluids A* **1990**, *2*, 1711–1722. [CrossRef]
2. Gidaspow, D. *Multiphase Flow and Fluidization*; Academic Press: Cambridge, MA, USA, 1994.
3. Fullmer, W.D.; Hrenya, C.M. The clustering instability in rapid granular and gas-solid flows. *Annu. Rev. Fluid Mech.* **2017**, *49*, 485–510. [CrossRef]
4. Jackson, R. *The Dynamics of Fluidized Particles*; Cambridge University Press: New York, NY, USA, 2000.
5. Tsao, H.K.; Koch, D.L. Simple shear flows of dilute gas-solid suspensions. *J. Fluid Mech.* **1995**, *296*, 211–245. [CrossRef]
6. Sangani, A.S.; Mo, G.; Tsao, H.K.; Koch, D.L. Simple shear flows of dense gas-solid suspensions at finite Stokes numbers. *J. Fluid Mech.* **1996**, *313*, 309–341. [CrossRef]
7. Wylie, J.J.; Zhang, Q.; Li, Y.; Hengyi, X. Driven inelastic-particle systems with drag. *Phys. Rev. E* **2009**, *79*, 031301. [CrossRef] [PubMed]
8. Heussinger, C. Shear thickening in granular suspensions: Interparticle friction and dynamically correlated clusters. *Phys. Rev. E* **2013**, *88*, 050201. [CrossRef]
9. Chamorro, M.G.; Vega Reyes, F.; Garzó, V. Non-Newtonian hydrodynamics for a dilute granular suspension under uniform shear flow. *Phys. Rev. E* **2015**, *92*, 052205. [CrossRef] [PubMed]
10. Saha, S.; Alam, M. Revisiting ignited-quenched transition and the non-Newtonian rheology of a sheared dilute gas-solid suspension. *J. Fluid Mech.* **2017**, *833*, 206–246. [CrossRef]
11. Alam, M.; Saha, S.; Gupta, R. Unified theory for a sheared gas-solid suspension: From rapid granular suspension to its small-Stokes-number limit. *J. Fluid Mech.* **2019**, *870*, 1175–1193. [CrossRef]
12. Saha, S.; Alam, M. Burnett-order constitutive relations, second moment anisotropy and co-existing states in sheared dense gas-solid suspensions. *J. Fluid Mech.* **2020**, *887*, A9. [CrossRef]
13. Garzó, V.; Tenneti, S.; Subramaniam, S.; Hrenya, C.M. Enskog kinetic theory for monodisperse gas-solid flows. *J. Fluid Mech.* **2012**, *712*, 129–168. [CrossRef]
14. Biben, T.; Martin, P.; Piasecki, J. Stationary state of thermostated inelastic hard spheres. *Physica. A* **2002**, *310*, 308–324. [CrossRef]
15. Gómez González, R.; Garzó, V. Kinetic theory of granular particles immersed in a molecular gas. *J. Fluid Mech.* **2022**, *943*, A9. [CrossRef]
16. Subramaniam, S. Multiphase flows: Rich physics, challenging theory, and big simulations. *Phys. Rev. Fluids* **2020**, *5*, 110520. [CrossRef]
17. Gómez González, R.; García Chamorro, M.; Garzó, V. Rheology of granular particles immersed in a molecular gas under uniform shear flow. *arXiv* **2023**, arXiv:2311.16717.
18. Hayakawa, H.; Takada, S. Kinetic theory of discontinuous rheological phase transition for a dilute inertial suspension. *Prog. Theor. Exp. Phys* **2019**, *2019*, 083J01. [CrossRef]
19. Grad, H. On the kinetic theory of rarefied gases. *Commun. Pure Appl. Math.* **1949**, *2*, 331–407. [CrossRef]
20. Ikenberry, E.; Truesdell, C. On the pressures and the flux of energy in a gas according to Maxwell's kinetic theory. *J. Rat. Mech. Anal.* **1956**, *5*, 1–54. [CrossRef]

21. Chapman, S.; Cowling, T.G. *The Mathematical Theory of Nonuniform Gases*; Cambridge University Press: Cambridge, UK, 1970.
22. Cercignani, C. *The Boltzmann Equation and Its Applications*; Springer: New York, NY, USA, 1988.
23. Ben-Naim, E.; Krapivsky, P.L. The Inelastic Maxwell Model. In *Granular Gas Dynamics*; Pöschel, T., Luding, S., Eds.; Springer: Berlin/Heidelberg, Germany, 2003; Volume 624, pp. 65–94.
24. Vega Reyes, F.; Garzó, V.; Santos, A. Granular mixtures modeled as elastic hard spheres subject to a drag force. *Phys. Rev. E* **2007**, *75*, 061306. [CrossRef]
25. Brown, E.; Jeager, H.M. Dynamic jamming point for shear thickening suspensions. *Rep. Prog. Phys.* **2014**, *77*, 046602. [CrossRef]
26. Gómez González, R.; Garzó, V. Simple shear flow in granular suspensiones: Inelastic Maxwell models and BGK-type kinetic model. *J. Stat. Mech.* **2019**, *2019*, 013206. [CrossRef]
27. Pelargonio, S.; Zaccone, A. Generalized Langevin equation with shear flow and its fluctuation-dissipation theorems derived from a Caldeira-Leggett Hamiltonian. *Phys. Rev. E* **2023**, *107*, 064102. [CrossRef]
28. Bird, G.A. *Molecular Gas Dynamics and the Direct Simulation Monte Carlo of Gas Flows*; Clarendon: Oxford, UK, 1994.
29. Evans, D.J.; Morriss, G.P. *Statistical Mechanics of Nonequilibrium Liquids*; Academic Press: London, UK, 1990.
30. Brilliantov, N.; Pöschel, T. *Kinetic Theory of Granular Gases*; Oxford University Press: Oxford, UK, 2004.
31. Garzó, V. *Granular Gaseous Flows*; Springer Nature: Cham, Switzerland, 2019.
32. Dufty, J.W.; Santos, A.; Brey, J.J.; Rodríguez, R.F. Model for nonequilibrium computer simulation methods. *Phys. Rev. A* **1986**, *33*, 459–466. [CrossRef] [PubMed]
33. Brey, J.J.; Dufty, J.W.; Santos, A. Kinetic models for granular flow. *J. Stat. Phys.* **1999**, *97*, 281–322. [CrossRef]
34. Bhatnagar, P.L.; Gross, E.P.; Krook, M. A model for collision processes in gases. I. Small amplitude processes in charged and neutral one-component system. *Phys. Rev.* **1954**, *94*, 511–525. [CrossRef]
35. Santos, A.; Astillero, A. System of elastic hard spheres which mimics the transport properties of a granular gas. *Phys. Rev. E* **2005**, *72*, 031308. [CrossRef] [PubMed]
36. Gross, E.P.; Krook, M. Model for collision processes in gases. Small amplitude oscillations of charged two-component systems. *Phys. Rev.* **1956**, *102*, 593–604. [CrossRef]
37. Garzó, V.; Santos, A. *Kinetic Theory of Gases in Shear Flows. Nonlinear Transport*; Springer: Dordrecht, The Netherlands, 2003.
38. Parmentier, J.F.; Simonin, O. Transition models from the quenched to ignited states for flows of inertial particles suspended in a simple sheared viscous fluid. *J. Fluid Mech.* **2012**, *711*, 147–160. [CrossRef]
39. Wang, T.; Grob, M.; Zippelius, A.; Sperl, M. Active microrhelogy of driven granular particles. *Phys. Rev. E* **2014**, *89*, 042209. [CrossRef]
40. Puglisi, A.; Loreto, V.; Marconi, U.M.B.; Petri, A.; Vulpiani, A. Clustering and non-Gaussian behavior in granular matter. *Phys. Rev. Lett.* **1998**, *81*, 3848–3851. [CrossRef]
41. Santos, A. Granular fluid thermostated by a bath of elastic hard spheres. *Phys. Rev. E* **2003**, *67*, 051101. [CrossRef] [PubMed]
42. Takada, S.; Hayakawa, H.; Santos, A.; Garzó, V. Enskog kinetic theory of rheology for a moderately dense inertial suspension. *Phys. Rev. E* **2020**, *102*, 022907. [CrossRef] [PubMed]

entropy

Article

Species Richness Net Primary Productivity and the Water Balance Problem

Allen G. Hunt [1,*], Muhammad Sahimi [2] and Erica A. Newman [3]

1 Department of Physics, Wright State University, Dayton, OH 45435, USA
2 Mork Family Department of Chemical Engineering and Materials Science, University of Southern California, Los Angeles, CA 90089, USA; moe@usc.edu
3 Department of Integrative Biology, University of Texas at Austin, Austin, TX 78712, USA; erica.newman@austin.utexas.edu
* Correspondence: allen.hunt@wright.edu

Abstract: Species energy theory suggests that, because of limitations on reproduction efficiency, a minimum density of plant individuals per viable species exists and that this minimum correlates the total number of plant individuals N with the number of species S. The simplest assumption is that the mean energy input per individual plant is independent of the number of individuals, making N, and thus S as well, proportional to the total energy input into the system. The primary energy input to a plant-dominated ecosystem is estimated as its Net Primary Productivity (*NPP*). Thus, species energy theory draws a direct correspondence from *NPP* to S. Although investigations have verified a strong connection between S and *NPP*, strong influences of other factors, such as topography, ecological processes such as competition, and historical contingencies, are also at play. The lack of a simple model of *NPP* expressed in terms of the principal climate variables, precipitation P, and potential evapotranspiration, *PET*, introduces unnecessary uncertainty to the understanding of species richness across scales. Recent research combines percolation theory with the principle of ecological optimality to derive an expression for *NPP*(P, *PET*). Consistent with assuming S is proportional to *NPP*, we show here that the new expression for *NPP*(P, *PET*) predicts the number of plant species S in an ecosystem as a function of P and *PET*. As already demonstrated elsewhere, the results are consistent with some additional variation due to non-climatic inputs. We suggest that it may be easier to infer specific deviations from species energy predictions with increased accuracy and generality of the prediction of *NPP*(P, *PET*).

Keywords: ecology; percolation theory; net primary productivity; species energy theory

1. Introduction

A long-standing problem in ecology is the quantification of the relative importance of various factors that influence species number, S, including energy input [1], competition [2,3], history [4–6], and heterogeneity [7]. This well-known problem has been discussed by scientists as early as von Humboldt, Darwin, and Wallace [8–10]. The relative importance of these factors may not be universal but instead may depend on latitude [11,12], spatial scale [7,13–16], or the particular kind of organism under consideration [13]. Here, the discussion will be confined to the terrestrial plant species, and in particular, trees, for which evidence points to the dominant role of species energy theory, at least for sufficiently large spatial scales [13,17]. The fundamental hypothesis of species energy theory is that the number of species is proportional to the energy input into an ecosystem that, particularly for ectotherms, such as reptiles, may be expressed directly in terms of solar energy [13]. However, for plants, it appears that the dominant input energy pathway that is relevant to species number is the process of photosynthesis [17]. Total photosynthesis is estimated by Net Primary Productivity, *NPP*, which is limited roughly equally by water and solar energy [17,18]. As it has also been pointed out that the effects of these limitations on *NPP* are most parsimoniously expressed through the variable evapotranspiration, *ET* [18], it has thus also been suggested (and verified) that among climatic variables, plant species richness is primarily explained by *ET* [13] or by $P + PET - PET^2$ (in both cases explaining 78% of variation) [13,17], where *PET* is potential evapotranspiration.

The purpose of this study is to determine whether or not new predictions of *NPP* based on percolation theory [19] can help to reduce what uncertainty remains in the role of species energy theory in species number *S* when *NPP* is represented as a function of climate variables. The basis of predicting plant ecosystem *NPP* from percolation theory is the success of predicting *NPP* ($R^2 = 0.97$) with one universal adjustable parameter across all climate regimes [19].

The theoretical expression for *NPP* is developed based on the scaling relations of percolation theory for vegetation growth and soil formation in terms of the fundamental pathways for water, that is, evapotranspiration and run-off, combined with an optimality hypothesis that plant ecosystems with the greatest *NPP*, subject to all relevant constraints (climatic, edaphic, and so on), will tend to dominate [19,20]. Based on its key inputs, we refer to this optimization as the percolation optimality hypothesis. Although expressed in terms of ecosystem productivity, there is an advantage conferred upon individuals from their ability to intercept solar energy and reproduce themselves, which then may explain community dominance patterns. Maximizing ecosystem productivity by increasing niche and functional diversity, insofar as this increase is also consistent with maximizing the number of species [21–23], has received recent support [24,25]. This support addresses the concern in [13] that maximizing *NPP* could be accomplished by one or two dominant species.

The percolation optimality hypothesis can predict several quantities of potential interest, including (1) the energy input into an ecosystem as a function of climate variables; (2) the number of plant species present in such an ecosystem; (3) by inference, the prediction that the number of species present is proportional to the number of individuals (i.e., species energy theory) [1,13], the More Individuals Hypothesis [26,27]; and (4) by inference, the prediction that the average metabolic rate per individual is constant across climate regimes, even when the average energy per species is not. It is therefore also broadly consistent with earlier work showing that maximizing plant productivity in an ecosystem is possible when plant diversity is maximized [24–32]. Even if species diversity does contribute to ecological niche diversity [21], this diversity need not universally translate to genetic diversity, however. Although further discussion of the linkages of such distinct measures of diversity is undoubtedly required in the field of ecology, these areas of ongoing research are not further addressed here.

Another potential application of the results presented in the present paper is the Maximum Entropy Theory of Ecology (METE). METE [28–32] exploits a minimum information concept (the maximization of information entropy) to derive distributions of species, individuals, and energy transformations in ecosystems. Although previous statistical models of ecological metrics have used Boltzmann entropy as a starting place, developments from Shipley et al. (2006) [33] moved the conversation in macroecology from a state of trying to more accurately measure and account for the variation of all climate and mechanistic variables across scales [14–16] to insights into how to use information entropy to better model complex ecosystems. METE similarly uses the more general Shannon entropy ($-\sum P_i \ln P_i$) to predict the distribution of microstates from macrostates in an ecosystem in a dynamic steady state. The details of the treatment are beyond the scope of the present paper, but important inputs to the METE framework include the total numbers of individuals and species, as well as the total metabolic energy expended by an ecosystem, which the current approach can deliver, provided a steady-state has been reached. An extension of METE proposes an ecological equation of state, which unifies the relationships of species richness, productivity, abundance, and biomass in ecosystems. This equation shows a positive relationship between productivity and species richness, as mediated by the biomass and the number of individuals in the area under consideration.

In the context of species richness, Currie [13] details the central role of evapotranspiration *ET* on *NPP*, and we therefore consider the relationship between *NPP* and *ET* in more depth. Rosenzweig [18] determined empirically that *NPP* was proportional to a power *s* of *ET*, whose value (1.69) was extracted from a global study across biomes. However, Hunt et al. (2024 and 2021) [19,34] demonstrated that the actual value should be related to the root mass fractal dimension. Assuming the roots follow fractal paths of least cumulative resistance, *NPP* should more accurately equal the relevant mass fractal dimension predicted by percolation theory, which quantifies the topology of such paths [35]. Since, according to [36–38], dominant rooting depths appear to be relatively shallow (less than half a meter, a meter, or 2 m, respectively), it was assumed that the relevant percolation mass fractal dimension should be its value in two dimensions (2D) [35], i.e., $91/48 = 1.896$ (obtained through exact calculations and exact scaling relationships). In order to predict the proper dependence of *NPP* (*ET*) on climate variables, *P* and *PET*, however, it becomes necessary to determine *ET*(*P*, *PET*), which is a problem known in hydrology as the water balance and viewed as a central problem of that field of study [39–41].

The water balance quantifies how much precipitation goes to evapotranspiration *ET* and how much runs off, equal to $P - ET$ [39–41]. Measurement of the water balance is often more easily obtained through measurement of the streamflow exiting a drainage basin than through direct measurement of *ET*; however, *ET* is not related conceptually to the specific scale of a drainage basin except in cases of significant relief [17]. We show how the water balance is constructed out of soil formation (see, e.g., [42]) and vegetation growth (e.g., [43]) rates and give an overview of how the proposed ecological optimality can be used to find the partitioning of surface water on the terrestrial Earth surface that leads to maximum productivity. The process is based on a scaling relationship for solute transport that describes soil depth evolution [42] and one for optimal paths that describes the distinction between root growth rates and root lateral extent [43]. The results have been shown to predict plant productivity as a function of climate variables [19] and are shown here to provide accurate predictions of species richness in terms of the same climate variables. It is also noteworthy that the result obtained for *ET* (*P*, *PET*) from maximizing ecosystem *NPP* is almost identical to one obtained from a dynamic process model that maximizes carbon profit [44].

Before continuing, two points should be mentioned. First, the formation of biomass from sugar involves a set of complex processes using sugar for energy (metabolism), which are not considered further here but help differentiate between carbon assimilation [19] and carbon profit [44]. Second, it is important to remember that the word evapotranspiration is constructed, in principle, from a sum of the effects of two rather different processes: evaporation of water from bare ground or plant surfaces and the process of transpiration by which water evaporated from the stomata of leaves, thereby drawing further water from the ground into and through the plant.

2. Theoretical Background

NPP is limited by multiple factors. In view of its dependence on the process of photosynthesis, sugar is constructed from water and carbon dioxide through the absorption of photons. Therefore, solar energy, atmospheric carbon, and water may all be limiting components, as well as soil nutrients. However, atmospheric carbon contents, though they change significantly over time, are essentially spatially uniform in comparison with the climatic inputs [17,18]. Spatial variability in the carbon cycle is critical at any particular time and is, thus, due in large part to the spatial variability of climate variables, precipitation *P*, and potential evapotranspiration, *PET* (a measure of solar energy). Because water drawn from the soil requires water evaporation from stomata, the indirect demand for plant-transpired water is higher than the direct demand from photosynthesis. However, an insufficient supply of either water or solar energy will suppress photosynthesis and, thus, plant productivity. It was proposed nearly 60 years ago [18] that both limiting components of the carbon cycle could be accounted for using a single variable, evapotranspiration, *ET*.

More recently, it has been proposed [19,34] that the soil depth should be a second input into the dependence of *NPP* on *ET*. The soil depth also relates to percolation theory.

Roots grow primarily within a thin soil layer, variously estimated as 0.5 m, 1 m, or 2 m [36–38]. Compared with typical drainage basins, or catchments, which are measured in kilometers or greater, this suggests that the root zone may, over a wide range of scales and conditions, be considered nearly two dimensional (2D). The 3D root zone has also been considered and is doubtless relevant at some scales [19]. In addition, for a variety of reasons, roots have been proposed to follow optimal paths [43]. In a model that treats such paths in the spirit of percolation theory, root masses occupying such paths have been shown [43] to form self-similar fractal structures with a fractal dimension d_f that is nearly equal to that of the largest 2D percolation cluster [35], $d_f = 91/48 = 1.9$.

It is argued [35] that flow in disordered porous media is dominated by the contribution from connected paths near the percolation threshold and, thus, universal power law scaling of percolation theory is relevant. The time required for solute introduced at a single source (such as from a reacting soil particle) at one side of a system to reach the other side was shown [45] to be proportional to the length of the system to a power equal to the fractal dimension of the percolation backbone, D_b, the multiply connected part of the percolation cluster in which fluid flow and solute transport occur (as the rest of the pores are dead ends). For conditions appropriate to weathering, i.e., either wetting soils or constant moisture content, D_b was determined [46] to be equal to 1.87. Validity of the advection-dispersion equation at the scale of an individual pore implies equality of water and solute velocities at that scale and requires a spatio-temporal scaling relationship for the solute's distance traveled, given by $x_s = x_0\ (t/t_0)^{1/D_b}$, where x_0 is a pore separation, or typical particle size, and $x_0/t_0 = v_0$ is, in natural ecosystems, a yearly mean pore-scale flow rate [43]. The result is

$$x = x_0\left(\frac{t}{t_0}\right)^{1/1.87} \tag{1}$$

Because solute transport tends to be the primary limiting factor in chemical weathering, which itself is the principal limitation to soil formation [47], Equation (1) gives the soil depth, while its time derivative, dx/dt, predicts the soil production function. The steady-state (constant) soil depth is defined by the equality of soil production and the rate of soil removal D_0. Equating dx/dt with D_0 yields [42]

$$x = x_0\left[\frac{P - ET}{1.87\phi D_0}\right]^{\frac{1}{D_b - 1}} \tag{2}$$

where porosity ϕ is necessary to convert a mean atmospheric velocity—or flux—of precipitation P less evapotranspiration ET to a water velocity in the porous medium. Since only water flowing through the soil into the subsurface can contribute to weathering and soil formation, the flux of water returned to the atmosphere through evapotranspiration ET does not contribute.

A broadly analogous argument provides a relationship for vegetation root growth [43] with the same form as Equation (1), but with one significant difference: the exponent in the power law is predicted from percolation theory to be $1/D_{opt}$, which, by considering the root zone to have shallow depth, is taken to be the value [46] appropriate for 2D, i.e., $D_{opt} = 1.21$. The explanation is shortened here. By applying a gradient in the soil water potential, the roots draw in water, which can intersect nutrient sources. Water flow toward the roots is dominated by the optimal paths of percolation. Accordingly, in a search for nutrients and water, root tip extension tends to follow optimal paths 'back up-gradient toward nutrient sources. Thus, the actual root length l is longer than the root radial extent, x_r, due to the tortuosity of the paths with the smallest cumulative resistance, and l =

$(x_r/x_0)^{Dopt}$. Since root tip extension rates tend to be a property of genetics plus nutrient and water availability [48], l has no obvious scale dependence, making $l = v_0\, t$, and

$$x_r = x_0 \left(\frac{t}{t_0}\right)^{1/1.21} \tag{3}$$

Although the length scale and flow rates for both soil formation and root growth are represented identically, the meaning of x_0 in Equation (3) is, instead of being a particle size, a typical plant xylem diameter, while v_0 is a little larger for plants, since they will be seen to take almost twice as much of the water available is left for flowing through the soil. Nevertheless, these particular distinctions have no effect on the vegetation optimality condition or the predictions of the percolation model described here. On an annual scale, the water evaporated off plant stomata is represented as the transpiration.

The production of biomass relates to the mass of an object, such as the roots. The root architecture can be modeled as a fractal (e.g., [37]). For consistency in approach, the choice of a 2D percolation exponent for the optimal paths' tortuosity requires the use of the 2D percolation exponent for the root mass fractal dimension. Using the argument that the root mass will depend on its lateral extent, it was then proposed [43] and references therein that *NPP* should be proportional to, ${x_r}^{1.9}$, i.e., the lateral spread raised to the power of $d_f = 1.9$. The result is the expression $NPP = c\, ET^{1.9}$, with c being a constant. Here, transpiration, which is typically [49] approximately 2/3 of evapotranspiration, *ET*, was approximated as *ET*. However, the total mass in the root system must also be expressed in terms of the vertical dimension. Thus, the yearly increase in biomass is proportional to the product of root depth and horizontal contribution to the biomass ($ET^{1.9}$).

Since *NPP* has an $ET^{1.9}$ dependence and soil depth x_s depends on P and ET as $(P\text{-}ET)^{1.15}$, it is straightforward to determine the maximum of *NPP* by differentiating the product $ET^{1.9}\,(P - ET)^{1.15}$ with respect to *ET* and setting the resulting equation to zero. Excluding the trivial endpoint solutions, $ET = 0$ and $ET = P$ (allowing division by *ET* and $ET - P$), the resulting equation is $1.9\, ET = (1.9 + 1.15)\, P$. The result [19] is a proportionality of *ET* to *P* with a proportionality constant involving only universal exponents from percolation theory [35] and a value 0.623; thus, $ET = 0.623\, P$. The second derivative of *NPP* with respect to *ET* is negative at $ET = 0.623\, P$, demonstrating that this particular value of *ET* generates a maximum; however, it is much easier to determine that the extremum is a maximum by applying physical arguments. Since annual *NPP* cannot be negative if plants even exist, *ET* can neither exceed *P* nor be negative. This argument is also consistent with a physical argument that *ET* can only take on values somewhere between 0 and *P*. But, at the endpoints of the domain, when $ET = P$, or when $ET = 0$, the above expression for *NPP* yields 0. Therefore, a maximum must exist for some *ET* value between these two limits.

The above optimization procedure must be extended when there is a deficit of either water or energy, i.e., $P \neq PET$, with *PET* defined explicitly as the depth of water that would evaporate off a free surface in one year. An index called the aridity index, $AI = PET/P$, has been proposed to distinguish between the water and energy limitations. For water-limited systems corresponding to $AI > 1$, the same optimization procedure was applied only to that fraction of the ground covered by vegetation (which is proportional to AI^{-1}), with *ET* elsewhere approximated as being equal to *P*, and for energy-limited systems, with $AI < 1$, the optimization procedure was applied only to that fraction of *P* that could be evaporated (i.e., equal to *PET*) [19], and the remaining *P* contributing to run-off.

The results of the optimization, including energy and water limitations for *ET* (*PET*,*P*), were then substituted back into $NPP = c\, ET^{1.9}$ to obtain two equations for aridity indices higher and lower than 1 [19].

$$NPP = c\,(0.623\, PET)^{1.9} \quad AI \le 1 \tag{4}$$

$$NPP = c\left[P - 0.377\frac{P^2}{PET}\right]^{1.9} \quad AI > 1 \tag{5}$$

For $P = PET$ ($AI = 1$) the two expressions are identical and equal to c $(0.623\ P)^{1.9}$. In Equations (4) and (5), a single unknown parameter, c, is used to convert units of precipitation in $mm^{1.9}\ yr^{-1.9}$ to units of productivity, $gC\ m^{-2}\ yr^{-1}$. Assuming the validity of species energy theory (supported by Refs. [13,17], with further evidence of a direct proportionality of S and NPP [28,50]), Equations (4) and (5) can now be utilized to predict species richness as well. In such a procedure, an additional multiplicative parameter of unspecified magnitude with units of $m^2\ yr\ gC^{-1}$ must be utilized in order to obtain results that are unitless. The two parameters may be combined into a single parameter, h, with units of $yr^{1.9}\ mm^{-1.9}$. Effectively, this gives the product of the number of species per gram of carbon biomass produced (per year per meter), times the biomass produced per unit of evapotranspiration. Doing so does, however, somewhat obscure the (assumed) central role of NPP in the determination of species number S.

It was shown in percolation optimality theory (Hunt et al. 2024) [19] that, by comparison with a data summary of Budyko (1974) [40], the constant c may be treated as universal across climate regimes with PET varying by a factor of 7, and for each separate PET value, the aridity index AI varying by 2 orders of magnitude. The direct comparison [19] of all the values in [40] with Equation (4) resulted in an R^2 value of 0.97, assuming a universally relevant value of c. However, the combined factor relating climate variables to S is different, depending on what subset of the total number of plant species is considered, with the largest value of the constant appropriate for all plant species, a smaller value for the number of vascular plants, and a yet smaller value if only tree species are meant. Because of their high production of biomass, expressions for tree species number S tend to require a larger NPP per species, with fewer individuals in a given area and a correspondingly smaller number of species. Other variation in h is expected to be traceable to other inputs to species richness, such as heterogeneity (e.g., topography) for a given scale, geologic or climatic history, or variation in scale.

3. Principal Data Sources

We chiefly use data from Currie, 1991 [13], who reported the dependence of tree species richness on climatic data across North America north of Mexico. He divided up this region into lat-lon quadrats of 2.5° by 2.5° south of 50° N and of the same width in longitude, but double the height in latitude, north of 50° N. Thus, not all regions were of the same size, which introduces an uncontrolled variation in S of up to a factor 2 that is not generated by climate, with the justification of increased accuracy in other regards. To count the number of species in each quadrat, maps of the ranges of tree species were overlaid with one another. For each quadrat, Currie determined annual mean values of P and PET, as well as ET (and other climate variables), referring to the methods of Budyko [40]. At 70° N, roughly tundra latitude, PET dropped to under 100 mm yr^{-1} at ca. 50 mm yr^{-1}, marking the highest latitude and smallest PET values where trees were found. Quadrats completely covered with tundra had no tree species. The largest value of PET was about 1800 mm yr^{-1}, while that of P was almost completely restricted to 1600 mm yr^{-1}. Currie [13] also calculated NPP but did not represent S in terms of NPP. Latham and Rickfels (1993) [5] augmented the data set of tree species numbers from [13] by including additional compatible data for South America, Eastern Asia, and Japan. The purpose of their investigation was to determine whether the greater continuity in climate in Eastern Asia through the glacial episodes of the Pleistocene could contribute significantly to a greater tree species richness, S. The data for South America were expressed as ranges of values, not strictly compatible with the data of [13], and we did not incorporate them.

4. Comparison with Data

Currie [13] carefully investigated climate severity, climate variability, glacial history, heterogeneity, and energy as potential influences on species richness and concluded that energy was the strongest predictor, with variations in ET accounting for 78% of the variability in S. He went on to discuss that, although correlations do not demonstrate causality, they do serve useful functions, i.e., "when a correlation predicted by a hypothesis is weaker than other observed correlations, one may conclude that a better hypothesis exists". Using this guidance, we tested Equations (4) and (5) with a constant c independent of climate variables to investigate the accuracy of the percolation optimality model of NPP for predicting the species number (Figure 1). Because Currie (1991) [13] does not give both P and PET values for the same site, it is not possible to compare predictions for each site directly without the original data and generate a statistical evaluation. In their comparison with data from [40], Hunt et al. [19] apply Equations (4) and (5) in terms of a family of curves NPP (PET/P, PET), each as a function of $AI = PET/P$ for different values of PET. The separate representations of $S(PET)$ and $S(P)$ in [13] require developing an alternate procedure with a coordinated comparison between the two. Thus, the value of the constant c is the same in both comparisons, while the range of PET values shown in $S(PET)$ must be used as the source of distinct PET values chosen for comparing with $S(P)$ and vice versa. Thus, like Currie, we investigate correlations between S with PET (Figure 1) and P (Figure 2) separately for the entire data set.

In addition to the paired test, it is possible to augment the visual test of the predictions for $S(PET)$ by separating different geographic scales. This is accomplished by focusing on the data for the major provinces of the Canadian North (Nunavut, Yukon, and Northwest Territories) since this region, with the southern boundary at 60 degrees N, accounts for the S values in Figure 1 with $PET < 400$ mm yr^{-1} (Figure 8 in Ref. [13]). In Statistique Canada (2022) [51], Average Annual Precipitation, by Ecoprovince (1979–2006), all P values in these three provinces fall within the range of 80 mm yr^{-1} to 510 mm yr^{-1}. Applying these values of P for $PET < 400$ mm yr^{-1} leads to the results shown in Figure 3.

The author [13] investigated various other organisms (birds, reptiles, and so on) as well but found a monotonic relationship with climate variables only for tree species. The author noted a strong correlation with any of the energy variables, but the strongest was with ET. The relationship with P was described as "weak". It is important to note, however, that the slope of the upper bound on species richness for $S(PET)$ is not a monotonic function of PET and that our prediction generates the observed change in curvature.

In Figures 1–3, the value of the constant h, which converts units of $ET^{1.9}$ to the unitless S, is maintained at 0.0055 yr$^{1.9}$ mm$^{-1.9}$.

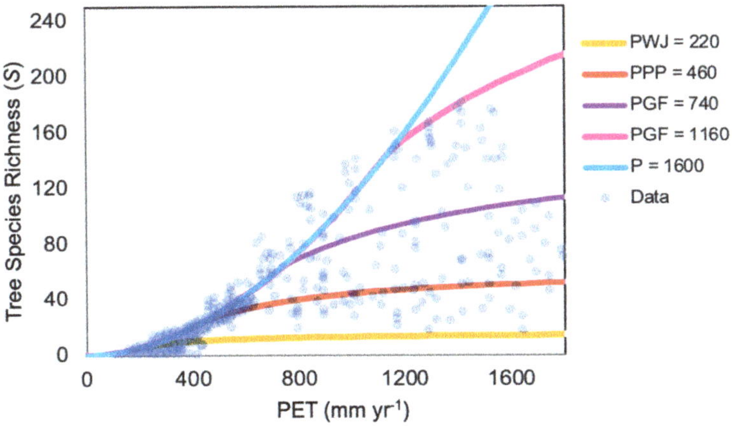

Figure 1. Comparison of predicted tree species number, $S(PET)$, for various values of P with observed numbers of tree species as a function of PET. The largest value of PET reported was 1800 mm yr^{-1}. PWJ 220 (mm yr^{-1}) refers to the lowest rainfall boundary where Western Junipers are reported [52] in Eastern Oregon, which approaches the lowest elevations along the valley floors [53]. PPP = 460 refers to the lowest P, where Ponderosa Pines are reported [52]. PGF = 740 refers to the smallest value of P for which Grand Fir is reported. PGF = 1160 refers to the largest value of P for which Grand Fir is

reported. $P = 1600$ refers to the upper precipitation limit on significant data (a few additional points are found in the vicinity of 2000 mm yr^{-1}). Because the Canadian North data are severely restricted in *PET*, the range $0 < PET < 400$ mm yr^{-1} is treated separately below, while the guidance given for Eastern Oregon is more appropriately restricted to $PET > 400$ mm yr^{-1}.

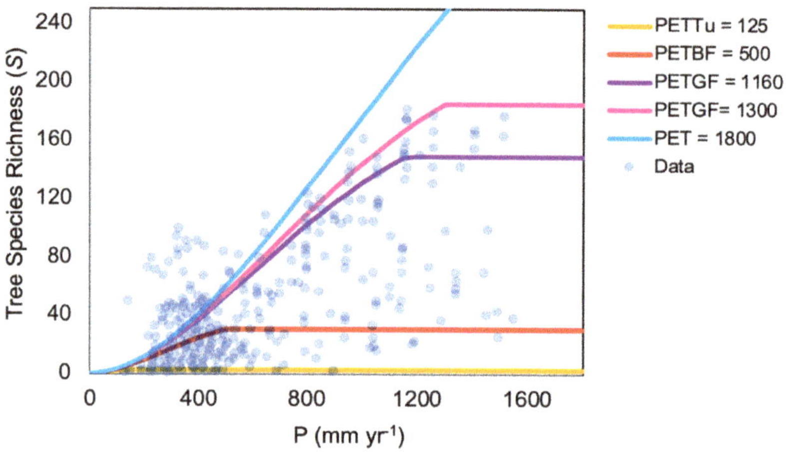

Figure 2. Comparison of predicted tree species number, S, as a function of P for various values of *PET*. PETTu = 125 corresponds to the Arctic Circle, a proxy for the onset of the tundra. *PET* 1800 corresponds to the largest *PET* values found in Figure 1. The other values are labeled analogously to Figure 1, with GF referring to the smallest and largest *PET* values bounding the habitat of the Grand Fir and the label BF meaning Boreal Forest.

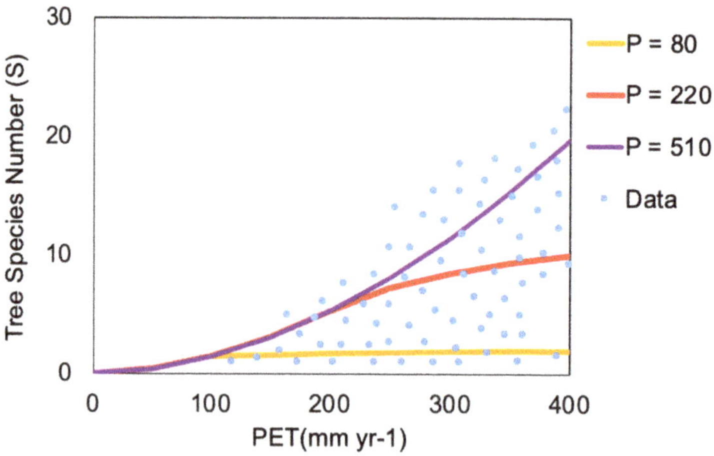

Figure 3. Comparison of predicted tree species number, *S*, as a function of *PET* for ranges of *P* and *PET* appropriate for the Canadian Northwest of Hudson Bay. The area between the yellow ($P = 80$) and the purple ($P = 510$) curves is where data are expected. The intermediate value (brown curve, $P = 220$) forms a lower bound for measured tree species at much larger values of *PET* in, e.g., Eastern Oregon and represents the driest climate there for Western Juniper but has no special relevance in the Canadian North, where *PET* does not exceed 400 mm yr^{-1}.

In Figure 1, the smallest *PET* with a non-zero *S* is *PET* = 117 mm yr^{-1} with *S* =1. For smaller *PET*, *S* appears to be 0. In Figure 2, the smallest *P* reported is 128 mm yr^{-1}, likewise with *S* = 1, and converging to 0 for smaller *P* (though one point at 138 mm yr^{-1} has *S* = 49). Equations (4) and (5) yields *S* = 1.4 for *PET* = 100 mm yr^{-1} and *P* = *PET*, and equally for *P* = 100 mm yr^{-1} and $PET \geq P$. For Figure 2, we chose *PET* = 125 mm yr^{-1} to denote the onset of the tundra for the reasons described next.

PET = 50 mm yr^{-1}, for which Equation (4) predicts $S = 0.4 < 1$, occurs at (70 °N) (Figure 8A in Ref. [13]), which skims the northern edge of Canada west of Hudson Bay (and marks the *S* = 0 contour in Figure 1A of [13]). At the Arctic Circle, values of *PET* range from about 50 mm yr^{-1} to 200 mm yr^{-1}, with a mean of 125 mm yr^{-1}, which generates *S* = 2, barely distinguishable from 0 in Figure 2. On the tree species map of Canada (Figure 1A of [13]), *S* drops below 10 near the Arctic Circle, but not to 2. However, for *P* and *PET* both 200 mm yr^{-1}, within the range of *PET* values shown in Figure 8A of Ref. [13], the calculated value of *S* is 5.

It is important to note that a *PET* of approximately 500 mm yr^{-1} (and for equal precipitation, calculated *S* = 30) occurs just north of the latitude 50° N [54], which corresponds approximately to the onset of the boreal forest. Starting at approximately this latitude, the contours of species follow latitudinal boundaries rather closely across Canada, with *S* between 30 and 40 at 50° N. As can be seen, for such latitudes with strong energy limitations, *PET* describes most of the variability in *S*, both in the figure and on the map. This characteristic is shared in a global data set of vascular plants addressed below.

In Figure 1, within the range $1160 < PET < 1300$, which corresponds to the range of the Grand Fir [52] in *PET*, PGF = 740, at the dry end of the range of Grand Fir in *P*, approximately matches the largest *S* values in the Western USA that varies from 80 to 100. However, PGF = 1160, which generates the upper limit of the *P* range of Grand Fir, corresponds to *S* = 160 or larger, almost double the maximum number of species in the west, and a better match for the largest values of S found in the Southeastern USA at similar *PET*.

Consider Figure 2 again. Since the *PET* ranges for the Western Juniper and the Ponderosa Pine range from 1300 mm yr^{-1} to 1610 mm yr^{-1}, but the *p* values vary from 220 mm yr^{-1} to 760 mm yr^{-1}, their ranges are compressed on this figure to lie between the purple and the light blue curves, where these curves are adjacent. The Grand Fir, however, is constrained to lie between the purple and the magenta curves on the left side of the plateau in the purple curve. Again, this puts the Grand Fir in the region where between about 130 and 160 species are expected, but only 80–100 are found.

Currie [13] explains the existence of the cloud of points outside the range of expected *S* values (above the 1600 curve) as tracing to the higher-than-expected species richness in the (arid to semi-arid) Southwestern USA. Strong heterogeneity in topography and water resources over relatively small spatial separations is likely the cause here, as was demonstrated through Currie's correlation between heterogeneity and richness. The interpretation of the overall results is also enhanced by an examination of Figure 1 in Currie's work [13], which shows local minima in $S < 30$ in the rain shadow of the Sierra Nevada (Great Basin) and in two regions along the Southern High Plains with $S < 20$. Juniper habitats are found in both of these relatively arid regions. Sufficient heterogeneity exists to place juniper habitat into the same quadrat with nearby habitats that support more (and less) trees, but not on as fine a scale as in California, where tree species numbers are everywhere above 40 and mostly above 60. Thus, topographic effects help to limit the number of tree species (Figure 1) at the lower range of WJ to 20 rather than 2 (which would reflect the local association with pinon pine). Thus, the map helps in understanding the cut-off in data at P of approximately 220 mm yr^{-1} for a wide range of *PET* with a fairly large *S* value (near 20).

Although we do not show the data from Kreft and Jetz (2007) [55] for the global vascular plant species number as a function of *PET*, the authors report that their data reflect a strong correlation of species number *S* with *PET* for $PET < 500$, but little predictability solely from that climate variable at larger *PET*, similar to the result found by

Currie (1991) [13], although, in that case, the strong correlation extends to $PET > 600$. These observations likely have the same cause, namely, that for sufficiently large *PET*, omission of consideration of the variability in *P* misses most of the predictability from climate variables.

Figure 4 demonstrates that similar climates (Eastern Asia [5] and North America-Europe [13]) with identical values of *ET* and, thus, to the lowest approximation of the same *NPP*, may have different numbers of tree species. The enhancement factor in Eastern Asia is approximately 2. The explanation for this discrepancy is given by Ref. [5] as related to natural history. Thus, North America and Europe, covered in the relatively recent geological past by ice sheets, may have fewer species since, in contrast to Eastern Asia, sufficient time has not elapsed for the ecosystem with the maximum number of species to develop. Currie [13] appears to have anticipated this argument, investigating North America for a correlation between *S* and Pleistocene glaciation, and found none. His interpretation was that a time span of over 10,000 years was adequate to remove the signs of past climate within the vegetation. However, Pleistocene climates affected strongly all of North America north of Mexico, so perhaps such an evaluation misses a critical input. If the interpretation of Ref. [5] is correct, the Shannon entropy associated with Asian ecosystems would be somewhat larger. Thus, maximizing Shannon entropy would, similar to conventional statistical mechanics, eliminate information that would help establish the history of a system. In the particular case discussed related to Holocene climate change, the simplest calculation, replacing each probability that a tree (in N. America) belongs to a given species by half its value (twice the value of *S* in E. Asia), would result in an increase in Shannon entropy of ln (2). In the event that this is a non-equilibrium value, that difference would then be expected to disappear with sufficient time. If, however, the difference is due to, e.g., a greater relief (heterogeneity), then the distinction would tend to remain.

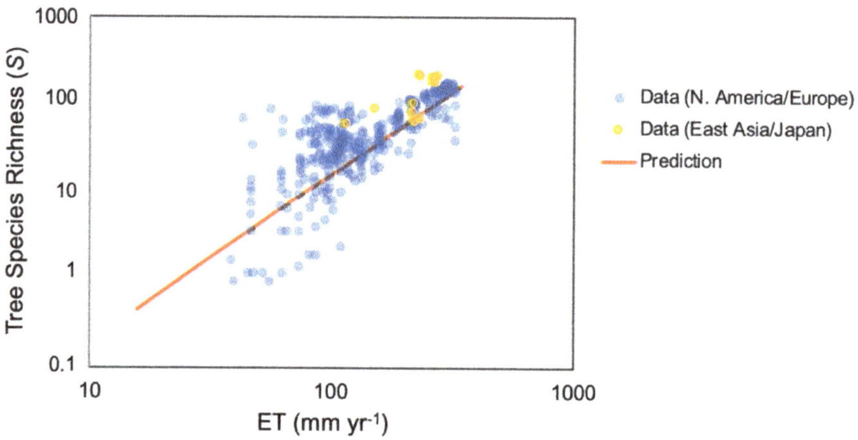

Figure 4. *NPP*, predicted as proportional to $ET^{1.9}$, is compared with species richness data from North America and Europe [13] and separately from Eastern Asia and Japan [5].

In any case, the slope of log *S* vs. log *ET* is compatible with the percolation prediction, 1.9 in both cases.

5. Discussion

Our predicted results for *NPP*, known to accurately forecast [19] plant productivity when compared with the classic data set [40], also track species richness rather closely, including the precipitation and potential evapotranspiration dependencies individually. Using a universal constant (for North America, at least), the observed regular dependence of tree species number on latitude over the northern half of the continent was predicted. In the US, representative species (Grand Fir, Ponderosa Pine, and Western Juniper) showed up in approximately their appropriate positions in the *NPP* (*P*, *PET*) space (equivalent to *S* (*P*, *PET*) space with a single constant of proportionality). The resulting proportionality of *S* to *NPP* is consistent with the species energy theory, which makes *S*, *N*, and energy input all proportional to each other.

The basis for the theoretical predictions lies in a new theory of the water balance [19] and the associated carbon cycle, itself developed using percolation scaling for growth, productivity, and soil formation, and the principle of ecological optimality. The fundamental basis for the optimality chosen here is similar to that proposed originally in Ref. [20] in its assumption that a relationship between *NPP* and water fluxes can be used as an objective function to optimize maximum productivity. However, in Ref. [20], the optimization was internal (i.e., within plants). Here, the optimality expresses a competition for the water between plants and soil together with the mostly symbiotic relationship between soil and plants. If the plants do not receive any water from precipitation, they cannot grow and enrich the soil with carbon and such nutrients as nitrogen, nor can the organic acids they otherwise produce initiate chemical weathering of the subsurface, and the water flux can at most transport carbonic acid from the atmosphere. Thus, a necessary precursor to soil development is completely missing in this particular scenario. On the other hand, if the plants take all the water from precipitation, none passes by to take the organic acids or nutrients down into the subsurface, and no soil is produced. Since most plants scarcely grow without soil, there must be a maximum in productivity somewhere in between, i.e., $0 < ET < P$ (a point summarized in a University of Texas Dallas Geonews video: https://www.youtube.com/watch?v=xv-n54NTd9M (accessed on 23 July 2024)).

Determining the optimum in the water balance (*ET*) through maximization of *NPP* allows the prediction of *NPP* as a function of climate variables. The maximization of *NPP* may require maximum diversity [22], which often implies the need for sufficient time under a stable climatic regime for all relevant ecosystem members to adapt or be transported into the region studied. However, disturbance regimes may strongly influence these patterns in the double role of potentially generating diversity over long time scales and taking up *NPP* (as through wildfire) [56]. Our approach is in general accord with the advice of Whittaker and Field (2000) [57]: "Predictable global patterns in species richness at the macro-scale can be accounted for by a simple climate model, based not on actual evapotranspiration (AET), which is an ambiguous variable, but on potential evapotranspiration (PET) and annual rainfall [...] Historical contingency provides an important part of the explanation for residual variation, left over when climate has been accounted for." We now check whether our results can help with the analysis of the role of historical contingency in the distinction between Asian and North American tree species richness.

Using [58], the maximum values of $PET = 2000$ mm yr^{-1} calculated by the Thornthwaite (1948) [59] or Penman–Monteith (1965) [60] method or $PET = 2200$ mm yr^{-1} calculated by the Priestley–Taylor (1972) [61] for tropical conditions (NDVI between 0.6 and 0.8), Equations (4) and (5) for $P \geq PET$ yields values of *S* equal to 418 and 501, respectively. Accounting for the ca. factor of 2 increase in *h* for Eastern Asia (Latham and Rickfels, 1993 [5]), as compared with North America, would generate maximum values of 836 and 1002 for *S*. Notably, these two values are close to the largest recorded values in the tropics ($1000 < S < 1500$) in both the Americas and Southeastern Asia (though at much smaller scales) in [62]. The greater continuity of the predictability of species richness in Eastern Asia as opposed to the significant disconnect in the Americas is consistent with the assertion of Latham and Rickfels (1993) [5] regarding the role of history ("tropical conservatism"), as well as the argument of Whittaker and Field [57] regarding the importance of an accurate representation of the primary, i.e., climate, input to species richness in determining the relative importance of secondary inputs. It is also possible that greater heterogeneity in topography in Eastern Asia, with its extreme relief, might have contributed to the greater diversity found in the Latham–Rickfels study [5].

6. Conclusions

We presented a proof of concept for applying the hydrological principles of water balance, the near-2D development of a soil layer across the landscape, the fractal-like development of root structure (following the logic of percolation theory), and the water requirements of plants to link potential evapotranspiration with species richness. We tested this prediction with previously studied data [13] and found realistic trends for the entire data set and for several individual species. In a later extension of the range of PET values investigated to values consistent with the maximum observed PET in the tropics, we discovered that a single formula (Equations (4) and (5)) generates values of S consistent with the observation from the Arctic to the Equator and helped to understand quantitatively the distinction between Asian and N. American tree species richness consistent with the tropical conservatism hypothesis.

This work aims to fill a gap in existing knowledge, that is, how species richness is mechanistically related to climate variables and water balance through constraints on water use by transpiration by plants and spatial constraints on their root growth. It also highlights important questions in biodiversity sciences as to what constrains overall species diversity. It provides an energetic justification/mechanism for the formation of richness that was already mentioned in previous work (e.g., [63]). Future work may delve more deeply into additional data sets or link this percolation optimality framework with the Maximum Entropy Theory of Ecology through the prediction of species richness. With such climate-related and mechanistic constraints, it may be possible, through the use of macroecological theory, to develop a much more complete picture of constraints on species diversity and productivity at local and global scales.

Important aspects missing from the present formulation include that it does not address the role of ecological networks [64], for which the data used in the present study have no temporal dependence. Without any means to address temporal dependences, our study therefore omits a priori potential causal inputs. Given the definition of geographic regions defined by the author [13] of the tree study in terms of a global coordinate system (latitude and longitude), it also seems to be impossible to expand this particular study to address the potential effects of drainage basin architecture and river corridor correlations in species numbers, even though the importance of such inputs is known [65,66]. In contrast to the impossibility of modifying the present study to address such important inputs into the temporal and spatial variability of species numbers, the results of our study may be relatively easily incorporated into (functional or geographical) network analyses of (plant and animal) species richness patterns and their development through time. For, as has long been known, animal species patterns should be linked to the plants with which they are associated [67].

Author Contributions: A.G.H. was responsible for the initial investigation, E.A.N. was responsible for making ecological concepts precise, and M.S. was responsible for the overall direction of the final manuscript, as well as contributing to the conceptual framework of solute transport used here. All authors have read and agreed to the published version of the manuscript.

Funding: This research received no external funding.

Data Availability Statement: Data used in this paper are available in the original articles in which they appeared.

Acknowledgments: A.H. is grateful to Wright State University for research support.

Conflicts of Interest: The authors declare no conflicts of interest.

References

1. Wright, D.H. Species–energy theory: An extension of species-area theory. *Oikos* **1983**, *41*, 496–506. [CrossRef]
2. Grime, J.P. *Plant Strategies and Vegetation Processes*; Wiley: New York, NY, USA, 1979.
3. Tilman, D. *Resource Competition and Community Structure*; Princeton University Press: Princeton, NJ, USA, 1982.
4. Prance, G.T. (Ed.) *The Biological Model of Diversification in the Tropics*; Columbia Univ. Press: New York, NY, USA, 1982.

5. Latham, R.E.; Rickfels, R.E. Global patterns of tree species richness in moist forests: Energy-diversity theory does not account for variation in species richness. *Oikos* **1993**, *67*, 325–333. [CrossRef]
6. Kerkhoff, A.J.; Moriarty, P.E.; Weiser, M.D. The latitudinal species richness gradient in New World woody angiosperms is consistent with the tropical conservatism hypothesis. *Proc. Natl. Acad. Sci. USA* **2014**, *111*, 8125–8130. [CrossRef]
7. Šímová, I.; Li, Y.M.; Storch, D. Relationship between species richness and productivity in plants: The role of sampling effect, heterogeneity and species pool. *J. Ecol.* **2013**, *101*, 161–170. [CrossRef]
8. Von Humbolt, A. *Ansichten der Natur mit Wissenschaftlichen Erlauterungen*; Cotta: Tubingen, Germany, 1808.
9. Darwin, C. *On the Origins of Species by Means of Natural Selection*; John Murray: London, UK, 1859.
10. Wallace, A.R. *Tropical Nature and Other Essays*; Macmillan: London, UK, 1878.
11. Simpson, G.G. Species densities of North American mammals. *Syst. Zool.* **1964**, *13*, 361–389. [CrossRef]
12. Cook, R.E. Variation in species density of North American birds. *Syst. Zool.* **1969**, *18*, 63–84. [CrossRef]
13. Currie, D.J. Energy and large-scale patterns of animal- and plant- species richness. *Am. Nat.* **1991**, *137*, 27–49. [CrossRef]
14. Stoms, D.M. Scale dependence of species richness maps. *Prof. Geogr.* **1994**, *46*, 346–358. [CrossRef]
15. Chase, J.M.; McGill, B.J.; Thompson, P.L.; Antão, L.H.; Bates, A.E.; Blowes, S.A.; Dornelas, M.; Gonzalez, A.; Magurran, A.E.; Supp, S.R.; et al. Species richness change across spatial scales. *Oikos* **2019**, *128*, 1079–1091. [CrossRef]
16. Whittaker, R.J.; Willis, K.J.; Field, R. Scale and species richness: Towards a general, hierarchical theory of species diversity. *J. Biogeogr.* **2001**, *28*, 453–470. [CrossRef]
17. O'Brien, E.M.; Field, R.; Whittaker, R.J. Climatic gradients in woody plant (tree and shrub) diversity: Water-energy dynamics, residual variation, and topography. *Oikos* **2000**, *89*, 588–600. [CrossRef]
18. Rosenzweig, M.L. Net primary productivity of terrestrial communities: Prediction from climatological data. *Am. Nat.* **1968**, *102*, 67. [CrossRef]
19. Hunt, A.G.; Sahimi, M.; Ghanbarian, B. Predicting ecosystem net primary productivity by percolation and optimality principle. *Water Resour. Res.* **2024**, *60*, e2023WR036340. [CrossRef]
20. Odum, E. *Fundamentals of Ecology*; W.B. Saunders: Philadelphia, PA, USA, 1959; Volume 21, p. 712.
21. Broadhurst, L.; Breed, M.; Lowe, A.; Bragg, J.; Catullo, R.; Coates, D.; Encinas-Viso, F.; Gellie, N.; James, E.; Krauss, S.; et al. Genetic diversity and structure of the Australian flora. *Divers. Distrib.* **2017**, *23*, 41–52. [CrossRef]
22. Messier, J.; Violle, C.; Enquist, B.J.; Lechowicz, M.J.; McGill, B.J. Similarities and differences in intrapopulation trait correlations of co-occurring tree species: Consistent water-use relationships amid widely different correlation patterns. *Am. J. Bot.* **2018**, *105*, 1477–1490. [CrossRef]
23. Cadotte, M.W.; Carscadden, K.; Mirotchnick, N. Beyond species: Functional diversity and the maintenance of ecological processes and services. *J. Appl. Ecol.* **2011**, *48*, 1079–1087. [CrossRef]
24. Oehri, J.; Schmid, B.; Schaepman-Strub, G.; Niklaus, P.A. Biodiversity promotes primary productivity and growing season lengthening at the landscape scale. *Proc. Natl Acad. Sci. USA* **2017**, *114*, 10160. [CrossRef]
25. Chen, S.; Wang, W.; Xu, W.; Wang, Y.; Wan, H.; Chen, D.; Tang, Z.; Tang, X.; Zhou, G.; Xie, Z.; et al. Plant diversity enhances productivity and soil carbon storage. *Proc. Natl. Acad. Sci. USA* **2018**, *115*, 4027–4032. [CrossRef] [PubMed]
26. Srivastava, D.S.; Lawton, J.H. Why More Productive Sites Have More Species: An Experimental Test of Theory Using Tree-Hole Communities. *Am. Nat.* **1998**, *152*, 510–529. [CrossRef]
27. Storch, D.; Bohdalková, E.; Okie, J. The More-Individuals Hypothesis Revisited: The Role of Community Abundance in Species Richness Regulation and the Productivity-Diversity Relationship. *Ecol. Lett.* **2018**, *21*, 920–937. [CrossRef]
28. Harte, J.; Brush, M.; Newman, E.A.; Umemura, K. An Equation of State Unifies Diversity, Productivity, Abundance and Biomass. *Commun. Biol.* **2022**, *5*, 874. [CrossRef] [PubMed]
29. Newman, E.A.; Brush, M.; Umemura, K.; Xu, M.; Harte, J. Defining an Ecological Equation of State: Response to Riera et al. (2023). *Ecol. Modell.* **2023**, *486*, 110532. [CrossRef]
30. Harte, J.; Zillio, T.; Conlisk, E.; Smith, A.B. Maximum entropy and the state-variable approach to macroecology. *Ecology* **2008**, *89*, 2700–2711. [CrossRef] [PubMed]
31. Harte, J. *Maximum Entropy and Ecology: A Theory of Abundance, Distribution, and Energetics*; Oxford University Press: Oxford, UK, 2011.
32. Harte, J.; Smith, A.B.; Storch, D. Biodiversity scales from plots to biomes with a universal species–area curve. *Ecol. Lett.* **2009**, *12*, 789–797. [CrossRef] [PubMed]
33. Shipley, B.; Vile, D.; Garnier, E. From plant traits to plant communities: A statistical mechanistic approach to biodiversity. *Science* **2006**, *314*, 812–814. [CrossRef] [PubMed]
34. Hunt, A.G.; Faybishenko, B.; Ghanbarian, B. Predicting characteristics of the water cycle from scaling relationships. *Water Resour. Res.* **2021**, *57*, e2021WR030808. [CrossRef]
35. Sahimi, M. *Applications of Percolation Theory*, 2nd ed.; Springer: Berlin/Heidelberg, Germany, 2023.
36. Gentine, P.; D'Dodorico, P.; Linter, B.R.; Sivandran, G.; Salvucci, G. Interdependence of climate, soil, and vegetation as constrained by the Budyko curve. *Geophys. Res. Lett.* **2012**, *39*, L19404. [CrossRef]
37. Lynch, J. Root architecture and plant productivity. *Plant Physiol.* **1995**, *109*, 7–13. [CrossRef]
38. Fan, Y.; Miguez-Macho, G.; Jobbágy, E.G.; Jackson, R.B.; Otero-Casal, C. Hydrologic regulation of plant rooting depth. *Proc. Natl. Acad. Sci. USA* **2017**, *114*, 10572–10577. [CrossRef]

39. Horton, R. The field, scope, and status of the science of hydrology. Eos, Trans. *Am. Geophys. Union* **1931**, *12*, 189.
40. Budyko, M.I. *Climate and Life*; International Geophysics Series 18; Academic Press: New York, NY, USA, 1974.
41. Dooge, J.C.I. Hydrology in perspective. *Hydrol. Sci. J.* **1988**, *33*, 61–85. [CrossRef]
42. Yu, F.; Hunt, A.G. An examination of the steady-state assumption in certain soil production models with application to landscape evolution. *Earth Surf. Process. Landf.* **2017**, *42*, 2599–2610. [CrossRef]
43. Hunt, A.G.; Faybishenko, B.; Powell, T.L. Test of model of equivalence of tree height growth and transpiration rates in percolation-based phenomenology for root soil interaction. *Ecol. Model.* **2022**, *465*, 109853. [CrossRef]
44. Nijzink, R.C.; Schymanski, S.J. Vegetation optimality explains the convergence of catchments on the Budyko curve. *Hydrol. Earth Syst. Sci.* **2022**, *26*, 6289–6309. [CrossRef]
45. Lee, Y.; Andrade, J.S.; Buldyrev, S.V.; Dokholoyan, N.V.; Havlin, S.; King, P.R.; Paul, G.; Stanley, H.E. Traveling time and traveling length in critical percolation clusters. *Phys. Rev. E* **1999**, *60*, 3425–3428. [CrossRef]
46. Sheppard, A.P.; Knackstedt, M.A.; Pinczewski, W.V.; Sahimi, M. Invasion percolation: New algorithms and universality classes. *J. Phys. A Math. Gen.* **1999**, *32*, L521–L529. [CrossRef]
47. Hunt, A.G.; Egli, M.; Faybishenko, B.A. (Eds.) Where are we and where are we going: Pedogenesis through chemical weathering, hydrologic fluxes, and bioturbation. In *Hydrogeology, Chemical Weathering, and Soil Formation*; AGU/Wiley Geophysical Monographs; American Geophysical Union: Washington, DC, USA, 2021.
48. Watt, M.; Silk, W.K.; Passioura, J.B. Rates of root and organism growth, soil conditions, and temporal and spatial development of the rhizosphere. *Ann. Bot.* **2006**, *97*, 839–855. [CrossRef]
49. Schlesinger, W.H.; Jasechko, S. Transpiration in the global water cycle. *Agric. For. Meteorol.* **2014**, *189*, 115–117. [CrossRef]
50. Gillman, L.N.; Wright, S.D.; Cusens, J.; McBride, P.D.; Malhi, Y.; Whittaker, R.J. Latitude, productivity and species richness. *Glob. Ecol. Biogeogr.* **2015**, *24*, 107–117. [CrossRef]
51. Statistique Canada. Average Annual Precipitation, by Ecoprovince, 1979 to 2016. 2022. Available online: https://www150.statcan.gc.ca/n1/pub/38-20-0001/2021001/l03-eng.htm#archived (accessed on 3 July 2024).
52. Berner, L.T.; Law, B.E. Water limitations on forest carbon cycling and conifer traits along a steep climatic gradient in the Cascade Mountains, Oregon. *Biogeosciences* **2015**, *12*, 6617–6635. [CrossRef]
53. Miller, R.; Chambers, J.C.; Evers, L.; Williams, C.J.; Snyder, K.A.; Roundy, B.A.; Pierson, F.B. *The Ecology, History, Ecohydrology, and Management of Pinyon and Juniper Woodlands in the Great Basin and Northern Colorado Plateau of the Western United States*; U.S. Department of Agriculture, Forest Service, Rocky Mountain Research Station: Fort Collins, CO, USA, 2019.
54. Kingston, D.G.; Todd, M.C.; Taylor, R.G.; Thompson, J.R.; Arnell, N.W. Uncertainty in the estimation of potential evapotranspiration under climate change. *Geophys. Res. Lett.* **2009**, *36*, L20403. [CrossRef]
55. Kreft, H.; Jetz, W. Global patterns and determinants of vascular plant diversity. *Proc. Natl. Acad. Sci. USA* **2007**, *104*, 5925–5930. [CrossRef] [PubMed]
56. Moritz, M.A.; Batllori, E.; Bolker, B.M. The role of fire in terrestrial vertebrate richness patterns. *Ecol. Lett.* **2023**, *26*, 563–574. [CrossRef]
57. Whittaker, R.J.; Field, R. Tree species richness modelling: An approach of global applicability? *Oikos* **2000**, *89*, 399–402. [CrossRef]
58. Fisher, J.B.; Whittaker, R.J.; Malhi, Y. ET come home: Potential evapotranspiration in geographical ecology. *Glob. Ecol. Biogeogr.* **2011**, *20*, 1–18. [CrossRef]
59. Thornthwaite, C.W. An approach toward a rational classification of climate. *Geogr. Rev.* **1948**, *38*, 55–94. [CrossRef]
60. Monteith, J.L. Evaporation and the environment. *Symp. Soc. Explor. Biol.* **1965**, *19*, 205–234.
61. Priestley, C.H.B.; Taylor, R.J. On the assessment of surface heat flux and evaporation using large scale parameters. *Mon. Weather Rev.* **1972**, *100*, 81–92. [CrossRef]
62. Davies, S.J.; Abiem, I.; Salim, K.A.; Aguilar, S.; Allen, D.; Alonso, A.; Anderson-Teixiera, K.; Andrade, A.; Arellano, G.; Ashton, P.S.; et al. ForestGEO: Understanding forest diversity and dynamics through a global observatory network. *Biol. Conserv.* **2021**, *253*, 108907. [CrossRef]
63. Harte, J. Maximum Entropy: A Foundation for a Unified Theory of Ecology. 2020. Available online: https://academic.oup.com/book/43963/chapter-abstract/369251208 (accessed on 5 July 2024).
64. Li, J.; Convertino, M. Inferring ecosystem networks as information flows. *Sci. Rep.* **2021**, *11*, 7094. [CrossRef]
65. Rinaldo, A.; Gatto, M.; Rodriguez-Iturbe, I. River Networks as Ecological Corridors: A Coherent Ecohydrological Perspective. *Adv. Water Res.* **2018**, *122*, 27–58. Available online: https://water.usask.ca/documents/dls-2021/dls-discussion_rinaldo.pdf (accessed on 5 July 2024). [CrossRef] [PubMed]
66. Rinaldo, A.; Gatto, M.; Rodriguez-Iturbe, I. *River Networks as Ecological Corridors*; Species, Populations, Pathogens; Cambridge University Press: Cambridge, UK, 2020.
67. MacArthur, R.H.; MacArthur, J.W. On bird species diversity. *Ecology* **1961**, *42*, 594–598. [CrossRef]

entropy

MDPI

Article

Cascades Towards Noise-Induced Transitions on Networks Revealed Using Information Flows

Casper van Elteren [1,2,*], Rick Quax [1,2] and Peter M. A. Sloot [1,2,3]

[1] Institute of Informatics, University of Amsterdam, 1098 XH Amsterdam, The Netherlands; r.quax@uva.nl (R.Q.); p.m.a.sloot@uva.nl (P.M.A.S.)
[2] Institute for Advanced Study, 1012 GC Amsterdam, The Netherlands
[3] Complexity Science Hub Viennna, 1080 Vienna, Austria
* Correspondence: c.vanelteren@uva.nl

Abstract: Complex networks, from neuronal assemblies to social systems, can exhibit abrupt, system-wide transitions without external forcing. These endogenously generated "noise-induced transitions" emerge from the intricate interplay between network structure and local dynamics, yet their underlying mechanisms remain elusive. Our study unveils two critical roles that nodes play in catalyzing these transitions within dynamical networks governed by the Boltzmann–Gibbs distribution. We introduce the concept of "initiator nodes", which absorb and propagate short-lived fluctuations, temporarily destabilizing their neighbors. This process initiates a domino effect, where the stability of a node inversely correlates with the number of destabilized neighbors required to tip it. As the system approaches a tipping point, we identify "stabilizer nodes" that encode the system's long-term memory, ultimately reversing the domino effect and settling the network into a new stable attractor. Through targeted interventions, we demonstrate how these roles can be manipulated to either promote or inhibit systemic transitions. Our findings provide a novel framework for understanding and potentially controlling endogenously generated metastable behavior in complex networks. This approach opens new avenues for predicting and managing critical transitions in diverse fields, from neuroscience to social dynamics and beyond.

Keywords: information theory; noise-induced transitions; metastability

1. Introduction

Multistability, a fundamental characteristic of complex systems [1,2], describes the capacity of a system to occupy multiple stable states and transition between them. This phenomenon is ubiquitous, manifesting in diverse domains from neural networks [3,4] to opinion dynamics [5] and ecosystems [6]. While state transitions are often attributed to external perturbations, we propose a novel perspective: in networked systems, noise-induced transitions can occur endogenously. These transitions emerge from local interactions that cascade through the network, triggering large-scale regime shifts in a process we term the "domino effect". This mechanism offers a new understanding of how complex systems can dramatically reconfigure without external forcing, challenging traditional views on system stability and change.

In nonlinear systems, such as interconnected neurons, noise plays a fundamental role in facilitating transitions between attractor states [7–9]. It enables the exploration of larger state spaces, allowing systems to escape local minima [10,11]. While multistability has historically been studied from an equilibrium perspective [10,12,13], recent research has revealed how network structure fundamentally affects the stability and transitions of complex systems [14–17].

Recent work has approached network control through algorithmic information theory, which measures the computational complexity of producing network states through controlled interventions [18–20]. While this provides powerful tools for steering networks through external manipulation, fundamental questions remain about how networks spontaneously transition between states through their internal dynamics. Our approach uses Shannon information theory to quantify the temporal correlations that emerge naturally as networks evolve, revealing how noise propagates through network structure to generate endogenous transitions. This complements algorithmic approaches by focusing on the statistical mechanisms underlying spontaneous state changes rather than the computational complexity of producing specific states.

Our study addresses a critical gap in understanding noise-induced transitions in networked dynamical systems out of equilibrium. We focus on systems where each node's state evolves according to the Boltzmann–Gibbs distribution, a framework applicable to various phenomena including neural dynamics [21], opinion formation, and ferromagnetic spins [22]. An example of a noise-induced transition in this model executed on a network is shown in Figure 1.

We introduce two novel concepts: *initiator* nodes that propagate noise and destabilize the system, and *stabilizing* nodes that maintain metastable states. To quantify the impact of short-term and long-term correlations in these transitions, we propose two information-theoretic measures: integrated mutual information and asymptotic information. These metrics, computable from observational data, provide powerful tools for analyzing metastable dynamics across different time scales.

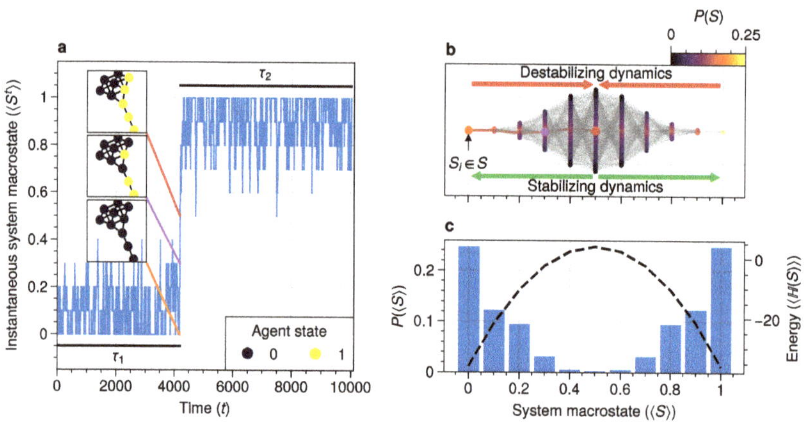

Figure 1. A dynamical network governed by kinetic Ising dynamics produces multistable behavior. (**a**) A typical trajectory is shown for a kite network for which each node is governed by the Ising dynamics with $\beta \approx 0.534$. The panels show system configurations $S_i \in S$ as the system approaches the tipping point (orange to purple to red). For the system to transition between attractor states, it must cross an energy barrier (**c**). (**b**) The dynamics of the system can be represented as a graph. Each node represents a system configuration $S_i \in S$ such as depicted in (**a**). The probability for a particular system configuration $p(S)$ is indicated with a color; some states are more likely than others. The trajectory from (**a**) is visualized. Dynamics that move towards the tipping point (midline) destabilize the system, whereas moving away from the tipping point are stabilizing dynamics. (**c**) The stationary distribution of the system is bistable. Crossing the tipping point requires crossing a high-energy state (dashed line). Transitions between the attractor states are infrequent and rare. For more information on the numerical simulations, see Appendix A.2.

Integrated mutual information captures the transient destabilization of the system, revealing the role of initiator nodes in triggering systemic transitions. Asymptotic information, on the other hand, quantifies the long-term memory encoded by stabilizer nodes, which ultimately reverses the domino effect and settles the network into a new stable attractor. By manipulating these roles, we demonstrate how targeted interventions can either promote or inhibit systemic transitions, offering a new approach to controlling critical transitions in complex networks.

Our computational method uncovers a network percolation process that facilitates noise-induced transitions without external parameter changes, offering a fresh perspective on tipping points in complex networks [23–26]. This approach bridges the gap between local equilibrium dynamics and global system behavior, providing insights into how network structure influences systemic transitions [14,15,27–29].

By revealing the domino-like mechanisms of endogenous state transitions, our work has broad implications for predicting and potentially controlling critical transitions in diverse, complex systems. From enhancing brain plasticity to anticipating ecosystem shifts, this framework provides a foundation for understanding and managing multistability in an interconnected world.

2. Methods

Our study focuses on dynamical systems where the state transitions of individual nodes are governed by the Boltzmann–Gibbs distribution. This distribution, fundamental in statistical mechanics, provides a probabilistic framework for describing the behavior of systems in thermal equilibrium. In our context, it determines the likelihood of a node transitioning from one state to another based on the energy difference between states and a global noise parameter. Specifically, the probability of a node transitioning from state s_i to state s_i' is given by:

$$P(s_i \rightarrow s_i') = \frac{1}{1 + \exp(-\beta \Delta E(s_i, s_i'))}, \tag{1}$$

where $\Delta E(s_i, s_i')$ represents the energy difference for the state transition, and β is the inverse temperature or noise parameter. This formulation captures the essence of how local interactions and global noise influence state changes in our networked system. Higher values of β correspond to lower noise levels, leading to more deterministic behavior, while lower β values introduce more randomness into the system's dynamics. This framework allows us to model a wide range of phenomena, from neural activity to opinion dynamics, within a consistent mathematical structure.

Fluctuations and their correlations at time τ are captured using Shannon's mutual information [30] shared between a node's state (s_i^t) at time t and the entire future system state ($S^{t+\tau}$), $I(s_i^\tau : S^{\tau+t})$. The time lag t is used to analyze two key features of information flows of a system: the area under the curve (AUC) of short-term information and the sustained level of long-term information.

The contribution of a node to the dynamics of the system will differ depending on the network connectivity of a node (Figure A3) [31,32]. The total amount of fluctuations shared between the node's current state and the system's short-term future trajectory is computed as the integrated mutual information.

$$\mu(s_i) = \sum_{t=0}^{\infty} (I(s_i^\tau : S^{\tau+t}) - \omega_{(s_i)})\Delta t. \tag{2}$$

Intuitively, $\mu(s_i)$ represents a combination of the intensity and duration of the short-term fluctuations on the (transient) system dynamics [31]. It reflects how much of the node state is in the "working memory" of the system.

The term $\omega(s_i) \in \mathbb{R}_{\geq 0}$ represents the system's long-term memory. As the system transitions between stable points, short-lived correlations evolve into longer-lasting ones, particularly among less dynamic nodes. When $\omega(s_i)$ is positive, it indicates a separation of time scales: ephemeral correlations dissipate, giving way to slower, more persistent fluctuations. These enduring fluctuations reflect the multiple attractor states accessible to the system, with fewer dynamic nodes becoming more aligned with future system states.

Near a stable attractor, the system primarily generates short-lived fluctuations. However, as it approaches a tipping point, longer-lasting correlations emerge. These persistent correlations facilitate the system's transition from one stable attractor to another, much like repeated nudges eventually push a ball over a hill. The asymptotic information, $\omega(s_i)$, quantifies this transition potential. Higher values of $\omega(s_i)$ indicate a greater likelihood of state transition, with the exact value reflecting each node's contribution to the tipping behavior.

Asymptotic information distinguishes itself from other early warning signals—such as increased autocorrelation, critical slowing down captured by Fisher information, changes in skewness or kurtosis, and increased variance—by specifically measuring the system's long-term memory and temporal correlation structure. While entropy captures the overall uncertainty or disorder in a system at a given moment, and mutual information quantifies the shared information between components at a particular time, asymptotic information focuses on the persistence of correlations over extended time periods. It reveals how past states influence future configurations, capturing aspects of the system's dynamics that are not explained by instantaneous or short-term pairwise measures.

Using these information features, each node can be assigned to a different *role* based on their contribution to the metastable transition. We denote nodes with short-lived correlations as *initiators* pushing nodes towards a tipping point. In contrast, nodes with longer-lived correlations are referred to as *stabilizers*. For these nodes, their dynamics are less affected by short-lived correlations, and they require a higher mixing state to transition from one state to another. The role assignment will be further discussed in Section 3.5.

We compute information flows using exact calculations on a randomly generated connected graph of $n = 10$ nodes. The states are grouped based on their distance to the tipping point, defined as the energy barrier between two locally stable states. For the Ising model, this corresponds to the collection of states where $\langle S \rangle = 0.5$. We evaluate the conditional distribution up to $\tau = 300$ time steps.

This computational process scales exponentially with the number of nodes, $O(n) = 2^n$, which limits its applicability to large-scale systems without employing variable reduction techniques such as coarse-graining. Extending this analysis to larger systems will be the focus of future research.

For detailed replication instructions, please refer to Appendix A.

3. Results

Our analysis reveals several key insights into the dynamics of metastable transitions and tipping points in complex networks. We observe a distinct *domino effect* where low-degree nodes initiate system destabilization. As the system approaches a tipping point, information flows shift from low-degree to high-degree nodes. We identify a rise in asymptotic information as a potential early warning signal for an impending tipping point. Finally, we uncover a division of roles among nodes, with some acting as *initiators* that propagate perturbations and others as *stabilizers* that influence the system's transition between attractor states.

In Figure 2, we visualize the information flows at different stages as the system approaches the tipping point. While we present detailed analysis using the kite graph for simplicity, these findings generalize to other network structures, as demonstrated in Figure 3 and further elaborated in the Appendix A.

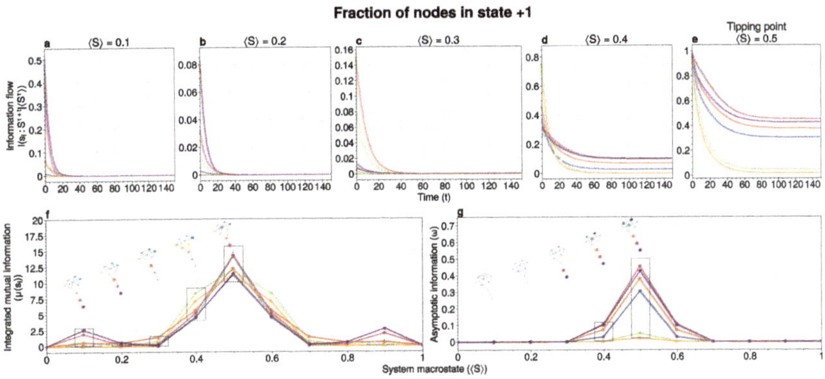

Figure 2. (**a–e**) Information flows as distance to a tipping point, where each line color corresponds to the matching-colored node in the kite graph inset. Far away from the tipping point, most information processing occurs in low-degree nodes (colored in blue/purple, tail of kite). As the system moves towards the tipping point, the information flows increase and shift towards higher-degree nodes (colored in red/orange, core of kite). (**f**) Integrated mutual information as a function of distance to the tipping point. The graphical inset plots show how noise is introduced far away from the tipping point in the tail of the kite graph (blue/purple nodes). As the system approaches the tipping point, the local information dynamics move from the tail to the core of the kite (red/orange nodes). (**g**) A rise in asymptotic information indicates that the system is close to a tipping point. At the tipping point, the decay maximizes as trajectories stabilize into one of the two attractor states. The color of each line consistently matches its corresponding node in the kite graph visualization.

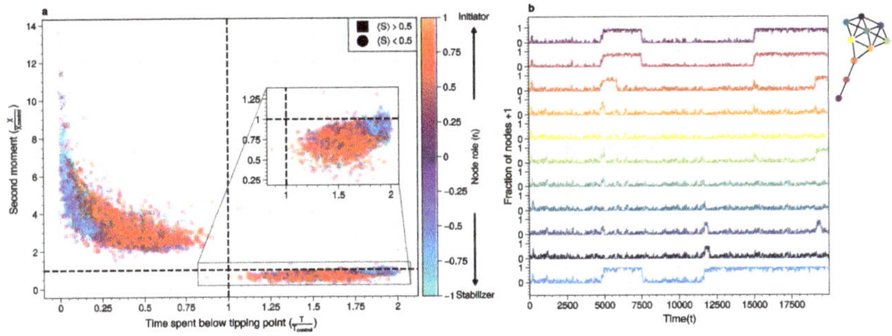

Figure 3. For system to cross a tipping point, two distinct types of nodes are essential: **stabilizers**, which contain information about the system's next attractor state and facilitate transitions between states; and **initiators**, which propagate noise through the system. (**a**) The effect of causal pinning interventions on node 0 states in Erdős–Rényi graphs ($N = 100$, 10 nodes each, $p = 0.2$, 6 seeds) is shown. Normalized system fluctuations (second moment) and time spent below the tipping point relative to the control are presented per network to indicate the effect of the pinning interventions. Pinning initiators increase tipping points while pinning stabilizers prevent tipping and increase noise above the tipping point. For more details on role approximation, see Section 3.5. (**b**) To exemplify the effect of the causal interventions in (**a**), typical system trajectories underpinning interventions on a node for the kite graph are shown. Colors reflect intervention on corresponding nodes in the inset kite graph. Initiator-based interventions remove fluctuations below the tipping point (<0.5) and increase fluctuations above, whereas stabilizer-based interventions stabilize tipping points while increasing noise.

3.1. Information Flow Dynamics and the Domino Effect

To decompose the metastable transition, we consider local information flows in a given system partition, $S_\gamma = \{S' \subseteq S | \langle S' \rangle = \gamma\}$ where $\gamma \in [0,1]$ represents the fraction of nodes that have state 1. This yields the conditional integrated mutual information:

$$\mu(s_i|\langle S \rangle) = \sum_{t=0}^{\infty} (I(s_i^\tau : S^{\tau+t}|\langle S^\tau \rangle) - \omega_{s_i})\Delta t. \tag{3}$$

Details about the estimation procedure can be found in Appendix A.5.

Two key observations emerge from Figure 2:

First, the tipping point is reached through a domino effect, with low-degree nodes acting as initiators early in the process. These nodes, being more susceptible to noise (see Figure A3), are more likely to pass fluctuations to neighbors—akin to pushing a ball up a hill. Far from the tipping point (Figure 2a), lower-degree nodes show higher integrated mutual information, $\mu(s_i|\langle S \rangle)$, than higher-degree nodes. This noise injection by lower-degree nodes increases the likelihood of a metastable transition.

Second, an increase in asymptotic behavior corresponds to the system transitioning between attractor states. As shown in Figure 2b,c, asymptotic information remains low far from the tipping point and steadily increases as the system approaches it. Nodes with higher asymptotic information possess greater predictive power regarding which side of the tipping point the system will settle on.

3.2. Path Analysis and Tipping Point Trajectories

To illustrate the information encoded in these flows, we computed trajectories from the attractor state $S = \{0, \ldots, 0\}$, simulated for $t = 5$ steps. Figure 4 shows a trajectory that maximizes:

$$\log p(S^{t+1}|S^t, S^0 = \{0, \ldots, 0\}, \langle S^5 \rangle = 0.5).$$

These trajectories reveal how the information flows measured in Figure 2c are generated by the sequence of flips originating from the tail of the kite graph. Tail nodes are uniquely positioned to pass on fluctuations to their neighbors, eventually causing a cascade of flips that reach the tipping point. This simple example illustrates how the network structure can influence the system's dynamics and the information flows that precede a metastable transition. Where noise pushes the system towards a tipping point, originating first in low-degree nodes for dynamics governed by the Boltzmann–Gibbs distribution.

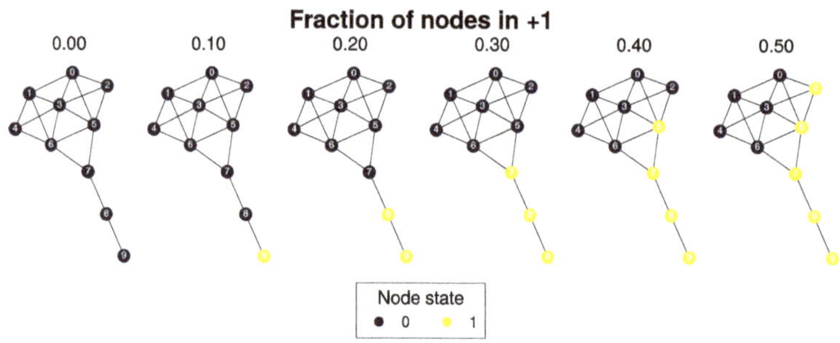

Figure 4. The tipping point is initiated from the bottom up. Each node is colored according to state 0 (black) and state 1 (yellow) Shown is a trajectory towards the tipping point that maximizes $\sum_{t=1}^{5} \log p(S^{t+1}|S^t, S^0 = \{0\}, \langle S^5 \rangle) = 0.5)$. As the system approaches the tipping point, low-degree nodes flip first and recruit "higher" degree nodes to further destabilize the system and push it towards a tipping point. In total, 30,240 trajectories reach the tipping point in 5 steps, and 10 trajectories have the same maximized values as the trajectory shown in this figure (see Figure A7 for the remaining trajectories and probabilities).

3.3. Network Structure and Node Roles in Metastable Transitions

The domino effect is not solely determined by node degree. As the system nears the tipping point, network effects become significant. For instance, in the kite graph, node 8 (degree 2) exhibits the highest integrated mutual information when 2 bits are flipped (Figure 2b). In contrast, node 3 (degree 6) shows low shared information prior to the tipping point but high shared information at the tipping point.

This transition highlights how the network structure as a whole contributes to a system's behavior. Local structural measures, such as degree centrality, may undervalue a node's contribution towards a tipping point and the eventual settlement in a new attractor.

3.4. Tipping Point Dynamics and Information Flow

At the tipping point, the system is most likely to either move to a new attractor state or relax back to its original state (Figure 4). Path analysis reveals that the most likely paths to the tipping point result in a configuration where a high-degree cluster of nodes must flip. This trajectory is less likely than reversing the path shown in Figure 4, explaining why most tipping points "fail" and relax back to the original attractor state (Figure 5b).

The increased information of node 8 around the tipping point can be understood by considering its predictive power about the system's future. As shown in Figure 5a, both node 3 and node 8 have low uncertainty about the future system state, but the nature of this certainty differs. Node 3 is more certain that the average system state will equal its state at the tipping point, while node 8 is more certain that the future system state will have the opposite sign to its state at the tipping point.

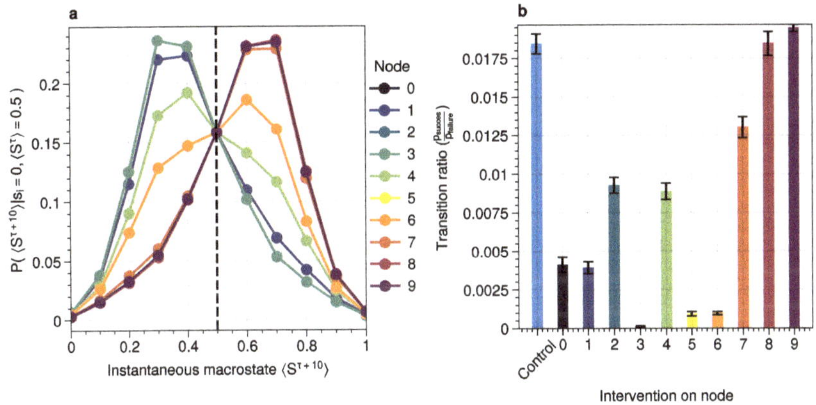

Figure 5. (**a**) Shown are the conditional probabilities at time $t = 10$ relative to the tipping point. The shared information between the hub node 3 and the tail node 8 is similar but, importantly caused through different sources. The hub (node 3) has high certainty that the system macrostate will be the same sign as its state. In contrast, node 8 has high certainty that the system macrostate will be opposite to its state at the tipping point. This is caused by the interaction between the network structure and the system dynamics whereby the most likely trajectories to the tipping point from the stable regime are mediated by the noise-induced dynamics from the tail to the core in the kite graph (see main text). (**b**) Successful metastable transitions are affected by network structure. Successful metastable transitions are those for which the sign of the macrostate is not the same prior to and after the tipping point, e.g., the system going from the 0 macrostate side to the +1 macrostate side or vice versa. Shown here are the number of successful metastable transitions for Figure 3 under control and pinning interventions on the nodes in the kite graph.

3.5. Role Division and Interventions in Tipping Behavior

We approximate the role of a node i using the difference between integrated mutual information and asymptotic information:

$$r_i = \max_{\langle S \rangle} \mu^*(s_i | \langle S \rangle) - \max_{\langle S \rangle} \omega^*(s_i) \in [-1, 1], \tag{4}$$

where μ^* and ω^* are normalized versions of μ and ω, respectively.

Nodes with role values close to 1 are classified as "initiators" with high predictive information about short-lived system trajectories. Nodes with values close to -1 are "stabilizers" with high long-term predictive information about future system states.

We validated these roles using simulated interventions (Figure 3). Pinning initiator nodes to the 0 state promotes tipping points while pinning stabilizer nodes is essential for stabilizing transitions between attractor states.

4. Discussion

Understanding how metastable transitions occur may help in understanding how, for example, a pandemic occurs or a system undergoes critical failure. In this paper, dynamical networks governed by the Boltzmann–Gibbs distribution were used to study how endogenously generated metastable transitions occur. The external noise parameter (temperature) was fixed such that the statistical complexity of the system behavior was maximized (see Appendix A.2).

The results show that in the network, two distinct node types could be identified: initiator and stabilizer nodes. Initiator nodes are essential early in the metastable transition. Due to their high degree of freedom, these nodes are more affected by external noise. They are instigators and propagate noise in the system, destabilizing more stable nodes. In contrast, stabilizer nodes have a low degree of freedom and require more energy to change state. These nodes are essential for the metastable behavior as they stabilize the system macrostate. During the metastable transition, a domino sequence of node state changes is propagated in an ordered sequence toward the tipping point.

This domino effect was revealed through two information features, unveiling an information cascade underpinning the trajectories toward the tipping point.

Integrated mutual information captured how short-lived correlations are passed on from the initiator nodes. In the stable regime (close to the ground state), low-degree nodes drive the system dynamics. Low-degree nodes destabilize the system, pushing the system closer to the tipping point. In most cases, the initiator nodes will fail to propagate the noise to their neighbors. On rare occasions, however, the cascade is propagated progressively from low degree to higher and higher degree. A similar domino mechanism was recently found in climate science [6,27]. Wunderling and colleagues provided a simplified model of the climate system, analyzing how various components contribute to the stability of the climate. They found that interactions generally stabilize the system dynamics. If, however, a metastable transition was initialized, the noise was propagated through a similar mechanism as found here, i.e., an initializer node propagated noise through the system, which created a domino effect that percolated through the system.

An increase in asymptotic information forms an indicator of how close the system is to a tipping point. Close to the ground state, the asymptotic information is low, reflecting how transient noise perturbations are not amplified, and the system macrostate relaxes back to the ground state. As the system approaches the tipping point, the asymptotic information increases. As the distance to the ground state increases, the system is more likely to transition between metastable states. After the transition, there remains a longer-term correlation. Asymptotic information reflects the long(er) timescale dynamics of the system. This "rest" information peaks at the tipping point as the system chooses its next state.

The information viewpoint uniquely offers an alternative view to understand how metastable transitions are generated by dynamical networks. Two information features were introduced that decompose the metastable transition in sources of high information processing (integrated mutual information) and distance of the system to the tipping point (asymptotic information). A domino effect was revealed, whereby low-degree nodes initiate the tipping point, making it more likely for higher-degree nodes to tip. On the tipping point, long-term correlations stabilize the system inside the new metastable state. Importantly, the information perspective allows for estimating integrated mutual information directly from data without knowing the mechanisms that drive the tipping behavior. The results highlight how short-lived correlations are essential to initiate the information cascade for crossing a tipping point.

5. Conclusions

Our information-theoretic approach offers an alternative view to understanding *how* metastable transitions are generated by dynamical networks. Two information features were introduced that decompose the metastable transition in sources of high information processing (integrated mutual information) and distance of the system to the tipping point (asymptotic information). A domino effect was revealed, whereby low-degree nodes initiate the tipping point, making it more likely for higher-degree nodes to tip. On the tipping point, long-term correlations stabilize the system inside the new metastable state. Importantly, the information perspective allows for estimating integrated mutual information directly from data without knowing the mechanisms that drive the tipping behavior. The results highlight how short-lived correlations are essential to initiate the information cascade for crossing a tipping point.

6. Limitations

Integrated mutual information was computed based on exact information flows. This means that for binary systems, it is necessary to compute a transfer matrix on the order of $2^{|S|} \times 2^{|S|}$. This reduced the present analysis to smaller graphs. It would be possible to use Monte-Carlo methods to estimate the information flows. However, $I(s_i^\tau : S^{\tau+t})$ remains expensive to compute. When using computational models, it is necessary to compute the conditional and marginal distributions, which are on order $\mathbb{O}(2^{|S|})$ and $\mathbb{O}(2^{t|S|})$, respectively. In Appendix A.11, we give a proof of principle of how the results presented here would generalize to larger systems.

In addition, the decomposition of the metastable transition depends on the partition of the state space. Information flows are, in essence, statistical dependencies among random variables. Here, the effect of how the tipping point was reached was studied by partitioning the average system state in terms of the number of bits flipped. This partitioning assumes that the majority of states prior to the tipping point are reached by having fraction $c \in [0,1]$ bits flipped. The contribution of each system state over time, however, reflects a distribution of different states; reaching the tipping point from the ground state 0 can be done at $t-2$ prior to tipping by either remaining in 0.4 bits or transitioning from 0.3 bits flipped to 0.4 and eventually to 0.5 in 2 time steps. The effect of these additional paths showed marginal effects on the integrated mutual information and asymptotic information.

Information flows conditioned on a partition is a form of conditional mutual information [33]. Prior results showed that conditional information produces synergy, i.e., information that is only present in the joint of all variables but cannot be found in any of the subsets of each variable. Unfortunately, there is no generally agreed-upon definition of how to measure synergy [34,35], and different estimates exist that may over or underestimate the synergetic effects. By partitioning, one can create synergy as, for a given partition, each spin has some additional information about the other spins. For example, by taking the states such that $\langle S \rangle = 0.1$, each spin "knows" that the average of the system equals 0.1. This creates shared information among the spins. Analyses were performed to estimate synergy using the redundancy estimation I_{min}[36]. Using this approach, no synergy was measured that affected the outcome of this study. However, it should be emphasized that synergetic effects may influence the causal interpretation of the approach presented here.

A general class of systems was studied governed by the Boltzmann–Gibbs distribution. For practical purposes, the kinetic Ising model was only tested, but we speculate that the results should hold (in principle) for other systems dictated by the Boltzmann–Gibbs distribution. We leave the extension to other Hamiltonian systems for future work.

The practical implementation of interventions based on our theoretical framework faces several real-world challenges. First, in actual complex systems, measuring and monitoring the complete state space in real time may be technically infeasible or prohibitively expensive. Second, the ability to perform precise, targeted interventions on specific components of the system may be limited by physical constraints or technological capabilities. Third, the assumption of perfect knowledge about system parameters and state transitions may not hold in real-world scenarios where noise, measurement errors, and external perturbations are present. Furthermore, the time scales at which interventions need to be implemented may be too rapid for practical human or automated response systems. These practical limitations suggest that while our framework provides valuable theoretical insights, its application may require significant adaptation and simplification for real-world implementation, potentially trading off optimal control for practical feasibility.

Author Contributions: Conceptualization, C.v.E.; Methodology, C.v.E.; Software, C.v.E.; Validation, C.v.E.; Formal analysis, C.v.E.; Investigation, C.v.E.; Resources, C.v.E.; Data curation, C.v.E.; Writing—original draft preparation, C.v.E.; Writing—review and editing, C.v.E., R.Q. and P.M.A.S.; Visualization, C.v.E.; Supervision, C.v.E., R.Q. and P.M.A.S.; Project administration, C.v.E.; Funding acquisition, C.v.E., R.Q. and P.M.A.S. All authors have read and agreed to the published version of the manuscript.

Funding: This research is supported by grant Hyperion 2454972 of the Dutch National Police.

Institutional Review Board Statement: Not applicable.

Data Availability Statement: The datasets generated and/or analyzed during the current study are available in the https://github.com/cvanelteren/metastability repository, accessed date (28 November 2024).

Conflicts of Interest: The authors declare no competing interests.

Appendix A

Appendix A.1. Background, Scope and Innovation

Noise-induced transitions produce metastable behavior that is fundamental for the functioning of complex dynamical systems. For example, in neural systems, the presence of noise enhances information processing through several mechanisms: stochastic resonance, where moderate levels of noise can amplify weak signals and improve signal detection; prevention of neural networks becoming stuck in local minima, therefore maintaining system flexibility; and enabling more efficient exploration of different neural states during computation [37–40]. These effects have been demonstrated both in experimental studies of neural circuits and theoretical models of neural computation.

Similarly, the relation between glacial ice ages and Earth's eccentricity has been shown to have a strong correlation. Metastability manifests itself through noise that can be of two kinds [9]. External noise originates from events outside the internal system dynamics [10,41]. Examples include the influence of climate effects, population growth, or random noise sources on transmission lines. External noise is commonly modeled by replacing an external control or order parameter with a stochastic process. Internal noise, in contrast, is inherent to the system itself and is caused by random interactions of elements within the system, e.g., individuals in a population or molecules in chemical processes. Both types of noise can generate transitions between metastable states. In this paper, we study the metastable behavior of internal noise in complex dynamical networks governed by kinetic Ising dynamics.

The ubiquity of multistability in complex systems calls for a general framework to understand *how* metastable transitions occur. The diversity of complex systems can be captured by interaction networks that dynamically evolve over time. These dynamics can be seen as a distributive network of computational units, where each unit or element of the interaction network changes its state based on input from its local neighborhood. Lizier demonstrated that the dynamic interaction of complex systems can be understood through their local information processing [42–44]. Instead of describing the dynamics of the system in terms of domain knowledge such as voltage over distance, disease spreading rate, or climate conditions, one can understand the dynamics in terms of *information dynamics*. In particular, the field of information dynamics is concerned with describing system behavior through its capacity to store, transmit, and modify information. By abstracting away the domain details of a system and recasting the dynamics in terms of *how* the system computes its next state, one can capture the intrinsic computations a system performs. The system behavior is encoded in terms of probability, and the relationships among these variables are explored using the language of information theory [45].

Information theory offers profound benefits over traditional methods used in metastability analysis, as the methods developed are model-free, can capture nonlinear relationships, can be used for both discrete and continuous variables, and can be estimated directly from data [30]. Shannon information measures, such as mutual information and Fisher information, can be used to study how much information the system dynamics shares with the control parameter [11,46].

Past research on information flows and metastable transitions focuses on methods to detect the onset of a tipping point [47–49]. It often centers around an observation that the system's ability to absorb noise reduces prior to the system going through a critical point. This critical slowing down can be captured as a statistical signature where the Fisher information peaks [50]. However, these methods traditionally use some form of control parameter driving the system towards or away from a critical point. Most real-world systems lack such an explicit control parameter and require different methods. Furthermore, detecting a tipping point does not necessarily lead to a further understanding of how the tipping point was created. For example, in a finite-size Ising model, the system produces bistable behavior. As one increases the noise parameter, the bistable behavior disappears. The increase in noise effectively changes the energy landscape, but little information is gained about how the metastable behavior initially emerged.

In this work, a novel approach using information theory is explored to study metastable behavior. The statistical coherence between parts of the system is quantified by the capability of individual nodes to predict the future behavior of the system [43]. Two information features are introduced: *Integrated mutual information* measures predictive information of a node on the future of the system, and *Asymptotic information* measures the long timescale memory capacity of a node. These measures differ from previous information methods such as transfer entropy [51], conditional mutual information under causal intervention [52], causation entropy [53], and time-delayed variants [54] in that these methods are used to infer the transfer of information between sets of nodes by possibly correcting for a third variable. Here, instead, we aim to understand how the elements in the system contribute to the macroscopic properties of the system. It is important to emphasize that information flows are not directly comparable to causal flows [33]. A rule of thumb is that causal flows focus on micro-level dynamics (X causes Y), whereas information flows focus on the predictive aspects, providing a holistic view of emergent structures [43]. In this sense, this work is similar to predictive information [55] where predictive information of some system (S) is projected onto its consistent elements ($s_i \in S$) and computed as a function of time (t).

Appendix A.2. Methods and Definitions

Appendix A.2.1. Model

To study metastable behavior, we consider a system as a collection of random variables $S = \{s_1, \ldots, s_n\}$ governed by the Boltzmann–Gibbs distribution:

$$p(S) = \frac{1}{Z} \exp(-\beta \mathcal{H}(S)),$$

where $\beta = \frac{1}{T}$ is the inverse temperature which controls the noise in the system, and $\mathcal{H}(S)$ is the system Hamiltonian which encodes the node-node dynamics. The choice of the energy function dictates what kind of system behavior we observe. Here, we focus on arguably the simplest models that show metastable behavior: the kinetic Ising model and the Susceptible-Infected-Susceptible model.

Temporal dynamics are simulated using Glauber dynamics sampling. In each discrete time step, a spin is randomly chosen, and a new state $X' \in S$ is accepted with probability

$$p(\text{accept } X') = \frac{1}{1 + \exp(-\beta \Delta E)}, \tag{A1}$$

where $\Delta E = \mathcal{H}(X') - \mathcal{H}(X)$ is the energy difference between the current state X and the proposed state X'.

Appendix A.2.2. Kinetic Ising Model

The traditional Ising model, originally developed to study ferromagnetism, is considered one of the simplest models that generate complex behavior. It consists of a set of binary distributed spins $S = \{s_1, \ldots, s_n\}$. Each spin contains energy given by the Hamiltonian:

$$\mathcal{H}(S) = -\sum_{i,j} J_{ij} s_i s_j - h_i s_i, \tag{A2}$$

where J_{ij} is the interaction energy of the spins s_i, s_j.

The interaction energy effectively encodes the underlying network structure of the system. Different network structures are used in this study to provide a comprehensive numerical overview of the relation between network structure and information flows (see Appendix A.2). The interaction energy J_{ij} is set to 1 if a connection exists in the network.

For sufficiently low noise (temperature), the Ising model shows metastable behavior (Figure 1c). Here, we aim to study *how* the system goes through a tipping point by tracking the information flow per node with the entire system state.

Appendix A.3. Information Flow on Complex Networks

Informally, information flows measure the statistical coherence between two random variables X and Y over time, such that the present information in Y cannot be explained by the past of Y but rather by the past of X. Estimating information flow is inherently difficult due to the presence of confounding factors, which potentially trap the interpretation in the "correlation does not equal causation" paradigm. Under some contexts, however, information flow can be interpreted as causal [31]. Let $S = \{s_1, \ldots, s_n\}$ be a random process, and S^t represent the state of the random process at some time t. The information present in S is given as the Shannon entropy:

$$H(S) = -\sum_{x \in S} p(x) \log p(x), \tag{A3}$$

where log is base 2 unless otherwise stated, and $p(x)$ is used as shorthand for $p(S = x)$. Shannon entropy captures the uncertainty of a random variable; it can be understood as the number of yes/no questions needed to determine the state of S. This measure of uncertainty naturally extends to two variables with Shannon mutual information. Let s_i be an element of the state of S, then the Shannon mutual information $I(S; s_i)$ is given as:

$$\begin{aligned} I(S; s_i) &= \sum_{S_i \in S, s' \in s_i} p(S_i, s') \log \frac{p(S_i, s')}{p(S_i) p(s')} \\ &= H(S) - H(S|s_i). \end{aligned} \tag{A4}$$

Shannon mutual information can be interpreted as the uncertainty reduction of S after knowing the state of s_i. Consequently, it encodes how much statistical coherence s_i and S share. Shannon mutual information can be measured over time to encode how much *information* (in bits) flows from state s_i^τ to $S^{\tau+t}$:

$$I(S^{\tau+t}; s_i^\tau) = H(S^{\tau+t}) - H(S^{\tau+t}|s_i^\tau). \tag{A5}$$

Prior results showed that the nodes with the highest causal importance are those nodes that have the highest information flow (i.e., maximize Equation (A5)) [31]. Intuitively, the nodes for which the future system "remembers" information from a node in the past are the ones that "drive" the system dynamics. Formally, these driver nodes can be identified by computing the total information flow between S^t and s_i, which can be captured with the integrated mutual information [31]:

$$\mu(s_i) = \sum_{\tau=0}^{\infty} I(s_i^{t-\tau}; S^t). \tag{A6}$$

In some contexts, the nodes that maximize (A6) are those nodes that have the highest causal influence in the system [31]. However, in general, information flows are difficult to equate to causal flows [33,43]. Here, the local information flows are computed by considering the integrated mutual information conditioned on part of the entire state space. This allows for mapping the local information flows between nodes and the system over time but does not guarantee that the measured information flows are directly causal. The main reason is that having predictive power about the future could be completely caused by the partitioning. In [31], the correlation measured considered all possible states, and the measures were directly related to a causal effect.

In addition, in [31], the shared information between the system with a node shifted over time ($I(S^\tau : s_i^{\tau+t})$) was considered. Applying this approach under a state partition $I(S^\tau : s_i^{\tau+t}|\langle S\rangle)$ causes a violation of the data processing inequality, as information may flow from a node at a particular time $t = t_1$ and then flow back to the node at $t = t_2$, where $t_2 > t_1$. To simplify the interpretation of the information flows and maintain the data processing inequality, the reverse $I(S^{t+\tau} : s_i^\tau|\langle S\rangle)$ was computed in the present study.

Appendix A.4. Noise Matching Procedure

The Boltzmann–Gibbs distribution is parameterized by noise factor $\beta = \frac{1}{kT}$, where T is the temperature and k is the Boltzmann constant. For high β values, metastable behavior occurs in the kinetic Ising model. The temperature was chosen such that the statistical complexity [56] was maximized. The statistical complexity C is computed as:

$$C = \bar{H}(S)D(S),$$

where $\bar{H}(S) = \frac{H(s)}{-\log_2(|S|)}$ is the system entropy, and $D(S)$ measures the distance to disequilibrium:

$$D(S) = \sum_i (p(S_i) - \frac{1}{|S|})^2.$$

A typical statistical complexity curve is shown in Figure A1. The noise parameter β is set such that it maximizes the statistical complexity using numerical optimization (COBYLA method in Scipy's `optimize.minimize` module) [57].

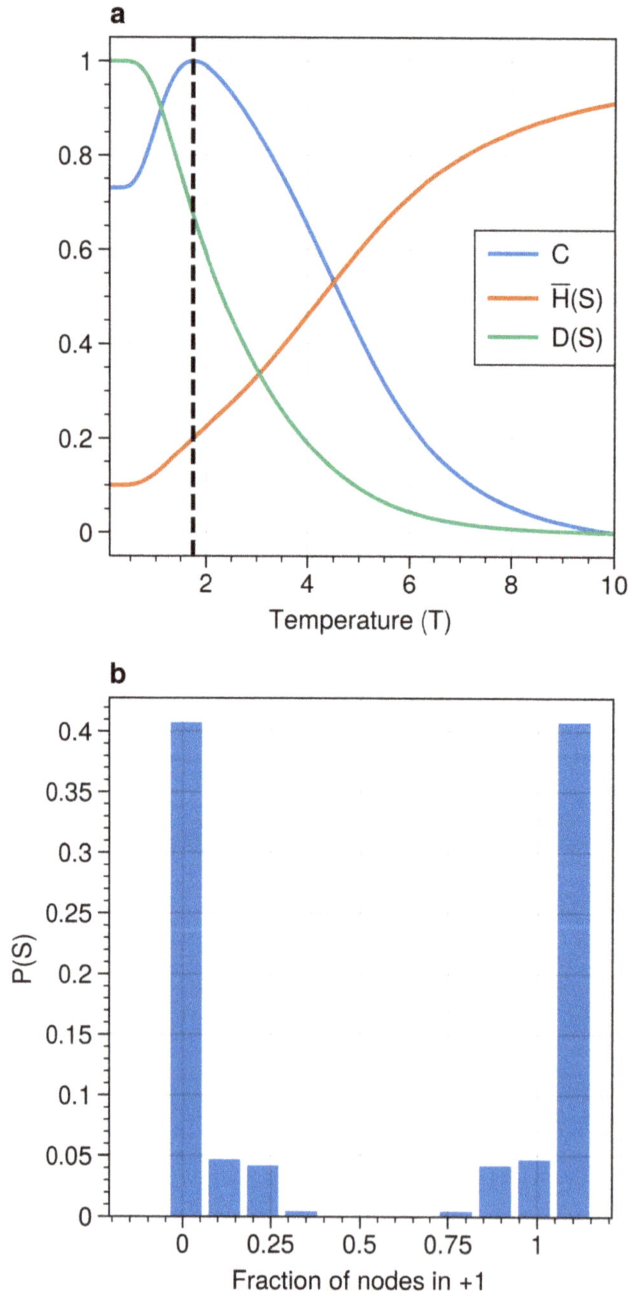

Figure A1. (**a**) Statistical complexity (C), normalized system entropy ($H(S)$) and disequilibrium ($D(S)$) as a function of the temperature ($T = \frac{1}{\beta}$) for Krackhardt kite graph. The noise parameter was set such that it maximizes the statistical complexity (vertical black line). The values are normalized between [0, 1] for aesthetic purposes. (**b**) State distribution $p(S)$ for temperature that maximizes the statistical complexity in (**a**) as a function of nodes in state 1.

Appendix A.5. Exact Information Flows $I(s_i^\tau; S^{\tau+t})$

In order to compute $I(s_i^\tau : S^{\tau+t})$, the conditional distribution $p(S^{\tau+t}|s_i^\tau)$ and $p(S^{\tau+t})$ need to be computed. This is achieved through direct computation on the partitions of the state consisting of the number of bits flipped—effectively encoding the distance towards or away from a tipping point.

We partition the state space based on the average magnetization $\langle S \rangle$, allowing us to track states that are n bit flips away from the tipping point.

For Glauber dynamics, the system S transitions into S' by randomly choosing node s_i to potentially flip. The transition matrix $p(S^t|s_i) = \mathbf{P}$ can be constructed by computing each entry p_{ij} as:

$$
\begin{aligned}
p_{ij,i\neq j} &= \frac{1}{|S|}\frac{1}{1+\exp(-\Delta E)}, \text{ with} \\
p_{ii} &= 1 - \sum_{j,j\neq i} p_{ij},
\end{aligned}
$$

where $\Delta E = \mathcal{H}(S_j) - \mathcal{H}(S_i)$ encodes the energy difference of moving from S_i to S_j.

For each partition $S_\gamma = \{S' \subseteq S | \langle S' \rangle = \gamma\}$, we:

1. Compute transition probabilities between states within the partition
2. Renormalize probabilities to ensure conservation within the partition
3. Evaluate $p(S^t|s_i)$ for all possible node states s_i in that partition

The marginal distribution $p(S^t)$ is then computed as:

$$
p(S^{\tau+t}) = \sum_{s_i} p(S^{\tau+t}|s_i^\tau)p(s_i^\tau).
$$

This procedure provides exact information flows for states at specific distances from the tipping point, allowing us to track how correlations evolve as the system approaches and moves away from the transition.

Extrapolation with Regressions

Exact information flows were computed per graph for $t = 500$ time steps. Using ordinary least squares, a double exponential was fit to estimate the information flows for longer t and estimate the integrated mutual information and asymptotic information.

Appendix A.6. Noise Estimation Procedure

Tipping point behavior under intervention was quantified by evaluating the level of noise on both sides of the tipping point. Let T_1 represent the ground state where all spins are 0, T_2 where all spins are 1, and let the tipping point TP be where the instantaneous macrostate $M(S^t) = 0.5$. Fluctuations of the system macrostate were evaluated by analyzing the second moment above and below the tipping point. This was achieved by numerically simulating the system trajectories under 6 different seeds for $t = 10^6$ time steps. The data were split between two sets (above and below the tipping point), and the noise η was computed as:

$$
\eta = \frac{1}{\alpha^2|S_w|}\sum_w {S_w^t}^2,
$$

where $w \in \{\langle S \rangle < 0.5, \langle S \rangle > 0.5\}$, and

$$
S_w^t = \begin{cases} S^t & \text{if } S^t < 0.5 \\ 1 - S^t & \text{if } S^t > 0.5 \end{cases} \tag{A7}
$$

is the instantaneous system trajectory for the system macrostate above or below the tipping point value. The factor α corrects for the reduced range the system macrostate has under interventions. For example, pinning a node s_i to state 0 reduces the maximum possible macrostate to $1 - \frac{1}{n}$, where n is the size of the system. The correction factor α is set such that for an intervention on 0 for a particular node, the range $S_{\langle S \rangle > 0.5}$, α is set to $\frac{n}{2} - \frac{1}{n}$.

Appendix A.7. Switch Susceptibility as a Function of Degree

First, we investigate the susceptibility of a spin as a function of its degree. The susceptibility of a spin switching its state is a function of both the system temperature T and the system dynamics. The system dynamics contribute to the susceptibility through the underlying network structure either directly or indirectly. The network structure produces local correlations, which affect the switching probability for a given spin.

As an initial approximation, we consider the susceptibility of a target spin s_i to flip from a majority state to a minority state, given the state of its neighbors, where the neighbors are not connected among themselves. Furthermore, we assume that for the instantaneous update of s_i, the configuration of the neighborhood of s_i can be considered to be the outcome of a binomial trial. Let N be a random variable with state space $\{0,1\}^{|N|}$, and let $n_j \in N$ represent a neighbor of s_i. We assume that all neighbors of s_i are i.i.d. distributed given the instantaneous system magnetization:

$$M(S^t) = \frac{1}{|S^t|} \sum_i s_i^t.$$

Let the minority state be 1, and the majority state be 0. The expectation of s_i flipping from the majority state to the minority state is given as:

$$\begin{aligned} E[p(s_i = 1|N)]_{p(N)} &= \sum_{N_i \in N} p(N_i) p(s_i = 1|N_i) \\ &= \sum_{N_i \in N} \prod_j^{|N_i|} p(n_j) p(s_i = 1|N_i) \\ &= \sum_{N_i \in N} \binom{n}{k} f^k (1-f)^{n-k} p(s_i = 1|f), \end{aligned} \tag{A8}$$

where f is the fraction of nodes in the majority states, n is the number of neighbors, and k is the number of nodes in state 0. As shown in Figure A3, this is computed as a function of the degree of spin s_i. As the degree increases, the susceptibility for a spin decreases relative to the same spin with a lower degree. This implies that the susceptibility to change due to random fluctuations is more likely to occur in nodes with fewer external constraints as measured by degree.

Appendix A.8. Additional Networks

The kite graph was chosen as it allowed for computing exact information flows while retaining a high variety of degree distribution given the small size. Other networks were also tested. In Figure A2, different network structures were used. Each node is governed by kinetic Ising spin dynamics.

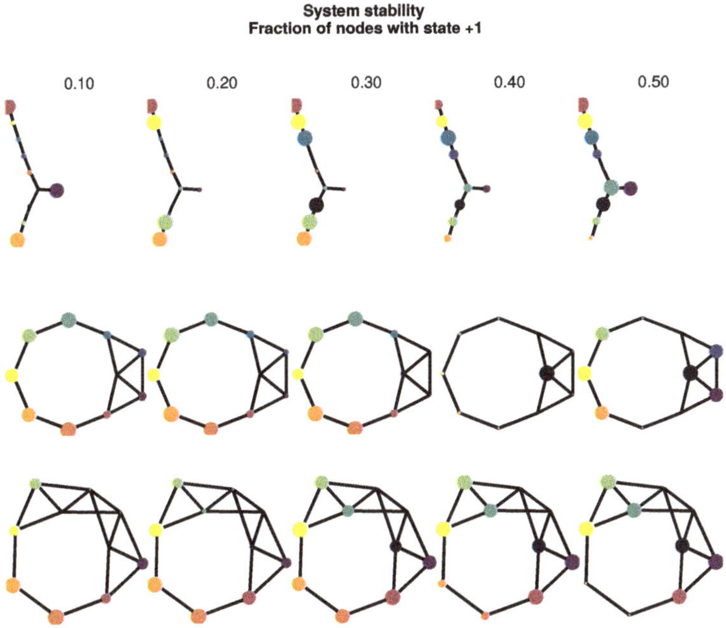

Figure A2. Adjusted mutual information for a random tree (**top**) and Leder–Coxeter–Frucht graphs (**middle,bottom**). Each node is governed by kinetic Ising spin dynamics. Far from the tipping point (fraction nodes 1 = 0.5), most information flows are concentrated on non-hub nodes. As the system approaches the tipping point (fraction = 0.5), the information flows move inwards, generating higher adjusted integrated mutual information for nodes with higher degrees.

Appendix A.9. Flip Probability per Degree

In Figure A3, the tendency for a node to flip from the majority to the minority state is computed as a function of the fraction of nodes possessing the majority state 1 in the system, denoted as N. Two things are observed. First, nodes with lower degrees are more susceptible to noise than nodes with higher degrees. For a given system stability, nodes with lower degrees tend to have a higher tendency to flip. This holds true for all distances of the system from the tipping point. In contrast, the higher the degree of the node, the closer the system must be to a tipping point for the node to change its state. This can be explained by the fact that lower-degree nodes have fewer constraints compared to nodes with higher degrees. For Ising spin kinetics, the nodes with higher degrees tend to be more "frozen" in their node dynamics than nodes with lower degrees. Second, for a node to flip with similar probability mass (i.e., $E[p(s_i)|N] = 0.2$), a node with higher degrees needs to be closer to the tipping point than nodes with lower degrees. In fact, the order of susceptibility is correlated with the degree; the susceptibility decreases with increasing degree and fixed fraction of nodes in state 1.

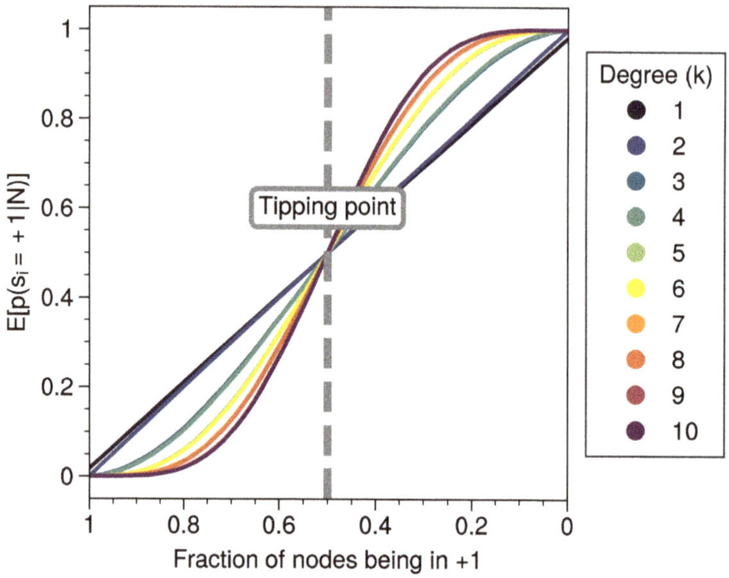

Figure A3. Susceptibility of a node with degree k switching from the minority state 0 to the majority state 1 as a function of the neighborhood entropy for $\beta = 0.5$. The neighborhood entropy encodes how stable the environment of a spin is. As the system approaches the tipping point, the propensity of a node to flip from the minority state increases faster for low-degree nodes than for high-degree nodes. Higher-degree nodes require more change in their local environment to flip to the majority state. See for details Appendix A.7.

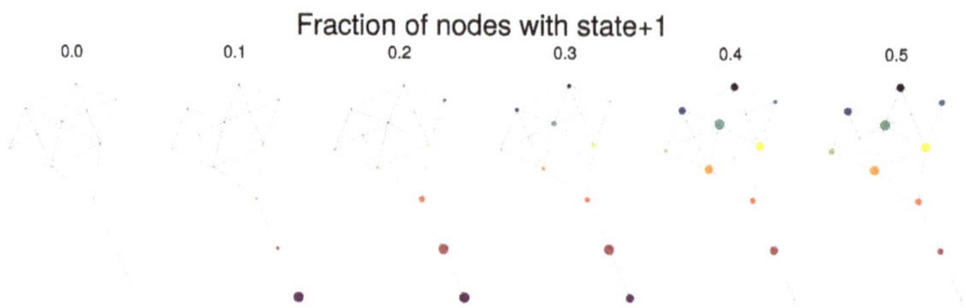

Figure A4. Shortest path analysis of the system ending up in the tipping point from the state where all nodes have state 0. The node size is proportional to the expectation value of a node that has state 1 ($E[s_i = 1]_{S^t, M(S^5)}$) as a function of the fraction of nodes that have state 1. The expectation values are computed based on 30240 trajectories; an example trajectory can be seen in Figure 4.

Appendix A.10. Synthetic Networks

For the synthetic graphs, 100 non-isomorphic connected Erdős–Rényi networks were generated with p = 0.2. Graphs were generated randomly and rejected if they did not contain a giant component or were isomorphic with already generated graphs. For each of the graphs, information curves were computed as a function of the macrostate $\langle S \rangle$.

Noise and Time Spent

Various network structures are generated in the synthetic networks. The variety of network structures has nonlinear effects on the information flows. The effect of intervention in Figure 3 is made relative to the control values for the graph and seed. The second moment (appendix: Appendix A.6) and the time spent below the tipping point are normalized with respect to the graph (Figure A5) and the seed. In total, 6 seeds are used (0, 12, 123, 1234, 123,456, 1,234,567).

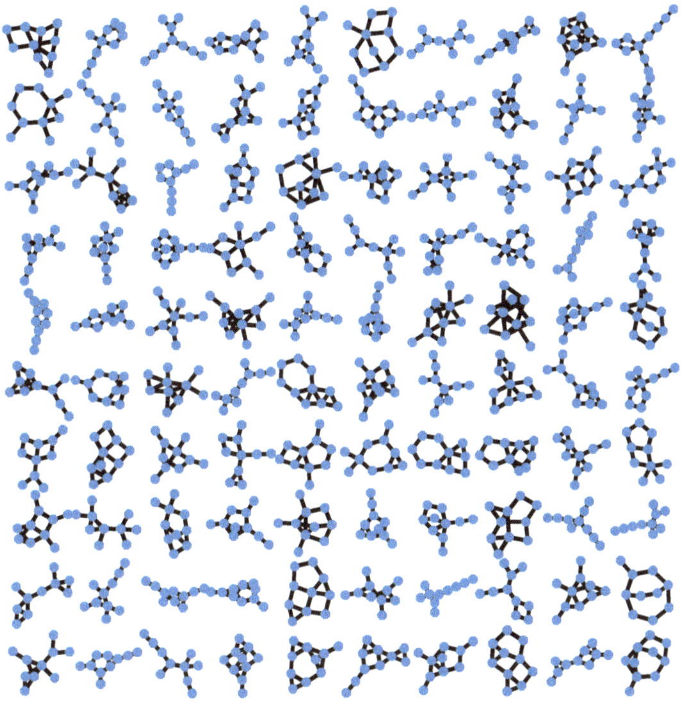

Figure A5. Erdős–Rényi graphs generated from seed = 0 to produce non-isomorphic connected graphs.

Appendix A.11. Case Study of a Larger System

In this section, we extend our analysis to a 15-node network to demonstrate the applicability of our findings to larger systems (see Figure A6). This case study serves to validate our theoretical insights derived from smaller networks and to illustrate how the fundamental mechanisms of metastable transitions are preserved as network size increases. Despite the increased computational complexity, our results indicate that the structural features driving these transitions in smaller networks are also evident in larger ones.

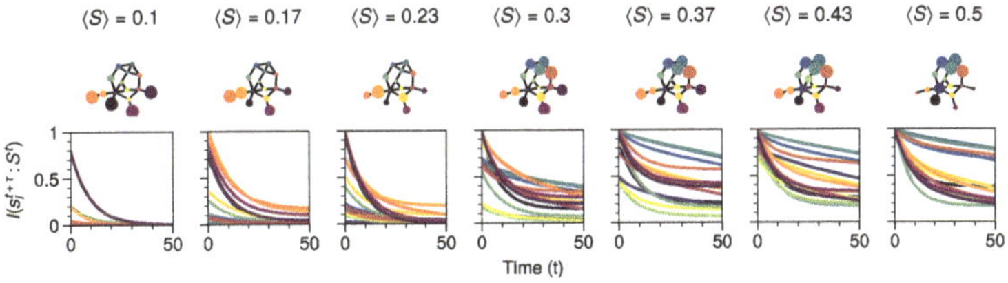

Figure A6. Example of tipping behavior in a system consisting of $N = 15$ nodes. The colors of the curves correspond to the nodes in the network. The information decay curves are bundled per degree. The transition from left to right increases the number of bits flipped until the tipping point. A wave can be seen where the integrated information flows from lower-degree nodes to higher ones as the number of bits flipped increases. The size of the nodes is proportional to the integrated mutual information.

As highlighted in Section 6, the state space of a network grows exponentially (2^n) with the number of nodes, making simulations of larger systems computationally demanding. Nevertheless, our analysis of the 15-node network supports our assertion that the foundational processes identified in our primary study can be extrapolated to more complex networks. Detailed results and discussion of this 15-node network analysis are provided to substantiate our approach and highlight the consistency of our findings across different network sizes.

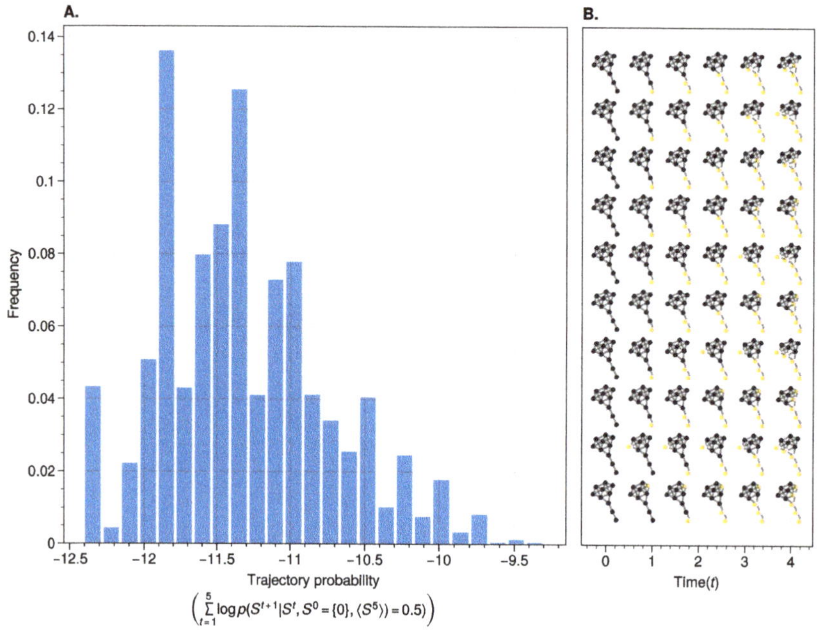

Figure A7. (**A**) Probability distribution of trajectories reaching the tipping point within 5 time steps, starting from an initial state where all nodes are set to 0. Out of all possible trajectories, 30,240 paths reached the tipping point. (**B**) Visualization of the highest-probability trajectories leading to system collapse. These equiprobable paths demonstrate the cascading failure mechanism, where specific initiator nodes trigger a domino effect throughout the network. The color gradient indicates the temporal progression of state changes, illustrating the sequential nature of the collapse process.

References

1. Ladyman, J.; Lambert, J.; Wiesner, K. What Is a Complex System? *Eur. J. Philos. Sci.* **2013**, *3*, 33–67. [CrossRef]
2. van Nes, E.H.; Arani, B.M.; Staal, A.; van der Bolt, B.; Flores, B.M.; Bathiany, S.; Scheffer, M. What Do You Mean, 'Tipping Point'? *Trends Ecol. Evol.* **2016**, *31*, 902–904. [CrossRef] [PubMed]
3. Kandel, E.R.; Schwartz, J.H.; Jessell, T.M. *Principles of Neural Science*, 4th ed.; McGraw-Hill Medical: New York, NY, USA, 2000.
4. Fries, P. Rhythms for Cognition: Communication through Coherence. *Neuron* **2015**, *88*, 220–235. [CrossRef] [PubMed]
5. Galam, S.; Cheon, T. Tipping Points in Opinion Dynamics: A Universal Formula in Five Dimensions. *Front. Phys.* **2020**, *8*, 566580. [CrossRef]
6. Wunderling, N.; Donges, J.F.; Kurths, J.; Winkelmann, R. Interacting Tipping Elements Increase Risk of Climate Domino Effects under Global Warming. *Earth Syst. Dyn.* **2021**, *12*, 601–619. [CrossRef]
7. Beggs, J.M.; Timme, N. Being Critical of Criticality in the Brain. *Front. Physiol.* **2012**, *3*. [CrossRef]
8. Mitchell, M.; Hraber, P.; Crutchfield, J.P. Revisiting the Edge of Chaos: Evolving Cellular Automata to Perform Computations. *arXiv* **1993**. [CrossRef]
9. Forgoston, E.; Moore, R.O. A Primer on Noise-Induced Transitions in Applied Dynamical Systems. *SIAM Rev.* **2018**, *60*, 969–1009. [CrossRef]
10. Czaplicka, A.; Holyst, J.A.; Sloot, P.M.A. Noise Enhances Information Transfer in Hierarchical Networks. *Sci. Rep.* **2013**, *3*, 1223. [CrossRef]

11. Nicolis, G.; Nicolis, C. Stochastic Resonance, Self-Organization and Information Dynamics in Multistable Systems. *Entropy* **2016**, *18*, 172. [CrossRef]
12. McNamara, B.; Wiesenfeld, K. Theory of Stochastic Resonance. *Phys. Rev. A* **1989**, *39*, 4854–4869. [CrossRef] [PubMed]
13. Kramers, H.A. Brownian Motion in a Field of Force and the Diffusion Model of Chemical Reactions. *Physica* **1940**, *7*, 284–304. [CrossRef]
14. Harush, U.; Barzel, B. Dynamic Patterns of Information Flow in Complex Networks. *Nat. Commun.* **2017**, *8*, 2181 . [CrossRef] [PubMed]
15. Gao, J.; Barzel, B.; Barabási, A.L. Universal Resilience Patterns in Complex Networks. *Nature* **2016**, *536*, 238–238. [CrossRef]
16. Dong, G.; Wang, F.; Shekhtman, L.M.; Danziger, M.M.; Fan, J.; Du, R.; Liu, J.; Tian, L.; Stanley, H.E.; Havlin, S. Optimal Resilience of Modular Interacting Networks. *Proc. Natl. Acad. Sci. USA* **2021**, *118*, e1922831118. [CrossRef]
17. Liu, Y.; Sanhedrai, H.; Dong, G.; Shekhtman, L.M.; Wang, F.; Buldyrev, S.V.; Havlin, S. Efficient Network Immunization under Limited Knowledge. *Natl. Sci. Rev.* **2021**, *8*, nwaa229. [CrossRef]
18. Zenil, H.; Kiani, N.A.; Marabita, F.; Deng, Y.; Elias, S.; Schmidt, A.; Ball, G.; Tegnér, J. An Algorithmic Information Calculus for Causal Discovery and Reprogramming Systems. *iScience* **2019**, *19*, 1160–1172. [CrossRef]
19. Zenil, H.; Kiani, N.A.; Zea, A.A.; Tegnér, J. Causal Deconvolution by Algorithmic Generative Models. *Nat. Mach. Intell.* **2019**, *1*, 58–66. [CrossRef]
20. Guo, C.; Yang, L.; Chen, X.; Chen, D.; Gao, H.; Ma, J. Influential Nodes Identification in Complex Networks via Information Entropy. *Entropy* **2020**, *22*, 242. [CrossRef]
21. Hopfield, J.J. Neural Networks and Physical Systems with Emergent Collective Computational Abilities. *Proc. Natl. Acad. Sci. USA* **1982**, *79*, 2554–2558. [CrossRef]
22. Glauber, R.J. Time-Dependent Statistics of the Ising Model. *J. Math. Phys.* **1963**, *4*, 294–307. [CrossRef]
23. Lenton, T.M.; Abrams, J.F.; Bartsch, A.; Bathiany, S.; Boulton, C.A.; Buxton, J.E.; Conversi, A.; Cunliffe, A.M.; Hebden, S.; Lavergne, T.; et al. Remotely Sensing Potential Climate Change Tipping Points across Scales. *Nat. Commun.* **2024**, *15*, 343. [CrossRef] [PubMed]
24. Peng, X.; Small, M.; Zhao, Y.; Moore, J.M. Detecting and Predicting Tipping Points. *Int. J. Bifurc. Chaos* **2019**, *29*, 1930022. [CrossRef]
25. Bury, T.M.; Sujith, R.I.; Pavithran, I.; Scheffer, M.; Lenton, T.M.; Anand, M.; Bauch, C.T. Deep Learning for Early Warning Signals of Tipping Points. *Proc. Natl. Acad. Sci. USA* **2021**, *118*, e2106140118. [CrossRef]
26. D'Orsogna, M.R.; Perc, M. Statistical Physics of Crime: A Review. *Phys. Life Rev.* **2015**, *12*, 1–21. [CrossRef]
27. Wunderling, N.; Stumpf, B.; Krönke, J.; Staal, A.; Tuinenburg, O.A.; Winkelmann, R.; Donges, J.F. How Motifs Condition Critical Thresholds for Tipping Cascades in Complex Networks: Linking Micro- to Macro-Scales. *Chaos Interdiscip. J. Nonlinear Sci.* **2020**, *30*, 043129. [CrossRef]
28. Yang, Y.; Motter, A.E. Cascading Failures as Continuous Phase-Space Transitions. *Phys. Rev. Lett.* **2017**, *119*, 248302. [CrossRef]
29. Yang, Y.; Nishikawa, T.; Motter, A.E. Small Vulnerable Sets Determine Large Network Cascades in Power Grids. *Science* **2017**, *358*, eaan3184. [CrossRef]
30. Cover, T.M.; Thomas, J.A. *Elements of Information Theory*; Wiley-Interscience: New York, NY, USA, 2005. [CrossRef]
31. van Elteren, C.; Quax, R.; Sloot, P. Dynamic Importance of Network Nodes Is Poorly Predicted by Static Structural Features. *Phys. A Stat. Mech. Its Appl.* **2022**, *593*, 126889. [CrossRef]
32. Quax, R.; Apolloni, A.; a Sloot, P.M. The Diminishing Role of Hubs in Dynamical Processes on Complex Networks. *J. R. Soc. Interface R. Soc.* **2013**, *10Q*, 20130568. [CrossRef]
33. James, R.G.; Barnett, N.; Crutchfield, J.P. Information Flows? A Critique of Transfer Entropies. *Phys. Rev. Lett.* **2016**, *116*, 238701 . [CrossRef] [PubMed]
34. Beer, R.D.; Williams, P.L. Information Processing and Dynamics in Minimally Cognitive Agents. *Cogn. Sci.* **2015**, *39*, 1–38. [CrossRef] [PubMed]
35. Kolchinsky, A. A Novel Approach to the Partial Information Decomposition. *Entropy* **2022**, *24*, 403. [CrossRef] [PubMed]
36. Williams, P.L.; Beer, R.D. Nonnegative Decomposition of Multivariate Information. *arXiv* **2010**, arXiv:1004.2515.
37. McDonnell, M.D.; Ward, L.M. The Benefits of Noise in Neural Systems: Bridging Theory and Experiment. *Nat. Rev. Neurosci.* **2011**, *12*, 415–425. [CrossRef]
38. Vázquez-Rodríguez, B.; Avena-Koenigsberger, A.; Sporns, O.; Griffa, A.; Hagmann, P.; Larralde, H. Stochastic Resonance at Criticality in a Network Model of the Human Cortex. *Sci. Rep.* **2017**, *7*, 13020. [CrossRef]
39. Roy, S.; Majumdar, S. The Role of Noise in Brain Function. In *Noise and Randomness in Living System*; Roy, S., Majumdar, S., Eds.; Springer: Singapore, 2022; pp. 99–110. [CrossRef]
40. Faisal, A.A.; Selen, L.P.J.; Wolpert, D.M. Noise in the Nervous System. *Nat. Rev. Neurosci.* **2008**, *9*, 292–303. [CrossRef]
41. Calim, A.; Palabas, T.; Uzuntarla, M. Stochastic and Vibrational Resonance in Complex Networks of Neurons. *Philos. Trans. R. Soc. A Math. Phys. Eng. Sci.* **2021**, *379*, rsta.2020.0236. [CrossRef]
42. Lizier, J.T.; Prokopenko, M.; Zomaya, A.Y. The Information Dynamics of Phase Transitions in Random Boolean Networks. In *Proceedings of the Eleventh International Conference on the Simulation and Synthesis of Living Systems (ALife XI)*, Winchester, UK, 5–8 August; 2008; pp. 374–381. p. 9.

43. Lizier, J.T.; Flecker, B.; Williams, P.L. Towards a Synergy-Based Approach to Measuring Information Modification. In Proceedings of the IEEE Symposium on Artificial Life (ALIFE), Singapore, 16–19 April 2013; pp. 43–51. [CrossRef]
44. Lizier, J.T.; Bertschinger, N.; Jost, J.; Wibral, M. Information Decomposition of Target Effects from Multi-Source Interactions: Perspectives on Previous, Current and Future Work. *Entropy* **2018**, *20*, 307. [CrossRef]
45. Quax, R.; Har-Shemesh, O.; Sloot, P.M. Quantifying Synergistic Information Using Intermediate Stochastic Variables. *Entropy* **2017**, *19*, 85. [CrossRef]
46. Lizier, J.T.; Prokopenko, M.; Zomaya, A.Y. Information Modification and Particle Collisions in Distributed Computation. *Chaos Interdiscip. J. Nonlinear Sci.* **2010**, *20*, 037109. [CrossRef] [PubMed]
47. Scheffer, M.; Bascompte, J.; Brock, W.A.; Brovkin, V.; Carpenter, S.R.; Dakos, V.; Held, H.; van Nes, E.H.; Rietkerk, M.; Sugihara, G. Early-Warning Signals for Critical Transitions. *Nature* **2009**, *461*, 53–59. [CrossRef] [PubMed]
48. Prokopenko, M.; Lizier, J.T.; Obst, O.; Wang, X.R. Relating Fisher Information to Order Parameters. *Phys. Rev. E* **2011**, *84*, 041116. [CrossRef] [PubMed]
49. Scheffer, M.; Carpenter, S.; Foley, J.A.; Folke, C.; Walker, B. Catastrophic Shifts in Ecosystems. *Nature* **2001**, *413*, 591–596. [CrossRef]
50. Eason, T.; Garmestani, A.S.; Cabezas, H. Managing for Resilience: Early Detection of Regime Shifts in Complex Systems. *Clean Technol. Environ. Policy* **2014**, *16*, 773–783. [CrossRef]
51. Schreiber, T. Measuring Information Transfer. *Phys. Rev. Lett.* **2000**, *85*, 461–464. [CrossRef]
52. Ay, N.; Polani, D. Information Flows in Causal Networks. *Adv. Complex Syst.* **2008**, *11*, 17–41. [CrossRef]
53. Runge, J.; Bathiany, S.; Bollt, E.; Camps-Valls, G.; Coumou, D.; Deyle, E.; Glymour, C.; Kretschmer, M.; Mahecha, M.D.; Muñoz-Marí, J.; et al. Inferring Causation from Time Series in Earth System Sciences. *Nat. Commun.* **2019**, *10*, 1–13. [CrossRef]
54. Li, C. Functions of Neuronal Network Motifs. *Phys. Rev. E* **2008**, *78*, 037101. [CrossRef]
55. Bialek, W.; Tishby, N. Predictive Information. *arXiv* **1999**. [CrossRef]
56. López-Ruiz, R.; Mancini, H.L.; Calbet, X. A Statistical Measure of Complexity. *Phys. Lett. A* **1995**, *209*, 321–326. [CrossRef]
57. Virtanen, P. SciPy 1.0: Fundamental Algorithms for Scientific Computing in Python. *Nat. Methods* **2020**, *17*, 15. [CrossRef]

entropy

MDPI

Article

Lattice Boltzmann Modeling of Additive Manufacturing of Functionally Graded Materials

Dmytro Svyetlichnyy

AGH University of Krakow, Faculty of Metals Engineering and Industrial Computer Science, al. Mickiewicza 30, 30-059 Kraków, Poland; svyetlic@agh.edu.pl; Tel.: +48-692-983-805

Abstract: Functionally graded materials (FGMs) show continuous variations in properties and offer unique multifunctional capabilities. This study presents a simulation of the powder bed fusion (PBF) process for FGM fabrication using a combination of Unity-based deposition and lattice Boltzmann method (LBM)-based process models. The study introduces a diffusion model that allows for the simulation of material mixtures, in particular AISI 316L austenitic steel and 18Ni maraging 300 martensitic steel. The Unity-based model simulates particle deposition with controlled distribution, incorporating variations in particle size, friction coefficient, and chamber wall rotation angles. The LBM model that simulated free-surface fluid flow, heat flow, melting, and solidification during the PBF process was extended with diffusion models for mixture fraction and concentration-dependent properties. Comparison of the results obtained in simulation with the experimental data shows that they are consistent. Future research may be connected with further verification and validation of the model, by modeling different materials. The presented model can be used for the simulation, study, modeling, and optimization of the production of functionally graded materials in PBF processes.

Keywords: functional graded materials; powder bed fusion; lattice Boltzmann method; modeling; simulation; selective laser melting

1. Introduction

Gradient materials (functionally graded materials, FGMs) are materials in which a continuous change in functional or construction properties has been achieved along at least one specific direction in a selected technological process. These materials have been known for many centuries, although their current name only appeared recently. FGMs allow for unique properties and behavior, tailored properties and multifunctional capabilities, thermal resistance, corrosion and wear resistance, improved mechanical performance, light weight, and high strength, and are widely used in aerospace, automotive, biomedical, energy, and other applications. Details of the concept of functionally graded materials, their use and fabrication processes can be found in [1]. Different aspects of FGM, state-of-the-art research, and development findings can also be found in [2].

Naebe and Shirvanimoghaddam [3] presented an overview of advances in FGM research. Materials selection, and the fabrication, characterization, analysis, and modeling of FGMs is the focus of the overview. Additionally, challenges in the fabrication of FGMs are discussed. Another overview of Kieback et al. [4] presents the achievements in the field of processing techniques involving metal melting and sintering. They considered modeling the formation of gradients in the microstructure and presented examples of numerical simulations. DebRoy et al. [5] reviewed the process, structure, and properties of additive manufacturing (AM) of metallic components. They described AM processes that consolidate materials such as powder, wire, or sheets into a dense metallic part by melting and solidifying with the aid of an energy source such as a laser, electron beam, or electric arc, or by the use of ultrasonic vibration in a layer-by-layer manner. One of the recent reviews [6] is a good guide on the current state of application areas, fabrication possibilities, and challenges in the field of FGMs. Yadav et al. [6] presented seven AM techniques that are used most frequently for the fabrication of FGM materials: vat photo polymerization, material extrusion, material jetting, sheet lamination, binder jetting, powder bed fusion (PBF), and direct energy deposition.

PBF is one of the most common AM technologies used to manufacture FGMs. PBF includes selective laser melting (SLM) and selective laser melting (SLS) processes, which allow for precise control of the composition and microstructure of FGMs. Mussatto [7] presented a review of several techniques of material deposition in the multi-material PFB process: conventional spreading, patterning drums, spreading plus suction, vibrating nozzle, hopper feeding, alternating, and electrophotographic.

Tian et al. [8] consider metallic FGMs and their processes. They noted that metallic FGMs can be classified in different ways: structural (porosity, grain size) and compositional, coating and bulk, and continuous and discontinuous. They also described 3D printing or AM techniques including PBF and experimental results on chemical composition and microstructure. Wu et al. [9] presented a review of the experimentally observed mechanical and microstructural characteristics of interfaces in multi-material laser PBF. They described mechanical spreading, nozzle-based, electrophotographic, and hybrid techniques that make it possible to achieve compositional gradients and discrete boundaries in two and three dimensions. They also show the challenges related to material transitions, such as defects, segregation, and phase separation.

Bandyopadhyay and Traxel [10] highlighted fundamental modeling strategies, considerations, results, and validation techniques using experimental data for application-optimized designs for metal additive manufacturing. They discussed the potential for modeling and how simulation would enhance the metal-AM optimization process.

Recent reviews on laser PBF modeling and simulations can be found in [11–13]. Multi-material AM processes modeled by Shinjo and Panwisawas [14] used the coupled level-set and volume-of-fluid method. They modeled the formation of the melt pool, the keyhole, and the flow for different elements. Küng et al. [15] presented a multi-material model for the lattice Boltzmann-based simulation of the PBF process of a binary alloy as a mixture of two elements. The heat capacities, latent heat, and other parameters are considered to be concentration-dependent. Melting and solidification were modeled according to the phase diagram. They simulated a single melt track in a multi-material powder mixture. Tang et al. [16] presented simulations of the multi-material laser PBF process using a volume-of-fluid model. They modeled two- and three-component powders with mixed titanium Ti, niobium Nb, vanadium V, and arsenic Ar particles.

Wang et al. [17] presented a coupled finite volume-discrete element model for SLM on powder scale. They modeled a multilayer, multitrack process. Fotovvati and Chou [18] presented a model based on the discrete element method and computational fluid dynamics for the simulation of the multilayer PBF process, creating 10 layers.

Recently, the author participated in the creation and development of the SLM and SLS processes platform, which allows for multistage multi-material simulations [19–21]. However, the scheme used in the previous simulations is based on the manufacturing process using metal alloys and bioactive glass alternately that cannot be applied for simulation of the PBF process of FGM fabrication.

It should be noted that it is very difficult to find publications with modeling of the process for FGMs. Among the various methods of material deposition that can be found in [7], the method proposed and presented in [22,23] was chosen for the simulation.

The objective of this paper is to present a modified PBF model that allows for the simulation of fabrication of the functionally graded material. To achieve this goal, the Unity-based model was adopted for the chosen deposition method, and the LBM-based model was extended with the diffusion model and concentration-dependent thermal and mechanical properties. The results of the simulation and comparison with the experimental data are also presented in the paper.

2. Model of PBF Processes

2.1. Holistic Model and Main Algorithm

The model of PBF processes is presented in Figure 1. First, the LBM-based model was developed for SLM. Details of the development, testing, verification, and validation of the SLM model and its submodels can be found in previous papers [19–21,24,25]. A platform was created for the 3D simulation of additive layer manufacturing processes characterized by changes in the state of matter (melting–solidification) [20]. Integration of the Unity-based powder bed generation model into the platform allowed for multistage multilayer multi-material simulations [19]. Then, it was shown that LBM allows for modeling not only of the SLM but also of the SLS process [21].

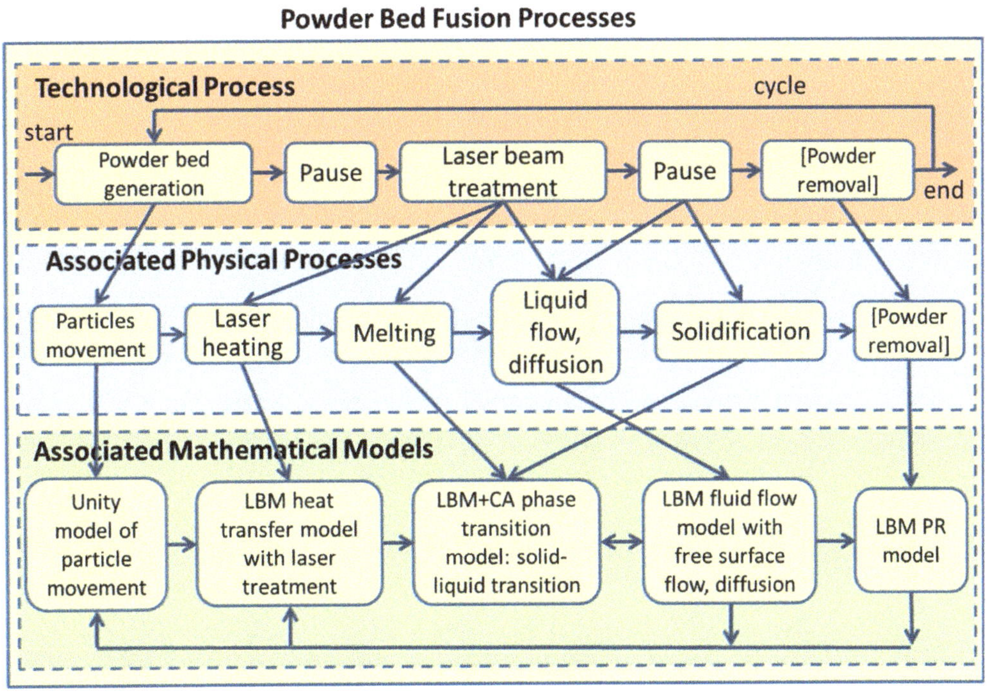

Figure 1. The PBF process with associated physical phenomena and extended submodels for modeling of mixture of two materials and diffusion.

The model presented in this paper is further developed in the direction of a simulation of a mixture of two materials. The previous multi-material model allowed for the simulation of two materials in a sequential way. The first material can be processed in several cycles with one or more layers. After that, the rest of the powder was removed and the powder of the second material was deposed. The second material can also be processed in several cycles with one or more layers. The first material (metals or alloys) with a higher melting temperature can be used as a matrix, and the second material (glasses or ceramics) can be used as a filler or coverage. The materials were not mixed.

The model is now extended with an additional diffusion model, which allows for the modeling of mixtures of different materials. Two materials with unlimited solubility to each other are chosen to simplify the model. They are AISI 316L austenitic steel and 18Ni maraging 300 martensitic steel. These materials can be used for the fabrication of functionally graded materials according to the scheme shown in [22,23].

The model contains the submodels presented in Figure 1, which are associated with the physical processes and phenomena that accompany the technological processes of PBF. Furthermore, the holistic model presented in [19,20] was supplemented with the possibility of modeling a mixture of two materials at the same time and diffusion. The powder removal remains in the model but is not used in calculations with mixed materials. The six basic processes and five submodels are considered at the second and third levels. Most of them are described in previous publications [19–21,24,25]; some elements are described in the following sections.

The main algorithm represented as a block diagram (Figure 2) was also modified compared to the previous algorithm [19,20]. The algorithm has two circuits. The first external circuit contains the powder bed generation (Unity-based model), cycle initialization, cycle modeling (LBM calculation module), and powder removal, which is not used here.

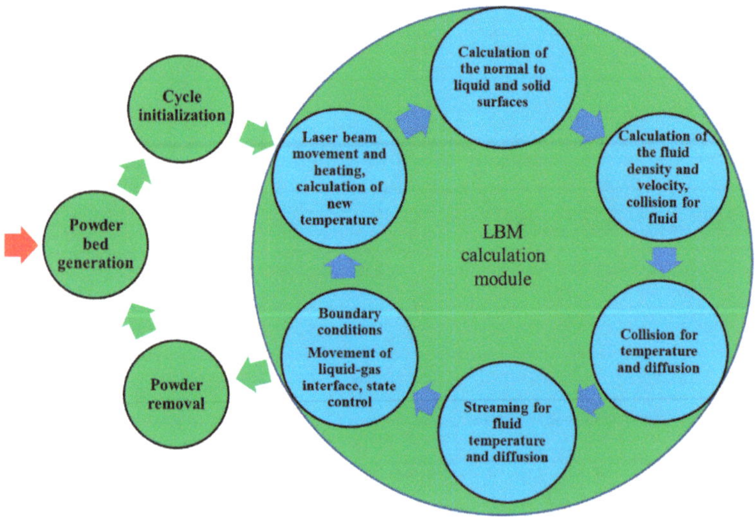

Figure 2. Graphical representation of the main algorithm.

The second internal circuit of the algorithm (shown in blue inside the 'LBM calculation module') performs one cycle of the PBF process. Operations are multiple repeated according to the presented sequence, which contains the laser beam movement, heating, and calculation of the new temperature, calculation of normal to the liquid and solid surfaces, calculation of the fluid density and velocity, collision for fluid, heat flow and diffusion, streaming for fluid, heat flow and diffusion, boundary condition, and movement of the liquid–gas interface.

The external cycle prepares the simulation of one layer, while the internal one realizes this simulation. The internal cycle is multiple repeated modeling one or many tracks, calculating small steps of laser movement and pauses after each track. The external cycle can be repeated as many times as many layers are modeled. In this paper, only a single track of the single layer is presented. The submodels, subroutines, and kernels of this circuit are essential for the proper functioning of the entire platform. Some of these models are described below; the others can be found elsewhere [19–21].

2.2. Powder Bed Generation Model

The Unity-based model was presented in [19,21]. This model was created by using the publicly available Unity game engine (https://unity.com, accessed on 20 December 2024). This model is modified to simulate the deposition of the powder of two materials. The modeling scene created in Unity is presented in Figure 3a. The primary elements of the scene are a powder chamber with parts separated by a thin wall with different powders, building platform, and coater. The Unity coordinates set is also presented. The thin wall can rotate around the vertical axes (axes Y) located in the center of the powder chamber (supply platform). A rotation angle is set before filling parts of the chamber with powder of different materials. The floor of the chamber can move vertically at the chosen level. The coater can then move horizontally along the Z axes, distributing the powder on the building platform. After that, the coater returns to the initial location. Finally, the floor of the building planform goes downward to the height of one layer, and the cycle with the movement of the floor of the powder chamber, the blade of the coater, and the floor of the building platform can be repeated. Two cases of filling of the powder chamber are also presented in Figure 3. The thin wall separating the parts of the chamber can either be rotated or not (Figure 3b). In the second case, this wall is rotated at an angle of 25° (Figure 3c). After rotation, the parts are filled with the appropriate powder material. The sizes of the powder chamber and the building platform are the same and equal to 25 × 25 mm in the Unity model, which represents the real domain of 1250 × 1250 μm. This means that the Unity model is created on a scale of 20:1 to respond to the real domain.

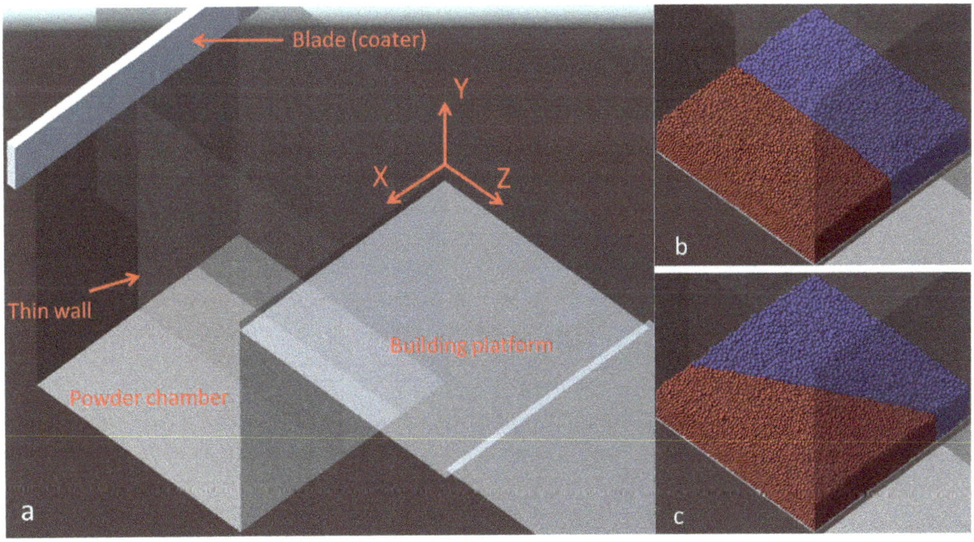

Figure 3. Modeling scene for simulation of the deposition of the powder of two materials shown in different colors (**a**) and two cases of filling of the powder chamber (**b**,**c**).

2.3. LBM Models of the PBF Process

LBM was developed from lattice gas automata [26] and is derived from the Boltzmann equation, a fundamental equation in statistical mechanics that can be considered as one of the computational fluid dynamics (CFD) methods. Several differential equations can be reduced to or approximated by the Lattice Boltzmann equation (LBE): Navier–Stokes, advection–diffusion, energy (heat conduction), Poisson, shallow water, elastic wave, Burgers, and magnetohydrodynamic. Details on how to approximate these equations can be found elsewhere [27,28].

LBM simulates the microscopic dynamics of particle distribution functions on a discrete lattice grid. These particle interactions and streaming processes yield macroscopic properties like density, velocity, temperature, enthalpy, concentration, etc. LBM uses discrete time, space, and velocity and calculates macroscopic properties through collision and streaming operations [27–29].

The approximations of different partial differential equations by LBE open up wide possibilities for multiphysics coupling and modeling. The algorithm of LBM is easy to implement due to the explicit streaming and collision steps and is highly suited for parallel computations on CPU and GPU. LBM operates effectively with complex boundaries. Thus, the model of the PFD process takes into account fluid flow with free surface (flow, surface tension, wettability, convection, etc.), heat flow (heat source, heat transfer, heat conduction, melting and solidification), and diffusion.

The LBM model of the PBF process is based on the previously developed model for the modeling platform, and most of the submodels can be found in detail in previous publications [19–21,24,25]. They contained two cooperative parts that are responsible for the fluid and heat flow processes. Fluid flow was considered for a single material.

In this paper, the flow of the mixture is simulated. To determine the composition of the mixture, a volume fraction of one material in the mixture is introduced. Fluid flow does not change the composition of the mixture. The composition of the mixture is changed by diffusion. Diffusion is simplified. Instead of considering the diffusion of many elements in different mixtures, a mixture is considered as a single homogeneous material in which both materials are diffused at the same rate, defined by a common diffusion coefficient.

The common algorithm of the LBM model is presented in Figure 2 as the internal circuit. Each of the three problems (fluid and heat flow and diffusion) is solved according to the same scheme: calculating macroscopic values based on the input distribution, determining the equilibrium distribution based on macroscopic values, determining the output distribution in the collision operation, and determining the input distribution based on the output distribution in the streaming operation by transferring the components of the distribution function to appropriate neighboring nodes, taking into account boundary conditions and other factors. Since all the mechanical and thermal properties of a material depend on chemical composition and temperature, all three problems are interdependent and can be solved simultaneously as much as possible. This explains the calculation sequence, which is described in detail in Section 2.3.4.

Such blocks as the calculation of macroscopic variables, collision and streaming operations for fluid and heat flow and diffusion are presented in the following Sections 2.3.1–2.3.3; the other blocks can be found in previous publications [19–21,24,25].

The D3Q19 velocity model is applied for all the tasks. Therefore, 19 components of the distribution functions that represent directions and velocities are used in the 3D models.

2.3.1. LBM Model of Fluid Flow

Macroscopic variables, that is, the density ρ of the mixture fluid and the velocity **v**, are calculated according to the following equation (vectors are indicated in bold):

$$\rho = \sum_{i=1}^{19} f_i^{\text{in}} \tag{1}$$

$$\rho \mathbf{v} = \sum_{i=1}^{19} f_i^{\text{in}} \mathbf{e}_i \tag{2}$$

where f_i^{in}—appropriate component of input distribution function of the liquid flow model, **e**—phase space variable—velocity (vector), set of discrete velocities, **v**—velocity vector $\{vx, vy, v_z\}$, and i—component of the velocity model.

The collision operation for fluid flow calculates the output distribution function f^{out} based on the input f^{in} and equilibrium f^{eq} distribution functions. Each distribution function contains 19 components. Then, the equilibrium f^{eq} and output distribution functions are defined as follows:

$$f_i^{\text{eq}} = w_i \rho \left[1 + 3\mathbf{e}_i \cdot \mathbf{v} + 4.5(\mathbf{e}_i \cdot \mathbf{v})^2 - 1.5\mathbf{v} \cdot \mathbf{v}\right] \tag{3}$$

$$f_i^{\text{out}} = f_i^{\text{in}} + \frac{\Delta t}{\tau_f}\left(f_i^{\text{eq}} - f_i^{\text{in}}\right) + F_i \tag{4}$$

where w_i—weights, F—external force, for example, gravity, $\Delta t = 1$—time step, and τ_f—relaxation time for fluid flow.

The streaming operation that defines input distribution function is performed according to:

$$f_i^{\text{in}}(\mathbf{x} + \mathbf{e}_i,\ t + 1) = f_i^{\text{out}}(\mathbf{x},\ t) \tag{5}$$

where **x**—node coordinates.

2.3.2. LBM Model of Heat Flow

The new temperature T is calculated as follows:

$$T = Q + \sum_{i=1}^{19} g_i^{\text{in}} \tag{6}$$

where g^{in}—distribution function of the heat flow model, Q—heat source.

The collision operation is carried out:

$$g_i^{\text{eq}} = w_i T \left[1 + 3\mathbf{e}_i \cdot \mathbf{v} + 4.5(\mathbf{e}_i \cdot \mathbf{v})^2 - 1.5\mathbf{v} \cdot \mathbf{v}\right] \tag{7}$$

$$g_i^{\text{out}} = g_i^{\text{in}} - \frac{1}{\tau_h}\left(g_i^{\text{in}} - g_i^{\text{eq}}\right) \tag{8}$$

where τ_h—relaxation time for heat flow.

And finally, the streaming operation is as follows:

$$g_i^{\text{in}}(\mathbf{x} + \mathbf{e}_i,\ t + 1) = g_i^{\text{out}}(\mathbf{x},\ t) \tag{9}$$

2.3.3. LBM Diffusion Model

This is a new part of the model. The fraction of the first material in the mixture (or its concentration) C is calculated as follows:

$$C = \sum_{i=1}^{19} c_i^{\text{in}} \tag{10}$$

where c—distribution function of the concentration flow.

The collision operation is as follows:

$$c_i^{\mathrm{eq}} = w_i C\left[1 + 3\mathbf{e}_i{\cdot}\mathbf{v} + 4.5(\mathbf{e}_i{\cdot}\mathbf{v})^2 - 1.5\mathbf{v}{\cdot}\mathbf{v}\right] \tag{11}$$

$$c_i^{\mathrm{out}} = c_i^{\mathrm{in}} - \frac{1}{\tau_d}\left(c_i^{\mathrm{in}} - c_i^{\mathrm{eq}}\right) \tag{12}$$

where τ_d—relaxation time for diffusion.

The streaming operation for diffusion is calculated as follows:

$$c_i^{\mathrm{in}}(\mathbf{x} + \mathbf{e}_i,\ t+1) = c_i^{\mathrm{out}}(\mathbf{x},\ t) \tag{13}$$

2.3.4. LBM Cycle and Model Interaction

The second internal circuit of the algorithm (Figure 2) begins with calculations of macroscopic variables such as temperature (6) and material fraction (10). However, the temperature model is more complex than (6); it takes into account melting and solidification, and its details are described in a previous publication [20]. Here, all the properties of the material that depend on temperature and material fraction are also calculated. Properties of the mixture are calculated as follows:

$$P = P_1 C + P_2(1 - C) \tag{14}$$

where P—properties of the mixture (density, melting point, specific heat capacity, thermal diffusivity, viscosity, etc.), P_1, P_2—properties of the first and second materials at the actual calculated temperature, and C—volume fraction of the first material.

The second block of the LBM algorithm is needed to define the interaction in different interfaces (liquid–gas, solid–liquid, and solid–liquid–gas), mainly used for fluid flow (surface tension, wettability) but also for heat transfer (additionally, solid-gas interface).

The third block calculates macroscopic variables for fluid flow: density (1) and velocity (from (2) and (1)). Macroscopic velocity is used for the calculation of the equilibrium distribution function not only for the fluid flow problem (3), but also for the heat flow (7) and diffusion (11) problems. The absence of velocity in (7) and (11) corresponds to the diffusion equation (Fourier), while the presence of velocity corresponds to the diffusion–advection equation.

Collision operations are performed in the third block for fluid flow (3) and (4) and in the fourth block for heat flow (7) and (8) and diffusion (11) and (12).

The fifth block performs a streaming operation for all three problems according to (5), (9), and (13). The components of the distribution functions are transferred to the appropriate neighboring nodes.

The last block gives the boundary condition and is responsible for the movement of interfaces connected with fluid flow (liquid–gas, solid–liquid–gas). The movements of interfaces connected with melting and solidification (solid–liquid–gas and solid–liquid) are considered in the first block. Details of the different interfaces used in the model can be found in [20].

A key parameter for proper simulation and agreement on kinetics of different processes is the relaxation time for three problems. They are calculated basing on simulation parameters (time step, lattice length, velocity model D3Q19) and material properties (kinematic viscosity ν, thermal diffusivity α, diffusion coefficient D) as follows:

$$\tau_f = \frac{3\nu\Delta t}{\Delta x^2} + 0.5 \tag{15}$$

$$\tau_h = \frac{3\alpha\Delta t}{\Delta x^2} + 0.5 \tag{16}$$

$$\tau_d = \frac{3D\Delta t}{\Delta x^2} + 0.5 \tag{17}$$

3. Simulation and Results

This stage of the research involves the development and verification of the model of the PBF process objected to the fabrication of grade materials. For the verification of the model, the data published in [22,23] were used. It explains the modeling scheme and some parameters of material distribution.

3.1. Materils

AISI 316L austenitic steel (material 1) and 18Ni maraging 300 martensitic steel (material 2) were chosen for the simulations. The physical properties of these materials are collected in Table 1. The comparison of the diffusion coefficient with the kinematic viscosity and thermal diffusivity shows the diffusion coefficient is several orders of magnitude smaller and could be neglected, and only advection and mixing can be taken into account. Since LBM considers the diffusion–advection equation, it allows us to model almost pure advection Taking into account the velocity in the calculation of the equilibrium distribution, Function (11) corresponds to advection (and mixing), while the relaxation time in the determination of the output distribution Function (12) and (17) determines the diffusion rate. Both materials are closely related to the movement (advection) of the fluid. When two streams with different fractions arrive at the same node (cell), the volume contents of both materials are summed to obtain the resulting fraction. When there is movement of the material in the liquid, the fraction can be changed. When the liquid is motionless, only diffusion is present.

Table 1. Physical properties of AISI 316L austenitic steel and 18Ni maraging 300 martensitic steel.

Properties	AISI 316L	18Ni Maraging
Density, kg/m^3	8000	8100
Melting point, K	1648–1673	1686
Specific heat capacity, J/(kg·K)	500	452
Thermal conductivity, W/(m K)	14.0–15.9	25.5
Thermal diffusivity, m^2/s	3.75×10^{-6}	5.5×10^{-6}
Convective heat transfer coefficient, W/(m^2 K)	50	50
Coefficient of thermal expansion, K^{-1}	$16–18 \times 10^{-6}$	11.3×10^{-6}
Dynamic viscosity, kg/(m s)	$2.4–3.3 \times 10^{-3}$	
Kinematic viscosity, m^2/s	$0.3–0.41 \times 10^{-6}$	
Surface tension, N/m $\times 10^{-4}$ [30]	1.86–1.54 (T-1811.15)	
Diffusion coefficient, m^2/s	$5 \times 10^{-9}–10^{-7}$.	
Relaxation times for liquid flow (15)	0.577	
Relaxation times for heat flow (16)	1.22	
Relaxation times for diffusion (17)	0.505	

3.2. Simulation of Powder Filling and Deposition

Figure 3 presents the simulation scheme and two examples of filling the powder chamber with a wall rotation angle equal to 0 and 25°. The simulations were performed for the following cases.

- Three rotation angles: (a) 0°, (b) 25°, and (c) 45°.
- Size of the particles: (a) equal to 40 μm, (b) material 1–35 μm, material 2–45 μm, (c) Gaussian distribution, material 1 average size 35 μm, material 2–45 μm, dispersion for both materials—5 μm.
- Friction coefficient: (a) 0, (b) 0.2, (c) 0.6.

The deposition process from the powder chamber to building platform with the coater is shown in Figure 4. The process is shown for a wall angle equal to 0° and ten layers with a friction coefficient of 0.2 and particles of 40 μm.

Figure 4. Powder deposition process (**a**) beginning of the first layer, (**b**) the end of the first layer, (**c**) results with 10 layers. Different colors presents different materials.

The deposition results for the case of different rotation angles, Gaussian distribution of the particle size, and coefficient of friction of 0.2 are shown in Figure 5. The white figures in Figure 5b represent an example of domains, which are transferred to the LBM model for further simulations. The location and sizes of the transferred domains are the same for all simulation cases.

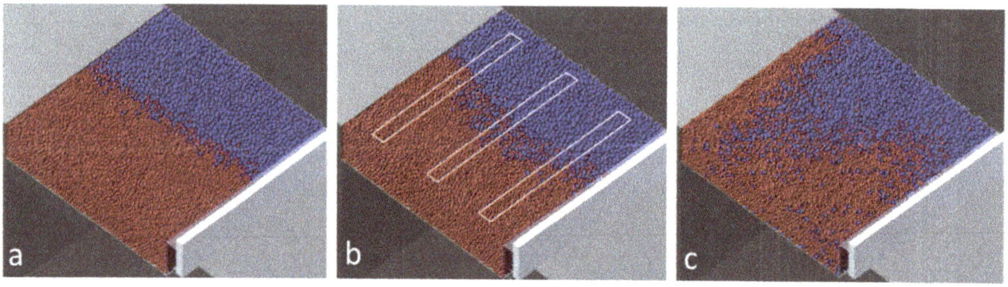

Figure 5. Particle powder deposition with Gaussian size distribution and different wall rotation angle: (**a**) 0°, (**b**) 25°, (**c**) 45°. Different colors present different materials.

3.3. *Simulation of the PBF Process with LBM*

The powder deposition results are transferred from the Unity-based model to the LBM model. In fact, the simulations in Unity and LBM are performed on different scales and different domains (Figure 5b). As mentioned above, the scale of the Unity-based model is 20:1 to respond to real sizes. LBM operates with dimensionless units of length and times. The lattice size in the LBM model, that is, the distance between the neighboring nodes, is equal to 1.25 μm. The entire domain is not simulated; only single tracks at different locations are. This allows us to reduce the simulation time.

Various simulations of the PBF process were performed, mainly at the angle of 0 and 25°, and the results of the distribution of the materials were compared with the distribution of the particles before the process. The simulation tracks are located on the axis of the building platform (1250 × 1250 μm) and at a distance of 375 μm from the axis symmetrically on both sides, as shown in Figure 5b. The size of the LBM modeling space is 960 × 160 × 80 μm or 768 × 128 × 64 nodes. The laser power is 300 W, the scan speed is 1000 mm/s, the laser movement distance is 850 μm (680 lattice lengths), and the laser spot diameter is 80 μm (64 lattice lengths).

The results of one of the simulation cases of the PBF process are presented in Figures 6–8. Figure 6 presents a process as the 3D perspective view with a rendering at five moments of time. The left snapshot presents the initial particle deposition before the PBF process transferred from the Unity-based model. The next three snapshots present three moments of time during the processing of one track—at the beginning, middle, and end of the track. The last snapshot demonstrates the track after the whole material is solidified. Unprocessed particles are shown in grades of red and blue. Solidified material is shown in dark colors, with the fractions of red and blue corresponding to the fractions of the first and second materials. Materials in the liquid state are shown in color as solidified materials but with the addition of green grade. The laser beam is shown in green.

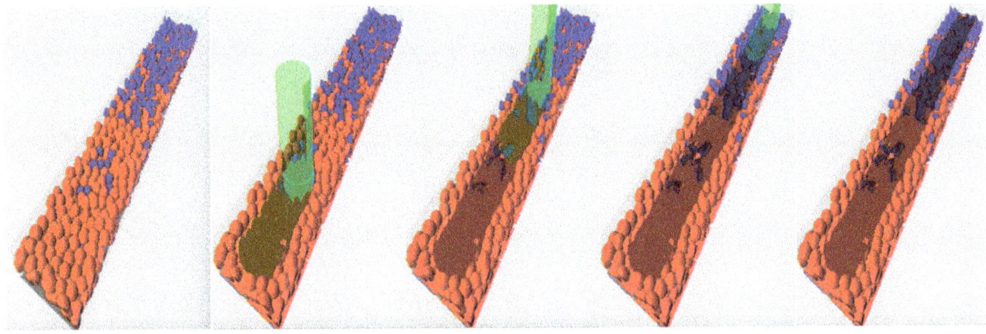

Figure 6. 3D view of several stages of the simulated PBF process with LBM. Colors are explained in the text.

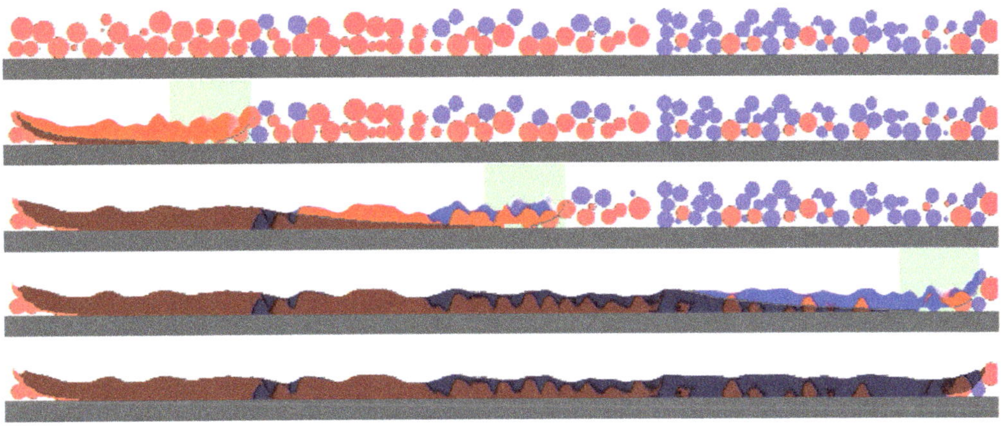

Figure 7. A longitudinal cross section of the particle of two materials before, during, and after the PBF process.

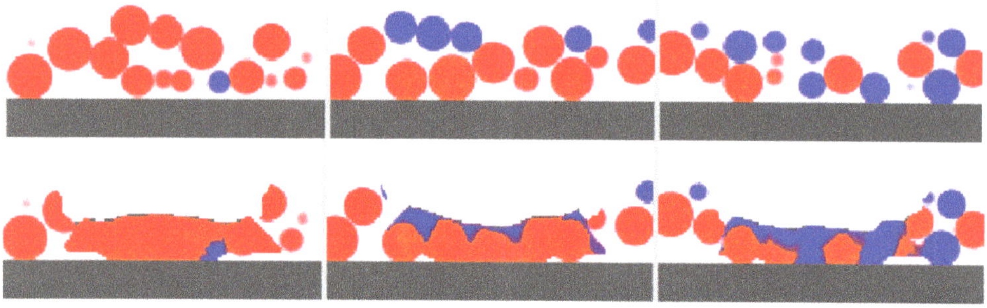

Figure 8. Transverse cross sections in three locations before (**a**) and after (**b**) the PBF process simulated with LBM.

Other results are presented in Figures 7 and 8. They show the materials before and after the PBF process as a longitudinal cross section on the track axis and three transverse cross sections located at the center of the modeled space and at the distance of 300 μm on both sides of the center. Unprocessed particles on the longitudinal cross sections (Figure 7) are shown in lighter colors; solidified materials are shown in deeper colors (red and deep blue); and a mixture is shown in an appropriate fraction of red and blue. Overheated molten material can melt some particles completely or partially and flow beneath them, filling the pores between unmolten particles. The flowing material initiates mixing, which also involves diffusion. However, for the single track on the microscale, in which the simulations with LBM were performed, highly mixed areas occupy only a small fraction, especially when cross sections are considered. Near the surface of the melting pool, the mixing is much higher than near its bottom. The higher mixing can be seen in the 3D view (Figure 6). This can be explained by the higher flow velocity and the additional fast stirring along with the slow diffusion. Near the bottom, the temperature is lower, the viscosity is lower, and the processes of stirring and diffusion are slower; the modeling results reflect this. Unprocessed and solidified materials are shown in the same colors on the transverse cross sections (Figure 8).

3.4. Analisys of Material Distribution

Some deposition results as a material distribution are presented in Figure 9. The left column represents the distribution before simulation of the PBF process. The volume fraction of the materials is in the range of 55–62%; the rest is occupied with gas. After the process, there is no gas and the total fraction of materials reaches 100%. The results before and after the process are almost the same; the error can be treated as a small random error. A transient zone from one material to another spreads from 0.25 mm for the angle of 0° through approximately 0.4–0.5 mm for the angle of 25° to the entire modeling space for the angle of 45°. The distribution is almost the same for different locations of the tracks for the rotation angle of 0° (Figure 9a); it can also be seen in Figures 4 and 5a. The increase in the rotation angle (Figure 9a,c) allows the transient zone to spread, but the difference in its location and length become uncontrolled. The heterogeneity of the distribution can also be seen in Figure 5b,c. There are several factors that influence the results obtained in the simulations. Even if the other conditions are the same, it depends on which powder is scraped first with the coater blade. In addition, smaller particles are deposited first; this is especially important when the average size of powder particles of different materials is different. The higher density particles are deposed first in one layer and go deeper when they are deposed in several layers without melting, but in the case of the materials presented in this paper, the density of the materials is almost the same, and this factor does not play a notable role. The influence of the friction coefficient on the deposition is also noticed, but it is difficult to formulate.

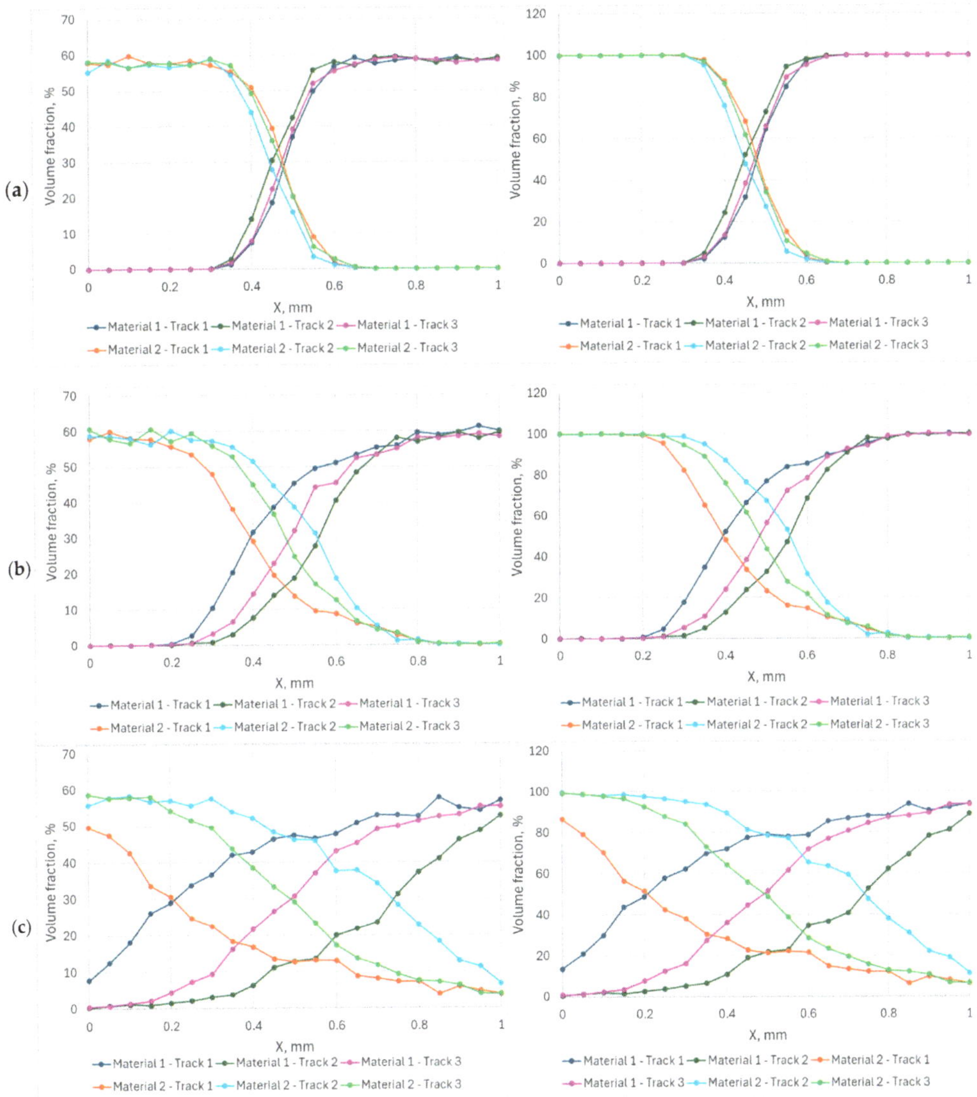

Figure 9. Distribution of two materials before (**left column**) and after (**right column**) the simulated PBF process for different rotation angles of the wall: (**a**) 0°, (**b**) 25° (**c**) 45°.

It is very difficult to perform a quantitative comparison with experimental data. The experimental results presented in [23] do not allow for the reproduction and repetition of the study in many details. That is why the comparison can be treated as qualitative only. The simulation can be considered qualitatively consistent with the results of the experimental studies presented in [23] and Figure 10. The experimental results show that diffusion can be neglected and either element can be chosen to define the fraction of the materials, while material fraction in simulations can be obtained directly [23]. The angle of the wall assumes the length of the transient gradient zone of 20 mm; perhaps the angle in the experimental studies was of 15–20°. The length of the powder chamber and the building platform was approximately 60–80 mm. The specimen sizes were 30 × 10 mm. Pairs of specimens were located symmetrically on a building platform at a distance of approximately 30 mm from each other. Two pairs differed in the order of the powders in the powder chamber. Many other parameters are unclear.

The results in [23] demonstrate significant differences in the distribution of different elements that define different materials. The distributions are different due to the different locations of the specimens and the order of the powder in the powder chamber. The lack of information does not allow the author to explain the differences obtained in experimental studies, but the studies presented in this paper can assume that these differences can be explained by the location of the specimens and the different sizes of the particles.

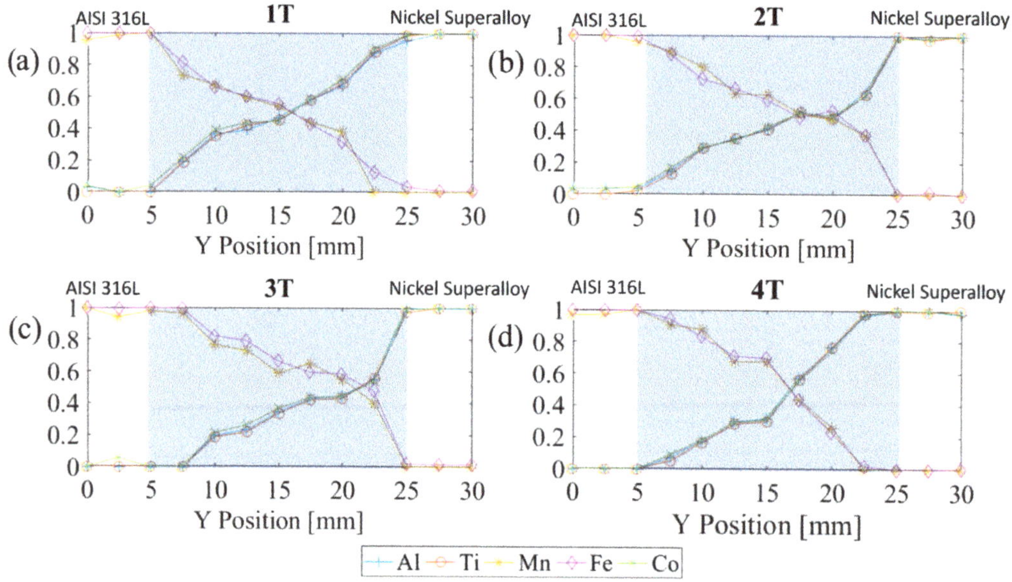

Figure 10. Results of chemical analysis performed on samples from different locations in [23]. Different orientation of the dividing wall (**a**,**b**)—left orientation, (**c**,**d**)—right orientation. AiSI 316L closer to the building platform (**b**,**c**), Ni superalloy closer to the platform (**a**,**d**).

4. Discussion

The results presented in the previous sections show the possibility of modeling a complex PBF process with the mixture of two materials to obtain graded materials. Analysis of the distribution of the materials in the final mixture and its comparison with experimental data allows us to conclude that the complex Unity–LBM model gives adequate results, at least qualitatively similar to the real PBF process. The analysis was performed not to show the effectiveness of the FGM fabrication method, but to demonstrate that the LBM-based model can be applied to the simulation of the fabrication of functionally graded material in additive manufacturing. The Unity-based model is of its own independent importance. Simulations by the Unity-based model of the deposition of materials with different physical and mechanical properties allow for the simulation, study, modeling, and optimization of deposition schemes and methods for graded materials because the PBF process that provides homogenization does not significantly influence the final material distribution. Future research may be connected with further verification and validation of the model, modeling different materials, and the optimization of the methods of manufacturing functionally graded materials in the PFD processes.

5. Conclusions

This paper presents the results of the further development of the platform for modeling multi-material multilayer PBF processes directed at the simulation of the fabrication of functionally graded materials. To achieve this goal, Unity- and LBM-based models were modified. The Unity-based model allows for the simulation of the deposition of materials with different mechanical properties to create input information for the LBM-based model. The LBM-based model was extended with several new elements. The diffusion phenomenon with material mixing was added; it led to the development and implementation of the diffusion model, which influenced the introduction of the third set of variables and distribution functions. In addition to diffusion, changes were implemented in the fluid and heat flow models. The fraction of the materials was introduced, which defines the physical, mechanical, and thermal properties of the mixture and the effect on the simulation parameters. The comparison of the results obtained in the simulation with the experimental data shows that they are consistent. The presented model can be used for the simulation, study, modeling, and optimization of the production of functionally graded materials in PBF processes.

Funding: This study was funded by the National Science Centre, Poland, grant number UMO-2018/31/B/ST8/00622, and by the Ministry of Science and Higher Education, Poland, Grant AGH University of Science and Technology no. 16.16.110.663.

Institutional Review Board Statement: Not applicable.

Informed Consent Statement: Not applicable.

Data Availability Statement: The original contributions presented in this study are included in the article material. Further inquiries can be directed to the corresponding author.

Conflicts of Interest: The author declares that there are no conflicts of interest. The funders had no role in the design of the study; in the collection, analysis, or interpretation of the data; in the writing of the manuscript; or in the decision to publish the results. The author is grateful to Dominik Grzegorz Sabat for his help with the modification and simulation of the Unity-based model.

References

1. Mahamood, R.M.; Akinlabi, E.T. *Functionally Graded Materials; In Topics in Mining, Metallurgy and Materials Engineering*, 1st ed.; Springer International Publishing: Cham, Switzerland, 2017; ISBN 978-3-319-53755-9.
2. Ebrahimi, F. *Advances in Functionally Graded Materials and Structures*; InTech: Houston, TX, USA, 2016; ISBN 978-953-51-2274-6.
3. Naebe, M.; Shirvanimoghaddam, K. Functionally Graded Materials: A Review of Fabrication and Properties. *Appl. Mater. Today* **2016**, *5*, 223–245. [CrossRef]
4. Kieback, B.; Neubrand, A.; Riedel, H. Processing Techniques for Functionally Graded Materials. *Mater. Sci. Eng. A* **2003**, *362*, 81–106. [CrossRef]
5. DebRoy, T.; Wei, H.L.; Zuback, J.S.; Mukherjee, T.; Elmer, J.W.; Milewski, J.O.; Beese, A.M.; Wilson-Heid, A.; De, A.; Zhang, W. Additive Manufacturing of Metallic Components—Process, Structure and Properties. *Prog. Mater. Sci.* **2018**, *92*, 112–224. [CrossRef]
6. Yadav, S.; Liu, S.; Singh, R.K.; Sharma, A.K.; Rawat, P. A State-of-Art Review on Functionally Graded Materials (FGMs) Manufactured by 3D Printing Techniques: Advantages, Existing Challenges, and Future Scope. *J. Manuf. Process.* **2024**, *131*, 2051–2072. [CrossRef]
7. Mussatto, A. Research Progress in Multi-Material Laser-Powder Bed Fusion Additive Manufacturing: A Review of the State-of-the-Art Techniques for Depositing Multiple Powders with Spatial Selectivity in a Single Layer. *Results Eng.* **2022**, *16*, 100769. [CrossRef]
8. Tian, X.; Zhao, Z.; Wang, H.; Liu, X.; Song, X. Progresses on the Additive Manufacturing of Functionally Graded Metallic Materials. *J. Alloys Compd.* **2023**, *960*, 170687. [CrossRef]

9. Wu, Z.; Wilson-Heid, A.E.; Griffiths, R.J.; Elton, E.S. A Review on Experimentally Observed Mechanical and Microstructural Characteristics of Interfaces in Multi-Material Laser Powder Bed Fusion. *Front. Mech. Eng.* **2023**, *9*, 1087021. [CrossRef]
10. Bandyopadhyay, A.; Traxel, K.D. Invited Review Article: Metal-Additive Manufacturing—Modeling Strategies for Application-Optimized Designs. *Addit. Manuf.* **2018**, *22*, 758–774. [CrossRef] [PubMed]
11. Chowdhury, S.; Yadaiah, N.; Prakash, C.; Ramakrishna, S.; Dixit, S.; Gupta, L.R.; Buddhi, D. Laser Powder Bed Fusion: A State-of-the-Art Review of the Technology, Materials, Properties & Defects, and Numerical Modelling. *J. Mater. Res. Technol.* **2022**, *20*, 2109–2172. [CrossRef]
12. Papazoglou, E.L.; Karkalos, N.E.; Karmiris-Obratański, P.; Markopoulos, A.P. On the Modeling and Simulation of SLM and SLS for Metal and Polymer Powders: A Review. *Arch. Comput. Methods Eng.* **2022**, *29*, 941–973. [CrossRef]
13. Zhou, R.; Liu, H.; Wang, H. Modeling and Simulation of Metal Selective Laser Melting Process: A Critical Review. *Int. J. Adv. Manuf. Technol.* **2022**, *121*, 5693–5706. [CrossRef]
14. Shinjo, J.; Panwisawas, C. Digital Materials Design by Thermal-Fluid Science for Multi-Metal Additive Manufacturing. *Acta Mater.* **2021**, *210*, 116825. [CrossRef]
15. Küng, V.E.; Scherr, R.; Markl, M.; Körner, C. Multi-Material Model for the Simulation of Powder Bed Fusion Additive Manufacturing. *Comput. Mater. Sci.* **2021**, *194*, 110415. [CrossRef]
16. Tang, C.; Yao, L.; Du, H. Computational Framework for the Simulation of Multi Material Laser Powder Bed Fusion. *Int. J. Heat Mass Transf.* **2022**, *191*, 122855. [CrossRef]
17. Wang, Z.; Yan, W.; Liu, W.K.; Liu, M. Powder-Scale Multi-Physics Modeling of Multi-Layer Multi-Track Selective Laser Melting with Sharp Interface Capturing Method. *Comput. Mech.* **2019**, *63*, 649–661. [CrossRef]
18. Fotovvati, B.; Chou, K. Multi-Layer Thermo-Fluid Modeling of Powder Bed Fusion (PBF) Process. *J. Manuf. Process.* **2022**, *83*, 203–211. [CrossRef]
19. Svyetlichnyy, D. Model of the Selective Laser Melting Process-Powder Deposition Models in Multistage Multi-Material Simulations. *Appl. Sci.* **2023**, *13*, 6196. [CrossRef]
20. Svyetlichnyy, D.S. Development of the Platform for Three-Dimensional Simulation of Additive Layer Manufacturing Processes Characterized by Changes in State of Matter: Melting-Solidification. *Materials* **2022**, *15*, 1030. [CrossRef] [PubMed]
21. Svyetlichnyy, D. Simulation of the Selective Laser Sintering/Melting Process of Bioactive Glass 45S5. *Simul. Model. Pract. Theory* **2024**, *136*, 103009. [CrossRef]
22. Campanelli, S.L.; Carone, S.; Casavola, K.; Errico, V.; Pappalettera, G.; Posa, P. Residual Stress Evaluation in Innovative Layer-Level Continuous Functionally Graded Materials Produced by Powder Bed Fusion-Laser Beam. *Int. J. Adv. Manuf. Technol.* **2024**, *134*, 511–527. [CrossRef]
23. Angelastro, A.; Posa, P.; Errico, V.; Campanelli, S.L. A Systematic Study on Layer-Level Multi-Material Fabrication of Parts via Laser-Powder Bed Fusion Process. *Metals* **2023**, *13*, 1588. [CrossRef]
24. Krzyzanowski, M.; Svyetlichnyy, D. A Multiphysics Simulation Approach to Selective Laser Melting Modelling Based on Cellular Automata and Lattice Boltzmann Methods. *Comput. Part Mech.* **2022**, *9*, 117–133. [CrossRef]
25. Dmytro, S.; Szymon, B.; Michal, K. An Extended Laser Beam Heating Model for a Numerical Platform to Simulate Multi-Material Selective Laser Melting. *Int. J. Adv. Manuf. Technol.* **2023**, *128*, 3451–3470. [CrossRef]
26. Hardy, J.; Pomeau, Y.; de Pazzis, O. Time Evolution of a Two-Dimensional Classical Lattice System. *Phys. Rev. Lett.* **1973**, *31*, 276–279. [CrossRef]
27. Succi, S. . *The Lattice Boltzmann Equation for Fluid Dynamics and Beyond*; Clarendon Press: Oxford, UK; Oxford University Press: Oxford, UK, 2001; ISBN 9780198503989.
28. Krüger, T.; Kusumaatmaja, H.; Kuzmin, A.; Shardt, O.; Silva, G.; Viggen, E.M. *The Lattice Boltzmann Method*; Springer International Publishing: Cham, Switzerland, 2017; ISBN 978-3-319-44647-9.
29. Mohamad, A.A. *Lattice Boltzmann Method*; Springer: London, UK, 2011; ISBN 978-0-85729-454-8.
30. Klapczynski, V.; Le Maux, D.; Courtois, M.; Bertrand, E.; Paillard, P. Surface Tension Measurements of Liquid Pure Iron and 304L Stainless Steel under Different Gas Mixtures. *J. Mol. Liq.* **2022**, *350*, 118558. [CrossRef]

entropy

MDPI

Article

Statistical Mechanics of Directed Networks

Marián Boguñá [1,2,*] and M. Ángeles Serrano [1,2,3,*]

1 Department of Condensed Matter Physics, University of Barcelona, Martí i Franquès 1, E-08028 Barcelona, Spain
2 University of Barcelona Institute of Complex Systems (UBICS), E-08028 Barcelona, Spain
3 Institució Catalana de Recerca i Estudis Avançats (ICREA), Passeig Lluís Companys 23, E-08010 Barcelona, Spain
* Correspondence: marian.boguna@ub.edu (M.B.); marian.serrano@ub.edu (M.Á.S.)

Abstract: Directed networks are essential for representing complex systems, capturing the asymmetry of interactions in fields such as neuroscience, transportation, and social networks. Directionality reveals how influence, information, or resources flow within a network, fundamentally shaping the behavior of dynamical processes and distinguishing directed networks from their undirected counterparts. Robust null models are crucial for identifying meaningful patterns in these representations, yet designing models that preserve key features remains a significant challenge. One such critical feature is reciprocity, which reflects the balance of bidirectional interactions in directed networks and provides insights into the underlying structural and dynamical principles that shape their connectivity. This paper introduces a statistical mechanics framework for directed networks, modeling them as ensembles of interacting fermions. By controlling the reciprocity and other network properties, our formalism offers a principled approach to analyzing directed network structures and dynamics, introducing new perspectives and models and analytical tools for empirical studies.

Keywords: complex networks; directed networks; maximum entropy; Fermi statistics; reciprocity

1. Introduction

A directed network [1] is a representation of a complex system that captures the asymmetry of interactions between its elements [2,3]. Directionality enriches network structure [4,5], and is essential for understanding how influence, information, or resources flow through a system [6], fundamentally distinguishing directed networks from undirected ones. This is critical across a wide range of domains, including neuronal systems, biological processes, transportation systems, and social networks. Moreover, directionality fundamentally influences the behavior of dynamical processes on networks [7–10].

To gain a deeper understanding of the principles shaping real directed networks, it is crucial to define models that accurately capture their essential characteristics and organization. In general, network models enable researchers to distinguish meaningful patterns from random fluctuations and provide principled explanations for the observed regularities. The family of network models derived by maximizing the entropy of graph ensembles subject to the constraints imposed by observations in real-world networks offer the least biased prediction for their properties [11,12]. However, designing maximum entropy models for directed networks is a challenging task. This difficulty arises from the need to account for the interplay between local node properties and global network structures.

https://doi.org/10.3390/e27010086

Specifically, key features in directed networks are in-degrees and out-degrees, accounting for the number of incoming and outgoing connected neighbors, their correlations, and reciprocity [11,13], or the tendency of pairs of nodes to form bidirectional connections. Reciprocity reflects the balance or imbalance of mutual interactions and serves as a critical indicator of the underlying structural and dynamical rules governing the system. Another key property is clustering, the tendency of pairs of neighbors to be connected, forming triangles in the network topology. In directed networks, triangles become multifaceted, splitting into seven distinct triangle motifs depending on the orientation of the arrows [14,15]. Despite the recent introduction of a directed network model [16] that simultaneously explains many features in directed networks, such as reciprocity, clustering, and other structural properties, a general theoretical approach based on the maximum entropy principle is still lacking.

In this paper, we introduce a statistical mechanics framework for directed networks, treating them as systems of interacting fermions. This approach leverages concepts from quantum statistics to describe directed networks in terms of ensembles, where network connections or fermions are constrained by conserved quantities and the entropy of the ensemble is maximized to fix its probability. By framing directed networks in this way, we provide a powerful theoretical tool for modeling their structure. Our framework not only offers new insights into the organization of real-world directed networks but also provides a principled basis for constructing models that respect key empirical properties.

2. General Formalism

The standard approach in network science treats the nodes of a network as the fundamental units of the system, with links representing the interactions between these units. This perspective naturally aligns with real-world systems, where nodes correspond to defined entities—countries in the world trade web, proteins or genes in biomolecular interaction networks, individuals in society, and so on—making focusing on nodes intuitive and practical. However, this node-centric viewpoint poses challenges when defining models using traditional tools from statistical mechanics, as it emphasizes the entities rather than the interactions.

In this work, we adopt a different perspective by shifting the focus from the nodes of the network to the links connecting them. In our approach, links are treated as fermionic "particles" that can occupy distinct energy states. The phase space of possible energy states is defined by the possible links between the N nodes of the network. This perspective is particularly intriguing for two reasons. First, links in a network are unlabeled, which makes them inherently indistinguishable. Second, in a simple network without multiple connections, only one link can occupy a given state, as no two identical links can exist between the same pair of nodes. These properties naturally lead to a statistical interpretation of links in a network as an ensemble of identical and independent fermions, obeying Fermi–Dirac statistics [11]. By reimagining directed networks in this manner, we not only provide a novel statistical framework for describing their structure but also lay the groundwork for constructing statistically rigorous principled models that capture the fundamental constraints of directed and undirected networks alike. For instance, fermionic mapping has been instrumental in the analytical study of different aspects of networks, from the explanation of structural correlations in scale-free networks [17] to a topological phase transition with divergent entropy involving the reorganization of network cycles [18].

2.1. Fermionic Approach to Directed Networks

Given a pair of nodes i and j, we define two distinct states, $i \to j$ and $j \to i$, which can be occupied by links, or fermions, pointing from i to j and from j to i, respectively, see Figure 1. Each state $i \to j$ has an associated energy ε_{ij}. The occupancy of these states is described by the asymmetric adjacency matrix $\{a_{ij}\}$, which equals 1 if the state $i \to j$ is occupied and 0 otherwise, analogous to the occupation number of states in systems of indistinguishable particles. All the topological properties of the network can be computed from the adjacency matrix. For instance, the number of incoming connections to a node, or the incoming degree, is

$$k_{in,i} = \sum_{j=1}^{N} a_{ji},$$

where N is the total number of nodes in the network. Analogously, the number of outgoing connections from a node, or the outgoing degree, is

$$k_{out,i} = \sum_{j=1}^{N} a_{ij}.$$

Reciprocity implies pairs of nodes with links pointing in both directions, as shown in the sketch at the bottom of Figure 1. In random network models, a certain default level of reciprocity is attained when links are independent, or fermions are non-interacting. However, higher or lower values require that links are correlated, or fermions are interacting. To account for this possibility, we assume that the energy of two links occupying the two states $i \to j$ and $j \to i$ simultaneously, that is, of mutual interactions, is $\tilde{\varepsilon}_{ij}$. In general, $\tilde{\varepsilon}_{ij}$ is different from $\varepsilon_{ij} + \varepsilon_{ji}$.

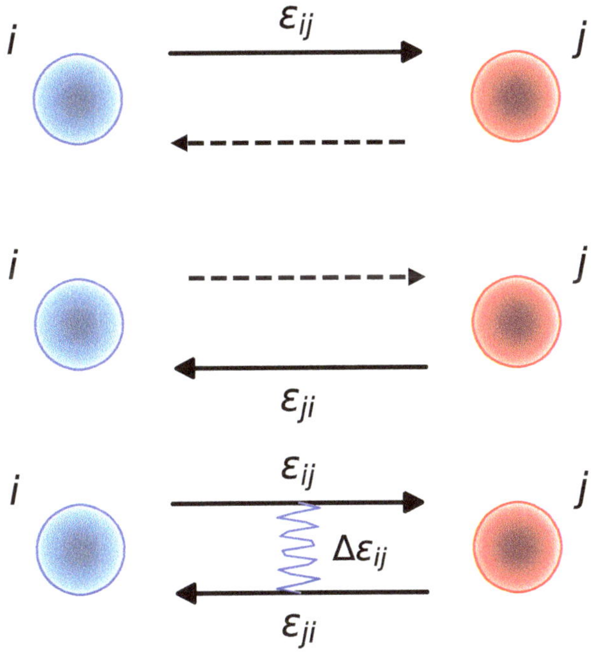

Figure 1. Possible fermionic states between a pair of nodes i and j, and their associated energies. The solid arrow indicates the presence of a directed link and the dashed arrow an empty state. When the two fermions simultaneously occupy the two states, $i \to j$ and $j \to i$, the total energy includes a correction $\Delta\varepsilon_{ij}$ added to the sum of the energies of the partially occupied states.

Due to the indistinguishability of links in a network, any directed network can be represented in the Fock space using the basis $\{|a\rangle \equiv \otimes_{i,j} |a_{ij}\rangle\}$ defining the number of particles/links occupying the set of possible single-particle states. Thus, the representation of the Hamiltonian of the network $\hat{H}$ in the basis of the Fock space defined by the adjacency matrix is

$$\langle a | \hat{H} | a \rangle = \sum_{i<j} \left[a_{ij}\varepsilon_{ij} + a_{ji}\varepsilon_{ji} + a_{ij}a_{ji}\Delta\varepsilon_{ij} \right], \tag{1}$$

where

$$\Delta\varepsilon_{ij} = \tilde{\varepsilon}_{ij} - \varepsilon_{ij} - \varepsilon_{ji}$$

is the correction due to the interaction of two fermions occupying the two states $i \to j$ and $j \to i$. When $\Delta\varepsilon_{ij} > 0$, the presence of two links connecting the same pair of nodes in opposite directions is energetically unfavorable, and thus the reciprocity is lower than in the random case. Conversely, when $\Delta\varepsilon_{ij} < 0$, the link reciprocity is higher than random case.

In analogy to the case of indistinguishable quantum particles, it is more convenient to work in the grand canonical ensemble, where the constraints are the following:

- the number of fermions (links) is fixed on average;
- The average energy is fixed as well.

In our formalism, this implies that the total number of links is a random variable that is fixed on average by the chemical potential μ. The grand partition function of the system is given by

$$\begin{aligned} Z &= Tr\left(e^{-\beta(\hat{H}-\mu\hat{N}_L)}\right) \\ &= \prod_{i<j}\left(1 + e^{-\beta(\varepsilon_{ij}-\mu)} + e^{-\beta(\varepsilon_{ji}-\mu)} + e^{-\beta(\tilde{\varepsilon}_{ij}-2\mu)}\right), \end{aligned} \tag{2}$$

where $\hat{N}_L$ is the number of links operator and the inverse temperature β controls the average energy of the network. The chemical potential μ fixes the average in-degree (and out-degree) through the relation

$$\langle k_{\text{in}} \rangle = \langle k_{\text{out}} \rangle = \frac{1}{N\beta}\left(\frac{\partial \ln Z}{\partial \mu}\right)_{\beta}, \tag{3}$$

and the entropy of the ensemble can be computed from the partition function as

$$S = \ln Z - \beta\left(\frac{\partial \ln Z}{\partial \beta}\right)_{\mu}. \tag{4}$$

Beyond these global thermodynamic properties, the probability of the ensemble generating a graph with adjacency matrix $\{a_{ij}\}$ is computed as the probability of a particular configuration of the system

$$\text{Prob}(\{a_{ij}\}) = \frac{1}{Z}\prod_{i<j} e^{-\beta\left[(\varepsilon_{ij}-\mu)a_{ij} + (\varepsilon_{ji}-\mu)a_{ji} + a_{ij}a_{ji}\Delta\varepsilon_{ij}\right]}. \tag{5}$$

The joint probability of the pair of states $i \to j$ and $i \leftarrow j$ between nodes i and j is

$$\text{Prob}(a_{ij}, a_{ji}) = \frac{e^{-\beta\left[(\varepsilon_{ij}-\mu)a_{ij} + (\varepsilon_{ji}-\mu)a_{ji} + a_{ij}a_{ji}\Delta\varepsilon_{ij}\right]}}{1 + e^{-\beta(\varepsilon_{ij}-\mu)} + e^{-\beta(\varepsilon_{ji}-\mu)} + e^{-\beta(\tilde{\varepsilon}_{ij}-2\mu)}}. \tag{6}$$

Finally, the probability of a directed link existing between nodes i and j, $p_{ij} \equiv \text{Prob}(a_{ij} = 1)$, is

$$p_{ij} = \frac{e^{-\beta(\varepsilon_{ij}-\mu)} + e^{-\beta(\tilde{\varepsilon}_{ij}-2\mu)}}{1 + e^{-\beta(\varepsilon_{ij}-\mu)} + e^{-\beta(\varepsilon_{ji}-\mu)} + e^{-\beta(\tilde{\varepsilon}_{ij}-2\mu)}}. \tag{7}$$

Equation (7) can be used to evaluate the average in- and out-degrees of individual nodes as

$$\kappa_{out,i} = \sum_j p_{ij}$$

and

$$\kappa_{in,i} = \sum_j p_{ji},$$

and the chemical potential as the solution of the equation

$$\langle k_{\text{in}} \rangle N = \sum_{i,j,j\neq,i} p_{ij}. \tag{8}$$

Finally, we can use these results to evaluate the reciprocity of the network r, defined as the ratio between the number of reciprocated links and the total number of links. Thus,

$$r = \frac{2\sum_{i<j} p_{ij}(1,1)}{\sum_{i,j\neq i} p_{ij}}, \tag{9}$$

where we have defined $p_{ij}(1,1) \equiv \text{Prob}(a_{ij} = 1, a_{ji} = 1)$. It is important to mention here that the freedom to chose the interaction energies $\Delta\epsilon_{ij}$ enables the possibility to adjust the level of reciprocity for particular sets of nodes or with specific topological properties.

2.2. Non-Interacting Fermions

When the links are independent or, equivalently, the fermions are non-interacting, $\Delta\varepsilon_{ij} = 0$ and the energy is $\tilde{\varepsilon}_{ij} = \varepsilon_{ij} + \varepsilon_{ji}$. In this situation, the connection probability p_{ij} of a directed link between nodes i and j takes the simple form

$$p_{ij}^{ni} = \frac{1}{1 + e^{\beta(\varepsilon_{ij}-\mu)}}. \tag{10}$$

The joint probability $\text{Prob}(a_{ij}, a_{ji})$ factorizes as $\text{Prob}(a_{ij}, a_{ji}) = p_{ij}^{ni} p_{ji}^{ni}$, and so does the partition function

$$Z = \prod_{i<j} \left(1 + e^{-\beta(\varepsilon_{ij}-\mu)}\right)\left(1 + e^{-\beta(\varepsilon_{ji}-\mu)}\right). \tag{11}$$

Finally, the reciprocity becomes

$$r = \frac{2\sum_{i<j} p_{ij}^{ni} p_{ji}^{ni}}{\sum_{i,j\neq i} p_{ij}^{ni}}, \tag{12}$$

which corresponds to the reciprocity expected by pure chance.

2.3. Interacting Fermions

The connection probability of the system without interactions, p_{ij}^{ni} in Equation (10), can be used to rewrite the connection probability for a directed link in the case of interacting fermions, p_{ij} in Equation (7), which leads to

$$p_{ij} = p_{ij}^{ni} \frac{1 - p_{ji}^{ni}(1 - e^{-\beta\Delta\varepsilon_{ij}})}{1 - p_{ij}^{ni} p_{ji}^{ni}(1 - e^{-\beta\Delta\varepsilon_{ij}})}. \tag{13}$$

In the case of weak interactions or high temperature, the term $\beta\Delta\varepsilon$ is small, leading to $p_{ij} \approx p_{ij}^{ni}$. Similarly, as seen from Equation (13), the connection probability remains unchanged by fermionic interactions in the limits $p_{ij}^{ni} \to 0$ or $p_{ij}^{ni} \to 1$, where $p_{ij} = p_{ij}^{ni}$ again holds. In these extreme situations, the lack or excess of bidirectional links leaves no room for the network to exhibit sensitivity to changes in the tendency for reciprocity. We will use this general property in the next section when dealing with specific models.

3. Specific Random Network Models

So far, we have not specified the energies of the states $\{\varepsilon_{ij}\}$, which ultimately define the particular model at hand. To illustrate the power of our approach, we focus on two different models within our formalism: the non-interacting Directed Soft Configuration Model (NI-DCM) [11], and the non-interacting Directed Geometric Soft Configuration Model (NI-DGCM) [16]. Furthermore, we also derive their maximum entropy interacting counterparts (I-DCM and I-DGCM).

A priori, our formalism works for an arbitrary number of fermions between 0 and $N(N-1)$. However, real complex networks are sparse, meaning that the average in- and out-degrees, $\langle k_{in} \rangle = \langle k_{out} \rangle$, are size-independent. In the rest of this paper, we consider ensembles of sparse networks.

3.1. Directed Configuration Model

To derive the probability of connection of the DCM [11,19] within our formalism, we make the simplest assumption that the energy of a directed link connecting nodes i and j comes from two sources: the energetic cost that node i incurs when creating an outgoing connection, $\varepsilon_{out,i}$, plus the energetic cost that node j incurs when accepting an incoming connection, $\varepsilon_{in,j}$. The total energy of the fermionic state is then

$$\varepsilon_{ij} = \varepsilon_{out,i} + \varepsilon_{in,j}. \tag{14}$$

Thus, each node in the network is characterized by an associated vector $(\varepsilon_{in}, \varepsilon_{out})$ accounting for incoming and outgoing connections. The distribution of such variables is given by the probability density function $\rho(\varepsilon_{in}, \varepsilon_{out})$, with marginal distributions for ε_{in} and ε_{out}, $\rho_{in}(\varepsilon_{in})$ and $\rho_{out}(\varepsilon_{out})$.

Although models in the DCM family are not fully realistic, they serve as prominent null models or baselines to evaluate whether observed features in real directed networks arise due to specific processes or simply by chance.

3.1.1. Non-Interacting Directed Configuration Model (NI-DCM)

Using Equation (8), and assuming that $\Delta\varepsilon_{ij} = 0$, and replacing sums with integrals, we can write

$$\langle k_{in} \rangle = Nz \int\int \frac{\rho_{in}(\varepsilon_{in})\rho_{out}(\varepsilon_{out})}{z + e^{\beta\varepsilon_{in}} e^{\beta\varepsilon_{out}}} d\varepsilon_{in} d\varepsilon_{out}, \tag{15}$$

where we have defined the fugacity in the standard way as $z \equiv e^{\beta\mu}$. Imposing sparsity in the thermodynamic limit of this particular model implies that the fugacity must scale with the system size as $z \sim N^{-1}$. This implies that the chemical potential takes the size-dependent form

$$\mu = \frac{1}{\beta} \ln \left[\frac{\langle k_{in} \rangle}{N \langle e^{-\beta\varepsilon_{in}} \rangle \langle e^{-\beta\varepsilon_{out}} \rangle} \right], \tag{16}$$

provided that $\langle e^{-\beta\varepsilon_{in}}\rangle$ and $\langle e^{-\beta\varepsilon_{out}}\rangle$ are bounded. In this case, the dependence between expected in- and out-degrees of nodes, κ_{in} and κ_{out}, and the in and out-energies, ε_{in} and ε_{out}, become

$$\kappa_{in} = \frac{\langle k_{in}\rangle}{\langle e^{-\beta\varepsilon_{in}}\rangle} e^{-\beta\varepsilon_{in}} \text{ and } \kappa_{out} = \frac{\langle k_{out}\rangle}{\langle e^{-\beta\varepsilon_{out}}\rangle} e^{-\beta\varepsilon_{out}}. \tag{17}$$

Substituting Equation (14) into Equation (10) and using Equations (16) and (17), the connection probability in Equation (10) becomes the one for the directed soft configuration model:

$$p_{ij}^{ni} = \frac{1}{1 + \frac{\langle k_{in}\rangle N}{\kappa_{out,i}\kappa_{in,j}}}. \tag{18}$$

Notice that, when the energies of states in Equation (14) are temperature-independent, the limit $\beta \to 0$ converges to the directed version of the classical Erdös-Rényi ensemble [20] because, in this limit, the expected degree of all nodes converges to the same value, as can be seen from Equation (17). In the opposite limit, when $\beta \gg 1$, the degree distribution becomes heavy-tailed and, depending on the distribution of energies, it may undergo a phase transition to a condensed phase where a finite fraction of nodes accumulate an extensive number of links, as shown in [21]. This effect will occur when the averages $\langle e^{-\beta\varepsilon_{in}}\rangle$ and/or $\langle e^{-\beta\varepsilon_{out}}\rangle$ diverge for $\beta > \beta_c$ for a critical inverse temperture β_c.

An alternative approach to Equation (14) is to fix the expected in- and out-degree distributions by defining temperature-dependent energy levels as

$$\varepsilon_{ij} = -\frac{1}{\beta}\ln\left(\kappa_{out,i}\kappa_{in,j}\right), \tag{19}$$

and the chemical potential as

$$\mu = -\frac{1}{\beta}\ln\left[\langle k_{in}\rangle N\right]. \tag{20}$$

These choices lead to the same connection probability Equation (18), with the difference that now the expected in- and out-degrees are temperature-independent and, thus, the degree distribution is fixed. Temperature-dependent energy levels appear in strongly interacting systems [22–25].

The entropy of the ensemble can be calculated using Equation (4), whose leading terms are

$$S = \langle k_{in}\rangle N(\ln\left[\langle k_{in}\rangle N\right] - 1) + O(\ln N), \tag{21}$$

recovering results in [12]. Notice that this expression does not depend on the ensemble temperature, only on the total number of links, which is a property that is fixed in the ensemble and does not depend on the degree distribution. This means that the same expression holds in the alternative definition of the model where the energy of the states is temperature-dependent.

Finally, the reciprocity of the ensemble can be evaluated using Equation (12), and reads

$$r = \frac{\langle k_{in}k_{out}\rangle^2}{N\langle k_{in}\rangle^3} - \frac{\langle k_{in}^2 k_{out}^2\rangle}{N^2\langle k_{in}\rangle^3} \approx \frac{\langle k_{in}k_{out}\rangle^2}{N\langle k_{in}\rangle^3}. \tag{22}$$

Thus, the reciprocity of the NI-SCM vanishes in the thermodynamic limit, even though it can become significant if the in- and out-degrees of nodes are positively correlated and their distributions heavy-tailed.

3.1.2. Interacting Directed Configuration Model (I-DCM)

The probability for a directed link in this model can be found by substituting Equation (14) into Equation (7), with $\Delta\varepsilon_{ij} \neq 0$, and imposing sparsity, which would again lead to Equations (16) and (17) if $\langle e^{-\beta\varepsilon_{in}}\rangle$ and $\langle e^{-\beta\varepsilon_{out}}\rangle$ are bounded. Alternatively, Equation (13), which relates the connection probabilities in the interacting and non-interacting formulations, provides a shortcut. The connection probability of the NI-DCM is size-dependent with p_{ij}^{ni} scaling as N^{-1}, hence approaching zero in the thermodynamic limit. At this extreme, Equation (13) indicates that $p_{ij} \approx p_{ij}^{ni}$, which implies that the energies ε_{in} and ε_{out}, along with β and μ, define the in- and out-degree distributions as in the non-interacting model.

In contrast, the joint probability $\mathrm{Prob}(a_{ij}, a_{ji})$ in the I-DCM does not factorize, thereby enabling the tuning of the reciprocity. The reciprocity can be calculated from Equation (9), using the probability of having a bidirectional connection between nodes i and j from Equation (6) after imposing the condition that, the two links are present simultaneously, $a_{ij} = a_{ji} = 1$. Using that

$$e^{\beta(\varepsilon_{ij}-\mu)} = \frac{N\langle k_{in}\rangle}{\kappa_{out,i}\kappa_{in,j}},$$

the reciprocity is

$$r = \frac{2}{N\langle k_{in}\rangle}\times \tag{23}$$

$$\times \sum_{i<j} \frac{\frac{\kappa_{out,i}\kappa_{in,j}}{N\langle k_{in}\rangle}\frac{\kappa_{out,j}\kappa_{in,i}}{N\langle k_{in}\rangle}e^{-\beta\Delta\varepsilon_{ij}}}{1+\frac{\kappa_{out,i}\kappa_{in,j}}{N\langle k_{in}\rangle}+\frac{\kappa_{out,j}\kappa_{in,i}}{N\langle k_{in}\rangle}+\frac{\kappa_{out,i}\kappa_{in,j}\kappa_{out,j}\kappa_{in,i}}{N\langle k_{in}\rangle)^2}e^{-\beta\Delta\varepsilon_{ij}}},$$

which, up to leading order in N, gives

$$r = \frac{1}{(N\langle k_{in}\rangle)^3}\sum_{i,j}\kappa_{out,i}\kappa_{in,i}\kappa_{out,j}\kappa_{in,j}e^{-\beta\Delta\varepsilon_{ij}}. \tag{24}$$

This result implies that reciprocity vanishes in the thermodynamic limit. The specific form in which $r \to 0$ as $N \to \infty$ depends on the form of the interaction energy. In all cases, when $\Delta\varepsilon_{ij} > 0$, reciprocity is energetically unfavorable, and thus lower than in the NI-SCM for the same temperature; conversely, when $\Delta\varepsilon_{ij} < 0$, the link reciprocity is higher.

For instance, a constant value independent of the specific pair of nodes, $\Delta\varepsilon_{ij} = \varepsilon$, leads to

$$r = \frac{e^{-\beta\varepsilon}}{N\langle k_{in}\rangle^3}\langle k_{in}k_{out}\rangle^2, \tag{25}$$

meaning that the interaction introduces temperature-dependent rescaling as compared to the reciprocity of the NI-SCM in Equation (22).

If, instead of a constant value, the nodes in the interaction have an additive contribution to the interaction correction energy, $\Delta\varepsilon_{ij} = \varepsilon_i + \varepsilon_j$, then

$$r = \frac{1}{N\langle k_{in}\rangle^3}\left(\sum_i \kappa_{out,i}\kappa_{in,i}e^{-\beta\varepsilon_i}\right)^2. \tag{26}$$

If ε_i is proportional to the temperature, $\varepsilon_i \propto 1/\beta$, the NI-SCM behavior is recovered with a temperature-independent constant rescaling. Additionally, it can incorporate dependencies on the hidden degrees of the corresponding node, for instance, $\varepsilon_i = -1/\beta \ln(\kappa_{out,i}\kappa_{in,i})$, and then

$$r = \frac{1}{N\langle k_{in}\rangle^3}\langle (k_{in}k_{out})^2\rangle^2. \tag{27}$$

Again, local correlations between the incoming and outgoing degrees of a node control the velocity of the reciprocity's decay. The results above also imply that a size-dependent negative interaction energy with intensity $|\varepsilon| \propto 1/\beta \ln N$ could counteract the decay of reciprocity in the SCM model and produce a finite value even in the thermodynamic limit.

3.2. Directed $\mathbb{S}^d$ Model

As we have seen in the previous section, reciprocity vanishes in the thermodynamic limit of the DCM even when fermions interact. Similarly, clustering also vanishes due to the size dependence of the connection probability. Finite reciprocity and clustering can be achieved in the framework of geometric networks [26,27], where nodes are distributed in an underlying metric space such that a distance x_{ij} can be defined between any pair of nodes. In this situation, we assume that the energies of sending out or accepting a link are supplemented with a cost associated with the distance between the nodes. Thus, the total energy of a link is

$$\varepsilon_{ij} = \varepsilon_{out,i} + \varepsilon_{in,j} + f(x_{ij}), \tag{28}$$

where $f(x)$ is a monotonically increasing function of the distance. An interesting choice is a logarithmic function, $f(x_{ij}) = \ln x_{ij}$, with nodes distributed in a d-dimensional Euclidean space, $\mathbb{R}^d$, according to a Poisson point process with a constant density, δ.

A prominent real system following this rationale is the Internet at the autonomous systems (AS) level, where link directions represent customer-to-provider relationships between autonomous systems [28]. A link from AS i to AS j indicates the flow of money when Internet traffic is routed through that connection. Notice that, in general, ASs can act as customers in some connections and as providers in others. In this system, the term $\varepsilon_{out,i}$ represents the cost that AS i must pay to maintain a connection as a customer. Similarly, $\varepsilon_{in,j}$ represents the cost that AS j must pay to maintain a connection as a provider. These two costs generally depend on the size and physical infrastructure of each AS. Finally, the term $f(x_{ij})$ represents the cost associated with the physical constraints of the connection, such as, for instance, its physical length. In the case of very large ASs, a negative interaction energy, $\Delta\varepsilon_{ij}$, represents the synergistic effect of having a bidirectional connection, typically between the tier 1 ASs that define the core of the Internet.

In a different domain, brain connectomes are well described by the hyperbolic geometry network framework encoded by the directed $\mathbb{S}^d$ model [16,29] and display an over-representation of reciprocal connections [30]. To form a synaptic connection, the costs of maintaining neurotransmitters and neuroreceptors are incurred, along with a cost associated with maintaining the neural fibers and the signal as it travels along the distance covered by them.

3.2.1. Non-Interacting Directed $\mathbb{S}^d$ Model (NI-DSM)

When $\Delta\varepsilon_{ij} = 0$, the expected out-degree of a node with energy ε_{out} , located, without a loss of generality, at the origin of coordinates, is given by

$$\langle k_{out}(\varepsilon_{out})\rangle = \delta \int \rho(\varepsilon_{in}) d\varepsilon_{in} \int_0^\infty \frac{V_{d-1} r^{d-1}}{1 + r^\beta e^{\beta(\varepsilon_{in}+\varepsilon_{out}-\mu)}} dr, \tag{29}$$

where $V_{d-1} = 2\pi^{d/2}/\Gamma(d/2)$ is the volume of a $(d-1)$-sphere. This expression can be rewritten for $\beta > d$ as

$$\langle k_{out}(\varepsilon_{out})\rangle = \delta V_{d-1} I(\beta, d) \langle e^{-d\varepsilon_{in}}\rangle e^{d\mu} e^{-d\varepsilon_{out}}, \tag{30}$$

where

$$I(\beta, d) = \int_0^\infty \frac{t^{d-1} dt}{1 + t^\beta} = \frac{\pi}{\beta \sin \frac{d\pi}{\beta}}. \tag{31}$$

Thus, if we redefine the expected out- and in-degrees as $\kappa_{out} \equiv e^{-d\varepsilon_{out}}$ and $\kappa_{in} \equiv e^{-d\varepsilon_{in}}$, with $\mu = -\frac{1}{d}\ln\left(\delta V_{d-1} I(\beta,d)\langle k_{in}\rangle\right)$, the connection probability becomes

$$p_{ij} = \frac{1}{1+\chi_{ij}^{\beta}} \quad \text{with} \quad \chi_{ij} \equiv \frac{x_{ij}}{(\hat{\mu}\kappa_{out,i}\kappa_{in,j})^{\frac{1}{d}}}, \tag{32}$$

and

$$\hat{\mu} = \frac{\beta\Gamma\left(\frac{d}{2}\right)\sin\left(\frac{\pi d}{\beta}\right)}{2\delta\pi^{1+\frac{d}{2}}\langle k_{in}\rangle}. \tag{33}$$

(The case $\beta < d$ can be analyzed as in [18]). This model can be immediately identified as the directed variant of the $\mathbb{S}^d$ model, first introduced in [16]. It represents a directed extension of the $\mathbb{S}^d$ model originally proposed in [31], along with its equivalent formulation in the hyperbolic plane, known as the $\mathbb{H}^2$ model [32]. Notably, numerous analytical results have been derived for the $\mathbb{S}^1/\mathbb{H}^2$ model, including studies on degree distribution [31–33], clustering [32–35], graph diameter [36–38], percolation [39,40], self-similarity [31], and spectral properties [41]. Moreover, this model has been extended to incorporate growing networks, weighted networks, multilayer networks, and networks with community structure, and it also serves as the foundation for defining a renormalization group for complex networks, see [26,27] and references therein. These analytical results and extensions of the undirected geometric model provide a guide for future studies of the Directed $\mathbb{S}^d$ Model.

Unlike the DCM, geometry implies that the connection probability is size-independent. In turn, this implies that the reciprocity and clustering are finite, as shown in [16]. Interestingly, this model undergoes a topological phase transition at the critical inverse temperature $\beta_c = d$ [18]. For $\beta > \beta_c$, clustering is finite in the thermodynamic limit, whereas it vanishes below this value. This phase transition is of a topological nature and involves the reorganization of cycles in the network; transitioning from being short-range in the clustered phase to long-range in the unclustered one. This transition is accompanied by an anomalous behavior of the entropy per link. From Equation (4), we can compute the entropy as

$$\frac{S}{N\langle k_{in}\rangle} = \frac{2\beta}{d}\left(1 - \frac{\pi d}{\beta}\cot\frac{\pi d}{\beta}\right). \tag{34}$$

Unlike standard continuous phase transitions, the entropy per link diverges at the critical temperature from below as

$$\frac{S}{N\langle k_{in}\rangle} \sim \frac{1}{\beta - d}, \tag{35}$$

whereas it diverges logarithmically at higher temperatures. The origin of this anomalous behavior is due to the fact that the number of available microstates per link at low temperatures is finite, primarily connecting pairs of nodes at bounded distances. However, once the temperature surpasses the critical temperature, the number of available microstates becomes that of the order of the number of nodes, as links can now connect pairs of nodes that are arbitrarily far apart.

3.2.2. Interacting Directed $\mathbb{S}^d$ Model (I-DSM)

When reciprocal links interact in the directed $\mathbb{S}^d$ model, the strategy applied for the I-DCM, based on using Equation (13) to relate the connection probabilities in the interacting and non-interacting formulations, cannot be used because p_{ij}^{ni} is independent of the system size and does not approach zero in the thermodynamic limit. The probability of a directed link in the I-DSM must be found by substituting Equation (28) into Equation (7), with $\Delta\varepsilon_{ij} \neq 0$, and imposing sparsity, which leads to new definitions of the chemical potential μ

and the relation between the expected in- and out-degrees of a given node and its in- and out-energies ε_{in} and ε_{out}. In particular, the connection probability can be written as

$$p_{ij} = \frac{\chi_{ji}^{\beta} + e^{-\beta\Delta\varepsilon_{ij}}}{\chi_{ij}^{\beta} + \chi_{ji}^{\beta} + \chi_{ij}^{\beta}\chi_{ji}^{\beta} + e^{-\beta\Delta\varepsilon_{ij}}}, \tag{36}$$

where

$$\chi_{ij} = x_{ij}e^{\varepsilon_{out,i}+\varepsilon_{in,j}-\mu}. \tag{37}$$

Using this expression, the average out-degree of a node with the in- and out-energies $\varepsilon_{in,i}$ and $\varepsilon_{out,i}$ can be written as

$$\begin{aligned} \langle k_{out}(\varepsilon_{in,i}, \varepsilon_{out,i})\rangle &= \delta V_{d-1}e^{d\mu}e^{-d\varepsilon_{out,i}} \times \\ &\times \int\int e^{-d\varepsilon_{in,j}}\rho(\varepsilon_{in,j}, \varepsilon_{out,j})d\varepsilon_{in,j}d\varepsilon_{out,j} \int_0^{\infty} \frac{t^{d-1}(q_{ij}t^{\beta} + e^{-\beta\Delta\varepsilon_{ij}})}{t^{\beta} + q_{ij}(1+t^{\beta})t^{\beta} + e^{-\beta\Delta\varepsilon_{ij}}}dt, \end{aligned} \tag{38}$$

where $q_{ij} \equiv e^{\varepsilon_{out,j}-\varepsilon_{out,i}+\varepsilon_{in,i}-\varepsilon_{in,j}}$. By integrating Equation (38) over the energies $\varepsilon_{in,i}$ and $\varepsilon_{out,i}$ and equating it to $\langle k_{in}\rangle$, we can obtain the value of the chemical potential μ from

$$e^{d\mu} = \frac{\langle k_{in}\rangle}{\delta V_{d-1}\langle e^{-d(\varepsilon_{out,i}+\varepsilon_{in,j})} \int_0^{\infty} \frac{t^{d-1}(q_{ij}t^{\beta} + e^{-\beta\Delta\varepsilon_{ij}})}{t^{\beta} + q_{ij}(1+t^{\beta})t^{\beta} + e^{-\beta\Delta\varepsilon_{ij}}}dt\rangle}, \tag{39}$$

where the average in the denominator is taken over the random variables $\varepsilon_{in,i}, \varepsilon_{in,j}, \varepsilon_{out,i}, \varepsilon_{out,j}$, and $\Delta\varepsilon_{ij}$. Using a similar approach, the reciprocity becomes

$$r = \frac{\langle e^{-d(\varepsilon_{out,i}+\varepsilon_{in,j})} \int_0^{\infty} \frac{t^{d-1}e^{-\beta\Delta\varepsilon_{ij}}}{t^{\beta} + q_{ij}(1+t^{\beta})t^{\beta} + e^{-\beta\Delta\varepsilon_{ij}}}dt\rangle}{\langle e^{-d(\varepsilon_{out,i}+\varepsilon_{in,j})} \int_0^{\infty} \frac{t^{d-1}(q_{ij}t^{\beta} + e^{-\beta\Delta\varepsilon_{ij}})}{t^{\beta} + q_{ij}(1+t^{\beta})t^{\beta} + e^{-\beta\Delta\varepsilon_{ij}}}dt\rangle}. \tag{40}$$

Equation (38) implies that the average in- or out-degree of a given node depends on both ε_{in} and ε_{out}, not only on one of them, as is the case for non-interacting fermions. This indicates that computing the degree distributions requires the explicitly solving Equation (38). However, in the particular case of fully correlated ε_{in} and ε_{out} and $\Delta\varepsilon_{ij} = \Delta\varepsilon$, the term $q_{ij} = 1$, and the average in- or out-degree becomes a function of ε_{in} or ε_{out} separately. Thus, as in the case of non-interacting fermions, we can write $\kappa_{out} \equiv e^{-d\varepsilon_{out}}$ and $\kappa_{in} \equiv e^{-d\varepsilon_{in}}$, with

$$\mu = -\frac{1}{d}\ln(\delta V_{d-1}\tilde{I}(\beta, d, \Delta\varepsilon)\langle k_{in}\rangle), \tag{41}$$

where

$$\tilde{I}(\beta, d, \Delta\varepsilon) = \int_0^{\infty} \frac{t^{d-1}(t^{\beta} + e^{-\beta\Delta\varepsilon})}{2t^{\beta} + t^{2\beta} + e^{-\beta\Delta\varepsilon}}dt, \tag{42}$$

and the reciprocity becomes

$$r = \frac{\int_0^{\infty} \frac{t^{d-1}e^{-\beta\Delta\varepsilon}}{2t^{\beta} + t^{2\beta} + e^{-\beta\Delta\varepsilon}}dt}{\tilde{I}(\beta, d, \Delta\varepsilon)}. \tag{43}$$

Figure 2 shows the results of the reciprocity in this case as a function of $\Delta\varepsilon$ for different values of β. The reciprocity converges to one in the limit $\Delta\varepsilon \to -\infty$ and approaches zero in the limit $\Delta\varepsilon \to \infty$, as expected. Furthermore, it increases as the temperature rises. Note that the convergence to 1 with very low temperatures and/or a highly negative $\Delta\varepsilon$ is only possible in the fully correlated case. In all other cases, the maximum possible value of the reciprocity is always less than one.

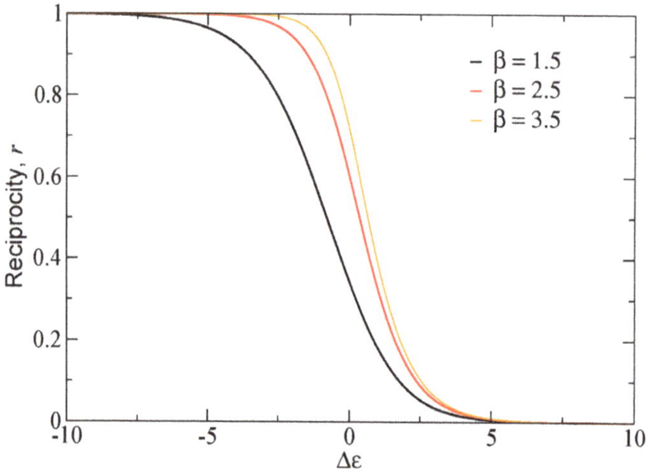

Figure 2. Reciprocity of the interacting directed $\mathbb{S}^d$ model for fully correlated in- and out-energies, as a function of $\Delta\varepsilon$. Different curves correspond to different temperatures β^{-1}.

4. Conclusions

The statistical mechanics framework for directed networks introduced in this work treats links as fermionic particles subject to constraints and interactions. This formalism allowed us to describe directed networks within a principled approach that incorporates the reciprocity and other structural properties, addressing the limitations of existing models. By leveraging concepts from quantum statistics, our methodology redefines network modeling, shifting the focus from node-centric descriptions to link interactions. Formulating directed networks within a grand canonical ensemble, we demonstrated how the chemical potential and key network features, such as the degree distribution and reciprocity, naturally emerge from the underlying statistical framework.

The versatility and analytical power of our formalism were illustrated through applications to specific cases, including the Directed Configuration Model and the Directed $\mathbb{S}^d$ model. The key results highlighted the influence of interactions on the reciprocity and clustering. In the non-interacting formulations, the reciprocity vanishes in the thermodynamic limit, whereas in the interacting models, the framework supports a tunable reciprocity that remains finite under specific conditions. The inclusion of a geometric component in the $\mathbb{S}^d$ model further showcased how spatial constraints shape the emergent properties of the network. This framework bridges theoretical advances with empirical applicability, providing a robust toolset for analyzing real-world directed networks. Additionally, it paves the way for exploring dynamical processes on directed topologies and designing models that better reflect the intricate balance of directed interactions. Future work could extend these principles to multilayer, temporal, or weighted networks, offering a deeper understanding of complex systems.

Author Contributions: M.B. and M.Á.S. both contributed equally to the conceptualization, methodology, formal analysis, investigation, writing—original draft preparation, writing—review and editing, and funding acquisition. All authors have read and agreed to the published version of the manuscript.

Funding: We acknowledge support from: Grant TED2021-129791B-I00 funded by MCIN/AEI/10.13039/501100011033 and by "European Union NextGenerationEU/PRTR"; Grant PID2022-137505NB-C22 funded by MCIN/AEI/10.13039/501100011033 and by "ERDF A way of making Europe"; and the

Generalitat de Catalunya grant number 2021SGR00856. M.B. acknowledges the ICREA Academia award, funded by the Generalitat de Catalunya.

Data Availability Statement: No new data were created.

Conflicts of Interest: The authors declare no conflicts of interest. The funders had no role in the design of the study; in the collection, analyses, or interpretation of data; in the writing of the manuscript; or in the decision to publish the results.

References

1. Newman, M.E.J. *Networks*; Oxford University Press: Oxford, UK, 2018.
2. Bianconi, G.; Gulbahce, N.; Motter, A.E. Local Structure of Directed Networks. *Phys. Rev. Lett.* **2008**, *100*, 118701. [CrossRef]
3. Asllani, M.; Lambiotte, R.; Carletti, T. Structure and dynamical behavior of non-normal networks. *Sci. Adv.* **2018**, *4*, eaau9403. [CrossRef]
4. Boguñá, M.; Serrano, M.Á. Generalized percolation in random directed networks. *Phys. Rev. E* **2005**, *72*, 016106. [CrossRef] [PubMed]
5. Serrano, M.Á.; De Los Rios, P. Interfaces and the edge percolation map of random directed networks. *Phys. Rev. E* **2007**, *76*, 056121. [CrossRef] [PubMed]
6. Serrano, M.Á.; Boguñá, M.; Vespignani, A. Patterns of Dominant Flows in the World Trade Web. *J. Econ. Interact. Coord.* **2007**, *2*, 111–124. [CrossRef]
7. Serrano, M.Á.; Klemm, K.; Vazquez, F.; Eguíluz, V.M.; San Miguel, M. Conservation laws for voter-like models on random directed networks. *J. Stat. Mech. Theory Exp.* **2009**, *2009*, P10024. [CrossRef]
8. Asllani, M.; Challenger, J.D.; Pavone, F.S.; Sacconi, L.; Fanelli, D. The theory of pattern formation on directed networks. *Nat. Commun.* **2014**, *5*, 4517. [CrossRef] [PubMed]
9. Muolo, R.; Carletti, T.; Gleeson, J.P.; Asllani, M. Synchronization dynamics in non-normal networks: The trade-off for optimality. *Entropy* **2020**, *23*, 36. [CrossRef]
10. Nartallo-Kaluarachchi, R.; Asllani, M.; Deco, G.; Kringelbach, M.L.; Goriely, A.; Lambiotte, R. Broken detailed balance and entropy production in directed networks. *Phys. Rev. E* **2024**, *110*, 034313. [CrossRef] [PubMed]
11. Park, J.; Newman, M.E.J. Statistical mechanics of networks. *Phys. Rev. E* **2004**, *70*, 066117. [CrossRef] [PubMed]
12. Bianconi, G. Entropy of network ensembles. *Phys. Rev. E* **2009**, *79*, 036114. [CrossRef] [PubMed]
13. Garlaschelli, D.; Loffredo, M.I. Patterns of link reciprocity in directed networks. *Phys. Rev. Lett.* **2004**, *93*, 268701. [CrossRef]
14. Fagiolo, G. Clustering in complex directed networks. *Phys. Rev. E—Stat. Nonlinear Soft Matter Phys.* **2007**, *76*, 026107. [CrossRef] [PubMed]
15. Ahnert, S.E.; Fink, T.M.A. Clustering signatures classify directed networks. *Phys. Rev. E* **2008**, *78*, 036112. [CrossRef]
16. Allard, A.; Serrano, M.Á.; Boguñá, M. Geometric description of clustering in directed networks. *Nat. Phys.* **2024**, *20*, 150–156. [CrossRef]
17. Park, J.; Newman, M.E.J. Origin of degree correlations in the Internet and other networks. *Phys. Rev. E* **2003**, *68*, 026112. [CrossRef] [PubMed]
18. Van der Kolk, J.; Serrano, M.Á.; Boguñá, M. An anomalous topological phase transition in spatial random graphs. *Commun. Phys.* **2022**, *5*, 245. [CrossRef]
19. Kim, H.; Del Genio, C.I.; Bassler, K.E.; Toroczkai, Z. Constructing and sampling directed graphs with given degree sequences. *New J. Phys.* **2012**, *14*, 023012. [CrossRef]
20. Erdös, P.; Rényi, A. On random graphs I. *Publ. Math.* **1959**, *6*, 290–297. [CrossRef]
21. Bianconi, G.; Barabási, A.L. Bose-Einstein Condensation in Complex Networks. *Phys. Rev. Lett.* **2001**, *86*, 5632–5635. [CrossRef]
22. Rushbrooke, G. On the statistical mechanics of assemblies whose energy-levels depend on the temperature. *Trans. Faraday Soc.* **1940**, *36*, 1055–1062. [CrossRef]
23. Landsberg, P. Statitical Mechanics of Teperature-Dependent Energy Levels. *Phys. Rev.* **1954**, *95*, 643.
24. Elcock, E.; Landsberg, P. Temperature dependent energy levels in statistical mechanics. *Proc. Phys. Soc. Sect. B* **1957**, *70*, 161. [CrossRef]
25. De Miguel, R.; Rubí, J.M. Statistical mechanics at strong coupling: A bridge between Landsberg's energy levels and Hill's nanothermodynamics. *Nanomaterials* **2020**, *10*, 2471. [CrossRef]
26. Boguñá, M.; Bonamassa, I.; De Domenico, M.; Havlin, S.; Krioukov, D.; Serrano, M.Á. Network geometry. *Nat. Rev. Phys.* **2021**, *3*, 114–135. [CrossRef]
27. Serrano, M.Á.; Boguñá, M. *The Shortest Path to Network Geometry: A Practical Guide to Basic Models and Applications*; Cambridge University Press: Cambridge, UK, 2022. [CrossRef]

28. Dimitropoulos, X.; Krioukov, D.; Fomenkov, M.; Huffaker, B.; Hyun, Y.; Claffy, K.; Riley, G. AS relationships: Inference and validation. *ACM SIGCOMM Comput. Commun. Rev.* **2007**, *37*, 29–40. [CrossRef]
29. Allard, A.; Serrano, M.Á. Navigable maps of structural brain networks across species. *PLOS Comput. Biol.* **2020**, *16*, e1007584. [CrossRef]
30. Lin, A.; Yang, R.; Dorkenwald, S.; Matsliah, A.; Sterling, A.R.; Schlegel, P.; Yu, S.c.; McKellar, C.E.; Costa, M.; Eichler, K.; et al. Network statistics of the whole-brain connectome of Drosophila. *Nature* **2024**, *634*, 153–165. [CrossRef]
31. Serrano, M.Á.; Krioukov, D.; Boguñá, M. Self-Similarity of Complex Networks and Hidden Metric Spaces. *Phys. Rev. Lett.* **2008**, *100*, 078701. [CrossRef]
32. Krioukov, D.; Papadopoulos, F.; Kitsak, M.; Vahdat, A.; Boguñá, M. Hyperbolic geometry of complex networks. *Phys. Rev. E* **2010**, *82*, 036106. [CrossRef] [PubMed]
33. Gugelmann, L.; Panagiotou, K.; Peter, U. Random Hyperbolic Graphs: Degree Sequence and Clustering. In *Automata, Languages, and Programming*; Czumaj, A., Mehlhorn, K., Pitts, A., Wattenhofer, R., Eds.; ICALP 2012, Part II, Lecture Notes in Computer Science; Springer: Berlin/Heidelberg, Germany, 2012; Volume 7392, pp. 573–585. [CrossRef]
34. Candellero, E.; Fountoulakis, N. Clustering and the Hyperbolic Geometry of Complex Networks. *Internet Math.* **2016**, *12*, 2–53. [CrossRef]
35. Fountoulakis, N.; van der Hoorn, P.; Müller, T.; Schepers, M. Clustering in a hyperbolic model of complex networks. *Electron. J. Probab.* **2021**, *26*, 1–132. [CrossRef]
36. Abdullah, M.A.; Fountoulakis, N.; Bode, M. Typical distances in a geometric model for complex networks. *Internet Math.* **2017**, *1*, 115–126. [CrossRef]
37. Friedrich, T.; Krohmer, A. On the Diameter of Hyperbolic Random Graphs. *SIAM J. Discret. Math.* **2018**, *32*, 1314–1334. [CrossRef]
38. Müller, T.; Staps, M. The diameter of KPKVB random graphs. *Adv. Appl. Probab.* **2019**, *51*, 358–377. [CrossRef]
39. Serrano, M.Á.; Krioukov, D.; Boguñá, M. Percolation in Self-Similar Networks. *Phys. Rev. Lett.* **2011**, *106*, 048701. [CrossRef]
40. Fountoulakis, N.; Müller, T. Law of large numbers for the largest component in a hyperbolic model of complex networks. *Ann. Appl. Probab.* **2018**, *28*, 607–650. [CrossRef]
41. Kiwi, M.; Mitsche, D. Spectral gap of random hyperbolic graphs and related parameters. *Ann. Appl. Probab.* **2018**, *28*, 941–989. [CrossRef]

MDPI AG
Grosspeteranlage 5
4052 Basel
Switzerland
Tel.: +41 61 683 77 34

Entropy Editorial Office
E-mail: entropy@mdpi.com
www.mdpi.com/journal/entropy

www.ingramcontent.com/pod-product-compliance
Lightning Source LLC
Chambersburg PA
CBHW061505210926
51850CB00008B/1830
* 9 7 8 3 7 2 5 8 4 6 3 3 7 *